P9-AFZ-393
LIBRARY
LEWIS-CLARK STATE COLLEGE
LEWISTON, IDAHO

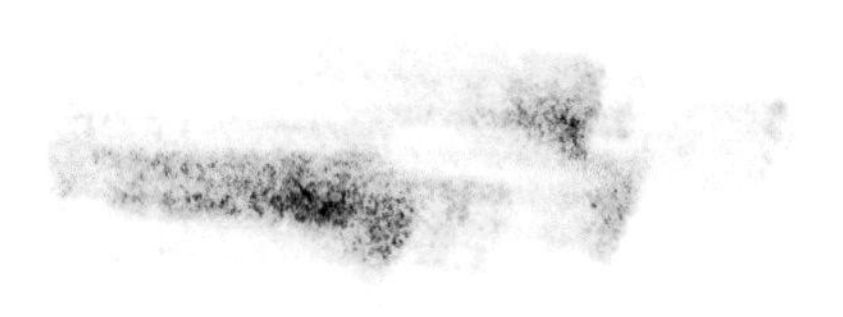

The End of Energy

The End of Energy

The Unmaking of America's Environment, Security, and Independence

Michael J. Graetz

The MIT Press
Cambridge, Massachusetts,
London, England

For information about special quantity discounts, please email special_sales@mitpress.mit.edu.

This book was set in Stone Sans and Stone Serif by Toppan Best-set Premedia Limited. Printed and bound in the United States of America.

Library of Congress Cataloging-in-Publication Data
Graetz, Michael J.
The end of energy : the unmaking of America's environment, security, and independence / Michael J. Graetz.
p. cm.
Includes bibliographical references and index.
ISBN 978-0-262-01567-7 (hbk. : alk. paper)
1. Energy policy—United States. 2. Energy resources development—United States. 3. Energy industries—United States. 4. United States—Economic policy. I. Title.
HD9502.U52G685 2011
333.7900973—dc22

2010040933

10 9 8 7 6 5 4 3 2 1

For my daughters
Casey, for her unflagging support and encouragement
Dylan, whose skepticism proved an inspiration
and
Sydney, for her estimable judgment and great good humor

Contents

Acknowledgments

I have had much generous and excellent help in writing this book. My idea to write about the development of energy and environmental policies emerged after several years of teaching a seminar on the 1970s for Yale law and history students with my colleague and friend Daniel Kevles. I am greatly indebted to him and the students in those courses for increasing my understanding of this transformative period not only for energy and the environment, but also for our country. As this project unfolded, several students at the Yale Law School contributed important research for various aspects of this book. I thank in particular Tomás Carbonell, David Chao, Miles Farmer, Steve Fisher, Brian Frazelle, Andrea Gelatt, Christopher Hurtado, Daniel Luskin, Jeffrey Tebbs, and Benjamin Zimmer. Brian Mahanna and Carrie Pagnucco were present from the inception of my efforts to transform inchoate ideas into this book and offered thoughtful comments and excellent research from beginning to end; they were more like collaborators than research assistants. John Nann of the Yale law library also provided valuable research. A number of my colleagues and friends from the Columbia and Yale law schools and elsewhere provided enormously helpful comments on a draft of the manuscript: Bruce Ackerman, Anne Alstott, Hannah Chang, Daniel Esty, Michael Gerrard, Linda Greenhouse, Doug Kysar, Daniel Markovits, Jerry Mashaw, Peter Merrill, Tom Merrill, Emma Neff, Alex Raskolnokov, and Ian Shapiro. Participants at the faculty workshop at Columbia Law School and a group of anonymous referees for MIT Press offered useful comments. I also learned much from conversations with Peter Merrill, Gilbert Metcalf, David Weisbach, and a number of other experts on the energy sector. My good friend Adam Haslett, an extraordinarily talented and successful writer, read early drafts of the manuscript and offered detailed comments and suggestions for how to make it better. His contributions were invaluable, and I marvel at his talents. Karen Williams began the typing of the manuscript and offered her enthusiastic support throughout. Margaret Symuleski typed, retyped, and retyped changes in the manuscript with great care and unflappable good humor. Yale Law School deans Harold Koh and Robert Post and Columbia Law School dean David Schizer provided both encouragement and financial support, as did Ian Shapiro, director of Yale's MacMillan Center. My agent, Wendy Strothman, helped shape the project and bring this book to fruition. I also thank my editor at MIT Press, John Covell, for his enthusiasm and comments and for shepherding the manuscript into print.

Finally, I thank my family—my wife, Brett Dignam, and my children, Lucas, Dylan, Jake, Sydney, and Casey—for their unwavering support and patience. Without them nothing would be possible—or any fun.

Michael J. Graetz
New York City
September 2010

Prologue: The Journey

It remains a basic fact of American life that, despite forty years of political fulminating, global conflict, and ever-increasing environmental awareness, most of us still take energy for granted. We take for granted that when we come home at night and flip on the light switch, the bulb will illuminate. We assume that when we turn up the thermostat, the heat will come on. And however acutely aware we may be of the price per gallon we pay, we take it as something close to a right of citizenship that when we drive an automobile up to one of the more than 100,000 gas stations in the United States, there will be fuel for our cars and trucks in the tanks beneath the asphalt. Without gasoline, the country would not run, and so there is gasoline, and barring extraordinary circumstances, there is plenty of it.

The fuels that let us take energy for granted come from all over our country and our planet: mines in West Virginia, wind farms in Texas, and nuclear power plants in California. But no doubt the most vexed fuel we use is petroleum, and although we now import it from more than 60 different foreign countries (Canada now first among them, Saudi Arabia second), it has been the need for supply from the Middle East that has exercised such an outsized influence on our foreign policy, our environmental politics, and our national security.

By the accidents of geology, the fossilized remains of prehistoric zooplankton and algae when heated over millennia form crude oil and are nowhere more plentiful than in a 174-mile long reservoir known as the Ghawar Field in Saudi Arabia. Ghawar lies along the eastern edge of the Arabian Peninsula, in the middle east of the Middle East, and buried beneath its sweltering sands is the biggest oil field in the world, having thus far produced 60 billion barrels of crude and providing to this day half the Saudi kingdom's daily output. For decades, Saudi Arabia has been able to act as a kind of central bank of oil, calming world markets at times of maximum scarcity by increasing output, as it did during the two gulf wars, and cutting supply when prices fall. Ghawar is what allows them to do this. It is their mother lode; in all of Iran, Iraq, Russia, and Nigeria, there is no deposit remotely its size.

The story of how the crude oil pumped out of the ground at Ghawar travels to the United States to become the gasoline, the jet fuel, and the ingredient in tens of thousands of plastics and consumer goods that together keep the American river of commerce flowing tells us much about the shape of our world today: about the vastly complex and tenuous system of global transport whose constant smooth functioning underwrites our complacency about the sprawl that requires two-hour commutes and makes sport utility vehicles (SUVs) even imaginable. It tells us about the consequences of failed states in blood and treasure. And it tells us something about the strange and constantly shifting balance we in the United States have tried to strike between unharnessing all of energy's potential to fuel economic growth and trying to limit its corrosive effects on our environment and our foreign policy. In short, to see how oil moves from Ghawar to our gas pumps depicts many of the most important issues this book confronts.

The journey begins innocuously enough, with oil sufficient to fill 5 million barrels (or 210 million gallons) a day moving from beneath the sands of Ghawar along a pipeline to the coast of the Persian Gulf, where it arrives at Ras Tanura, the largest oil terminal in the world. Already by this point, however, security is an issue. Given Al Qaeda's stated goal of overthrowing the Saudi monarchy and its desire to disrupt Western economies, every yard of pipeline is a potential target. To guard against the dangers of attack and the supply disruptions it would cause, the Saudis have the largest inventory of spare pipeline parts of any nation, much of it stored remotely along the pipeline route itself so that repair teams can be flown in by helicopter, access the needed materials on site, and get the oil flowing again as quickly as possible.

The terminal at Ras Tanura, which processes 10 percent of the global output of crude, is itself a fortress with multiple security checkpoints along the highway leading to it and on the grounds itself. Access to this command center of Saudi Aramco, the largest oil company in the world, is highly restricted. Storage tanks at the terminal house 50 million barrels worth of crude. Its original port lies along a small peninsula jutting into the Persian Gulf, but because of modern supertankers' need for deep water the Saudis have installed artificial islands in the nearby waters to allow for faster loading.

Once the crude has been transferred onto a tanker, the ship begins its journey westward by passing through the Strait of Hormuz. Only 34 miles across at its narrowest point, with Iran on the western shore and the United Arab Emirates and Oman on the eastern, this strait is the most important choke point in the world oil supply. Roughly 15 tankers a day pass through it, representing 40 percent of waterborne oil shipments. If Iran—attacked perhaps by the United States or Israel—ever decided to shut down the Persian Gulf oil business, this is where it would do so.

For a variety of legal, diplomatic, and logistical reasons, the great majority of tankers are not owned by oil companies or sovereign entities, but by independent shipping firms. A tanker might be built in South Korea, owned by a firm in Texas, fly the Liberian flag, and transport oil bought by the highest bidder. As an environmental matter, the most obvious risk the tanker poses in its 40- to 50-day passage from Ras Tanura to the Gulf of Mexico is the danger of a major spill caused by a storm, poor navigation, or an improperly maintained hull. But incidents such as the 1989 *Exxon-Valdez* spill in Alaska or the *Prestige* spill off the coast of Spain in 2002, which was nearly double the size of the *Valdez* spill, are only the most spectacular form of tanker damage. These ships also discharge large quantities of oil-contaminated ballast water, or bilge, as well as fuel oil, which, when all the ships in the tanker fleet are factored in, adds up to five *Exxon-Valdez* spills each and every year. Oil is a dirty business, not the least for our oceans.

Once a tanker has moved through the Strait of Hormuz and crossed the Gulf of Oman, it enters what, over the past decade, have become the most dangerous waters in the world—the Gulf of Aden. The ancient art of piracy, never entirely eradicated, but for most of the twentieth century more of an irritant than a threat to global supply chains, now thrives off the coast of the Horn of Africa. And what it thrives on is the ransom paid for richly laden ships. Supertankers display the three characteristics pirates like best: they are slow moving; they have a low freeboard—the distance from water to deck; and they carry precious cargo. The pirates operate from the failed state of Somalia, a country eviscerated by decades of armed conflict involving neighboring states, warlords, Islamic extremists, and botched superpower interventions. Their success has raised shipping insurance costs tenfold in recent years.

To elude the danger, some tankers now avoid the quicker passage through the Suez Canal and navigate around the Cape of Good Hope, extending the journey to the United States up to 95 days. Even this change in route, however, has not made them safe. In 2008, the *Sirius Star*, a Saudi-owned tanker following this route with its cargo of $100 million worth of crude was captured 520 miles off the coast of Kenya, becoming the largest ship ever hijacked. Making the short-term calculation that a $3 million ransom was worth it to get their tanker, crew, and oil back, the owners paid up, filling the pirates' coffers for future operations. American and British vessels have begun to patrol the area to ward off the pirates, but now that these increasingly sophisticated gangs have extended their reach to a million square miles of the Indian Ocean, Western authorities admit there is no sea-based enforcement that can stop them. Among the less obvious contributions to the price of the gas in your tank, then, is the fact that a large East African state has no functioning government.

Once through the Suez Canal or past the tip of Africa, a U.S.-bound ship typically makes its long transatlantic journey to the American facility most capable of unloading supertankers. It is called the Louisiana Offshore Oil Port (LOOP), and it lies 18 miles off of Port Fourchon in 110 feet of water. The tanker approaches one of the three floating buoys, which it attaches to via very large hoses that pump the oil into a 48-inch-wide undersea pipeline. It can take 36 to 48 hours to unload a 2-million-barrel supertanker.

The oil began its journey under the jurisdiction of a monarchy in Saudi Arabia; while at sea it was under the control of maritime and international law; and now here, in the Gulf of Mexico, as the black fuel moves through the pipeline to the Marathon Petroleum Company's eight underground salt caverns at Clovelly, Louisiana, which store some 50 million barrels, it enters for the first time the web of U.S. law and regulation. And quite a web it is. At this early stage of the process, the crude falls under the Hazardous Liquid Pipeline Safety Act, the Clean Air Act, the Outer Continental Shelf Lands Act, the National Environmental Policy Act, the Oil Pollution Act, and the Clean Water Act's National Pollutant Discharge Elimination System as well as its Spill Prevention Control and Countermeasure requirements, to name only some of the federal statutes involved. As the crude continues down the supply chain, it will encounter dozens of other state and federal controls and standards.

And yet these laws are only as effective as their credible enforcement. The responsibility to oversee more than 2 million miles of oil and gas pipelines, which lined up end to end would wrap the globe 88 times, falls to the U.S. Transportation Department's Office of Pipeline Safety, which employs 55 inspectors and hardly ever imposes a fine for violations, even for explosions and the death of pipeline workers. (And as we learned from the deep-well disaster in the Gulf of Mexico in the spring of 2010, the Office of Pipeline Safety is not the only federal agency performing too little oversight.)

In short, as soon as the oil reaches our waters and makes its way landward, it enters a different kind of conflict zone, where slow-moving bureaucratic battles between private corporations, government regulators, affected cities and towns, and environmental organizations have been fought continuously since at least the late 1960s. From its origin, the petroleum has acted to control and in many ways distort our foreign policy; now it enters domestic politics.

It will take roughly fourteen days for the vital liquid to make its way from LOOP via Capline, a major oil thoroughfare that parallels the Mississippi River, to, say, a refinery in southern Illinois, traveling at about four miles an hour, roughly walking speed. There it will spend another four to eight days being processed into gasoline, diesel, kerosene, jet fuel, and various other chemical distillates.

On its journey across the ocean, the shipment has already entered into the planning and formulas of various oil companies' supply-chain management systems, which constantly attempt to maximize profit by fine-tuning what quantities of oil will be processed into which products depending on daily changes in prices and supplies across the region and the country. Major refineries, the biggest of which handle more than half a million barrels a day, constantly adjust their product mix as the market changes, until finally the individual products are sent down the next set of pipelines to reach what in the industry is called "the rack"—that is, one of the fifteen hundred oil terminals spread throughout the country— triggering, as they travel, the attention of different environmental and taxing authorities with every state line they cross.

After spending less than a week on the rack, the gasoline or diesel is loaded into an oil truck, which typically drives no more than a couple of hours to reach the convenience stores and retail gas stations that dot our landscape. Depending on the sea route and refining time, it has taken anywhere from nine weeks at the shortest to more than four months at the longest for the fossilized zooplankton beneath the sands of Ghawar to travel halfway around the globe and end up in the tank of your automobile. Strangely enough, barring spills or accidents, no human eyes have seen it. And nor will they ever see it because it burns in engines and enters the atmosphere as invisible carbon dioxide and is not visible in the products it helps produce.

For most of the twentieth century, the process described here did not occur. We produced domestically the oil we used. And importing the relatively small quantities we sometimes needed to top off our domestic supply was far less complex as a technical and logistical matter—and given the remnants of colonialism, a much less tense geopolitical matter. That was another time. The decade that changed all that was the 1970s. It was then that the journey became the problem. Energy in America has never been the same since.

This book is about the problems, policies, and politics of energy in America, beginning with the crises of the 1970s, the varied responses to which continue to shape our current predicaments. It is about all the major forms of energy—oil, natural gas, coal, nuclear, hydro, solar, and wind—and how our government's attempts to control and decontrol, subsidize and command, legislate and repeal over the past four decades have produced a system and economy of energy production and consumption that fails to well serve our needs or those of our environment. The book is, then, in one sense a story of failure, but a story from which a great deal may be learned about how our democratic society might go about making better decisions for its energy future.

As this story unfolds—in all its complexity—many villains will come to the fore, including, no doubt, the Organization of Petroleum Exporting Countries (OPEC) cartel and in particular some of its members. At home, we have suffered from poor political leadership from both ends of Pennsylvania Avenue, where short-term political expediency has trumped sensible long-term policies. Key legislators far too frequently have elevated parochial regional interests over our national needs and have been overly responsive to the potential for short-term partisan gains. Our leaders have also commonly been seduced by sweet visions of technological silver bullets. Environmental organizations have sometimes insisted on unrealistic goals, now and then forged inapt alliances, and been used to further elite not-in-my-backyard (NIMBY) agendas. Energy companies have frequently underestimated risks and shifted to taxpayers those costs that the companies themselves should properly bear.

Amid all the currents and crosscurrents, however, one character plays a particularly central role: price. Although our government has enacted thousands of pages of energy legislation since the 1970s, it has never demanded that Americans pay a price that reflects the full costs of the energy they consume. Nothing that we did or might have done has had as much potential to be as efficacious as paying the true price. The contrast with tobacco, for example, where taxes have been used over time both to reduce its consumption and to help finance some of the costs it imposes on public budgets and society at large can hardly be more stark. This book makes clear that this failure, alongside many others, accounts for the state of affairs we face today.

1 A "New Economic Policy"

If you turned on your television set at 9:00 p.m. on Sunday evening, August 15, 1971, as millions of Americans did every week to follow the travails of the Cartwright family in the enormously popular Western *Bonanza,* you might have been surprised to see the somber visage of Richard M. Nixon, the thirty-seventh president of the United States. Nixon had come down from the mountaintop—down from Camp David, the presidential retreat nestled in the Catoctin Mountains of Maryland—to unveil a dramatic and far-reaching set of economic policy changes. His address, entitled "The Challenge of Peace," was rather short, given its dramatic content. By 9:20, *Bonanza* had resumed without a hitch, but our nation still lives with the aftermath of what Richard Nixon said and did that night.

The policies Nixon announced had been outlined to him a few weeks earlier by his charismatic Treasury secretary John Connally, and the president had already agreed to the essential elements. The details, however, were developed over the previous weekend by his economic advisers, who had been sequestered at Camp David since Friday afternoon. They had been cut off from any communication with the outside world to avoid leaks. The *New York Times* had published the heretofore secret *Pentagon Papers* only a month earlier, and Nixon was determined to keep a tight lid on his plans for the economy. Moreover, as Treasury Undersecretary Paul Volcker told the group, billions might be made with advance knowledge of what the president was about to do. The weekend at Camp David was more than a charade, but considerably less than a decision-making conclave.

Nixon began his speech by suggesting, as he so often had, that a Vietnam peace was at hand. He described his plan as essential "to create a new prosperity without war." He insisted the nation's economy was moving from war to peace—anticipating an end to Vietnam, which would not come for another four years. The "New Economic Policy" Nixon announced that evening was intended to address three pressing problems: rising unemployment, ongoing inflation, and the weakness of the dollar.

The measures to stimulate employment were relatively noncontroversial standard fare: a tax credit for business investments, elimination of an excise tax on automobiles, and a middle-class income tax cut. The wrinkle, if there was one, was Nixon's call to offset the fiscal costs of these tax cuts by reduced spending in order not to increase the federal government's deficit. But in classic Nixonian fashion, "reduced spending" was merely a feint. The president called for a postponement in spending that he knew Congress was never going to buy. He also expected Congress to add more tax cuts into his package. The combination of tax cuts and spending restraint he proposed, however, allowed him to describe these proposals as "reordering our budget priorities so as to concentrate more on achieving our goal of full employment."

The second prong of the New Economic Policy was far more dramatic, catching everyone who heard it by surprise. Even in retrospect, it seems positively unthinkable. A Democratic Congress, anxious to deflect the blame for inflation away from itself and onto the White House, had a year earlier handed Nixon an extraordinary grant of presidential power: to regulate wages and prices throughout the economy. The Democratic congressional leadership was convinced that this Republican president would never use this power. But John Connally had urged Nixon not to reject it, but instead to keep this authority in a closet in case he wanted to use it some day. On this summer Sunday, Nixon opened that closet. He told the country he was ordering a 90-day freeze on all wages and prices in the United States, "backed by government sanctions." He also announced that he was creating a new federal agency to regulate prices and allocate products, the Cost of Living Council, to maintain price stability after the freeze expired. This freeze came at a time when inflation was at 4 percent. Nixon had pulled out a sledgehammer, not a stick. Perhaps hoping that his audience had not fully comprehended what he had just said, Nixon had the gall to add: "Working together, we will break the back of inflation, and we will do it without the mandatory wage and price controls that crush economic and personal freedom." But his freeze on wages and prices was hardly voluntary, and price controls would last far longer than anyone expected.

Wage and price controls were a terribly blunt instrument, thought to be appropriate only in extraordinary wartime circumstances. That a Republican president would institute such governmental interference over what Nixon insisted was a peacetime economy was truly remarkable. Ironically, Nixon himself had worked in the Office of Price Administration during World War II, and having seen a price control bureaucracy up close, he knew how difficult such controls were to enforce. He also personally abhorred, for both practical and philosophical reasons, the idea of government trying directly to control wages and prices. So did virtually his entire team of economic advisers.

Nixon had often stated that he would never adopt this kind of an "incomes policy." But he had been tempted to embrace such controls not only by ongoing calls for them from key members of Congress, the public, and the press, but also by Arthur Burns, who had served as President Eisenhower's chairman of the Council of Economic Advisers and was in 1971 chairman of the Federal Reserve. George Shultz, Nixon's director of the Office of Management and Budget (who would subsequently succeed John Connally at the Treasury and serve as secretary of state under Ronald Reagan), had asked the critical question about wage and price controls that weekend at Camp David: "How do you stop it when you start?" An answer to that vexing question, such as it was, would have to wait. For energy prices, the wait would last a decade.

The clinching argument in favor of the wage-price freeze was political, not economic, and it had been made privately to the president by the politically savvy Connally, a man whom William Safire, Nixon's chief speechwriter, would later say Nixon had fallen in love with. The nation had first come to know Connally, a lifelong Democrat, when he was seriously wounded in November 1963 while riding as Texas governor in the motorcade during John F. Kennedy's assassination in Dallas. In February 1971, over some serious objections from Republican colleagues, Nixon brought Connally into his cabinet to serve as Treasury secretary and his key economic spokesman. Like Nixon, Connally was a lawyer and a politician, not an economist. Together, he and Nixon focused their attention on the 1972 election. As Herb Stein, chairman of Nixon's Council of Economic Advisers, gently put it later, Nixon "tended to worry exceedingly about his reelection prospects and so felt impelled to extreme measures to assure his reelection." Stein added, "He had a great longing for the dramatic gesture, for which he found a perfect supporter in John Connally."

The wage and price controls Nixon imposed on the nation that August night would subsequently prove to be very bad economic policy. By the time the freeze expired 90 days later, a large regulatory system had been created in an attempt to keep prices in check, and nearly two years later, in June 1973, Nixon ordered a repeat price freeze to keep inflation from raging. Prices had been held down to the point that companies lost more by marketing some products than by destroying them. By then, the public had witnessed scenes of chickens being drowned rather than grown for market and had become used to seeing empty grocery shelves. Looking back, Herb Stein confessed, "no one involved in the decision to impose controls foresaw how long they would last or how rigorous they would be." Except for energy, price controls were generally abolished in April 1974, when the president simply allowed his authority to impose them to expire.

The period of controls was followed by a large price explosion and ultimately contributed to the combination of high unemployment and high inflation—the dread "stagflation"— that would haunt the country later in the 1970s. Even when the controls were in place, they were frequently ineffective. From mid-1971, when the freeze was announced, until the end of 1974, when virtually all controls were eliminated, prices rose by an average annual rate of 6.6 percent, largely due to increases in energy prices. By the end of the decade, inflation was so out of control that short-term interest rates hit 21 percent. And, as we shall see, price controls on oil and gas remained in place long after other controls had expired. Nixon's New Economic Policy made fashioning sensible energy policy very difficult indeed.

A deep recession coupled with very tight monetary policy in the 1980s would ultimately wring out of the nation's economy most of the lingering aftermath of Nixon's wage and price controls—energy policy aside. But the wage–price freeze and the controls that followed, as bad as they proved to be economically, well served their purpose of helping Nixon politically. Congress cheered the wage–price freeze, the press applauded, the electorate was exuberant. On the Monday following the president's announcement, the Dow-Jones Average enjoyed the biggest one-day increase it had ever had. Polls showed that nearly three-fourths of the American people were glad to see Nixon institute the freeze. The public welcomed strong and decisive presidential action and was comforted by a faith that if their government would just adopt the right economic policies, they could always enjoy high employment with little or no inflation. It was, after all, a time when the planned economies, especially that of the Soviet Union, appeared to be doing rather well and when the directive "industrial policies" of Japan seemed to offer the best path to prosperity with price stability.

The economic havoc that price controls would ultimately wreak was pushed down the road, past Nixon's 1972 reelection. That election took place—just as Nixon and Connally planned—at a moment when the economy had strengthened and the controls seemed to be holding inflation in check.

The third and equally profound prong of the New Economic Policy that the president described that Sunday night unilaterally extinguished the world's international monetary arrangements. The headlines said that Nixon had "closed the gold window." What this meant was that the United States, which had long stood ready to exchange dollars for gold at the rate of $35 per ounce, would no longer routinely do so at any price. Exactly what abandoning the gold standard meant for the relationships among the values of the dollar and other currencies was not immediately clear, although it was apparent to anyone who understood the situation that the dollar would be devalued, especially vis-à-vis the Japanese yen and the German mark.

Nixon claimed, in a portion of the August 15 speech he had written himself in the wee hours that Sunday morning at Camp David, that his actions would "lay to rest the bugaboo of what is called devaluation." He insisted that it was necessary to protect the American people from the "international money speculators," who, he claimed, had "been waging an all-out war on the American dollar." As Watergate and its aftermath would subsequently reveal all too clearly, Nixon had a particularly wide streak of paranoia and a perverse ability to identify enemies or villains. So it is difficult to credit this claim—a difficulty magnified by the fact that if anyone had precipitated this crisis, if in fact it was a crisis, that person was a representative of the British national bank, who during the preceding week, according to the Treasury, had asked to convert $3 billion into gold.

Nevertheless, in a few short paragraphs President Nixon swept away the international monetary arrangements that had been agreed to nearly three decades earlier in July 1944 by delegates from 44 Allied nations at Bretton Woods, New Hampshire. The Bretton Woods agreements had set up a system that required each country to maintain the exchange rate of its currency within one percent of a specified value of gold. There was no consensus at Camp David about what would happen to the dollar after the president's announcement. Not that such a consensus would have mattered.

The underlying economic problem Nixon faced had been caused by large U.S. trade deficits, which had sent billions of dollars abroad in exchange for foreign goods. Our accumulated trade deficits had become so large that there were more than enough dollars in circulation around the world to empty all the gold from Fort Knox. The outflow of dollars in exchange for foreign goods had resulted in foreigners' exchanging an unprecedented number of dollars for gold. To address the trade issue directly, Nixon also announced that he was temporarily imposing a 10 percent tax—a tariff—on all goods imported into the United States.

After several months of hurried international diplomacy attempting to assuage hurt feelings, especially in European capitals and Japan, and to relieve the shock from Nixon's actions, a multilateral agreement about exchange rates, devaluing the dollar against other major world currencies, was reached in a meeting of the so-called Group of Ten nations in December 1971 at the Smithsonian Institution in Washington, D.C. In that agreement, there was some effort to maintain fixed currency exchange rates, albeit with a wider band of values, 2.25 percent. But this agreement was not to last: by the mid-1970s, the values of most of the currencies of the world would be allowed to float against one another. On December 18, 1971, when the "temporary" Smithsonian agreement was announced, the price of gold was said to be only $38 an ounce; a few weeks later it reached $44 an ounce. In 1972, it hit more than $70 an ounce and was still climbing.

Taken together, the dramatic policy changes Nixon announced on August 15, 1971—the wage and price freezes, the 10 percent import surcharge, and the revamping of worldwide currency arrangements—would profoundly affect U.S. energy policy. Herb Stein would later call the impact on energy the most "irksome legacy of the Nixon price-wage controls." Yet at Camp David that weekend there had been no mention of energy. Nor did Nixon mention it in his speech. Even John Connally, the former governor of Texas—then America's greatest oil-producing state—had managed to ignore it in the development of his plan for the president.

The fact was that in August 1971 the United States was complacent about energy policy. The president and his advisors, giving no particular thought to the effect of price controls on oil, expected the freeze to last only 90 days and only a short thawing period to follow. But that is not the way things turned out—especially for oil and gas: it took a 118-page Cost of Living Council report simply to describe the four phases of rules and regulations and some of the effects of price controls on petroleum products from the August 1971 freeze through 1974—nearly seven years before the controls would actually be lifted. The government's effort to supersede the market's allocation of oil and gas prices was byzantine and perverse. There were numerous tiers of prices for oil and petroleum products, including thirty-two different prices for natural gas. Buyers scrambled to qualify for the lowest possible prices, sellers for the highest. Producers turned their oil and gas spigots on or off depending on what prices they would receive. The Cost of Living Council's rules and regulations determined who got which supplies and at what price. Virtually all the prices were in some sense inappropriately low. Local product shortages became common. Price controls contributed to shortages of home heating oil; reduced production of domestic oil, coal, and natural gas; inspired hoarding and black market transactions; produced uncertainty throughout the energy industry and among energy users; and bestowed favorable and disfavorable treatment on categories of buyers and sellers completely unrelated to considerations of fairness or efficiency—to name just a few of its unfortunate and unintended consequences.

Energy policy complacency had settled in because for decades oil prices had been remarkably stable or even declining. Inflation in energy products had been lower than inflation generally. From the end of World War II through the late 1960s, the United States not only had produced the vast bulk of oil consumed domestically, but had also been able to play the role of emergency supplier to the rest of the free world. Large U.S. and British oil companies had long maintained control over vast oil reserves in the Middle East. Before the 1970s, our nation's policy struggle over oil had been over how to respond to abundant supplies.

But dramatic change was afoot. Due principally to strong economic growth and rising incomes, total world energy consumption more than tripled between 1949 and 1972 as Europe shifted from a coal-based economy to one fueled by cheap oil from the Middle East. Worldwide oil demand more than quintupled during this postwar period. But because of vast new discoveries and production of oil, especially in the Middle East, prices had remained remarkably stable. The public in both the United States and Europe regarded abundant, inexpensive oil as a birthright. Through the end of the 1960s, the problem, if there was one, was how to limit excess production. In the United States, stable prices were achieved largely through allocations of production and supply by state regulatory commissions, the most important among them the Texas Railroad Commission. Indeed, from the mid-1950s through the mid-1960s there was such a glut of oil that prices actually declined. To keep their revenues growing, the oil-producing nations of the Middle East were constantly increasing their oil output. Henry Kissinger later reflected, "In 1969, when I came to Washington, I remember a study on the energy problem that proceeded from the assumption that there would always be an energy surplus. It wasn't conceivable that there would be a shortage of energy."

In August 1971, a barrel of oil (42 gallons) in the United States cost only about $3, and even this price was about $1 higher than the worldwide market price due to a mandatory quota on imported oil imposed by the United States to encourage domestic exploration. Domestic U.S. oil production had peaked in 1970, but domestic demand continued to rise at a rate of about 5 percent a year. At the beginning of that decade, only about one-fifth of U.S. oil consumption was supplied by imports. A presidential task force convened by Nixon to study the import quota predicted in a 1970 report that by 1980 the United States would be importing only 27 percent of its oil and that prices would remain stable for the next decade. As it turned out, by 1980 we were importing about half of our oil and at a price more than ten times that of 1970. Today, imports account for nearly two-thirds of the oil we consume.

In the 1950s and 1960s, however, when domestic production was robust and protected from foreign competition by the quota as well as allocated by state regulatory systems, U.S. oil prices had remained remarkably stable: $2.90 a barrel in 1959 and $2.94 in 1968, 60 to 70 percent higher than the price of Middle East imports at East Coast ports. Due to increasing demand, domestic oil production was nearly 30 percent higher at the end of the 1960s than in the 1950s. Prices were considerably lower elsewhere, reaching, for example, a low of $1.20 to $1.30 per barrel for Iranian oil in Europe in 1967. Along with the U.S. economy, the European and Japanese economies were becoming ever more dependent on cheap foreign oil—mostly from the Middle East.

Our unwavering belief that plentiful, cheap oil was a permanent fact of life became abundantly clear in May 1969 when the shah of Iran visited the United States to attend President Dwight Eisenhower's funeral. While here, he offered to sell the United States one million barrels of oil a day for 10 years at a price of $1 per barrel. He also proposed that we establish a petroleum stockpile to protect us against any supply interruption. After short study, our nation's oil experts rejected the shah's offer. That October a second offer by the shah to sell us cheap oil was also turned away. In May 1971, when Saudi Arabia's King Faisal visited Nixon, oil was not even discussed. These decisions we would soon come to regret.

Nixon's price freeze and the energy price controls that followed it were imposed just as oil markets were undergoing a dramatic transformation. By the end of 1973, oil shortages would become obvious to every American. Insisting on keeping oil and gas prices from rising as surpluses turned into shortages would prove a fool's errand.

Dramatic upward pressure on worldwide prices resulting from upheaval in the Middle East would be felt in the United States first in the fall of 1973. But even before that Richard Nixon's New Economic Policy caused substantial disruptions. The original problem resulted from the president's timing and the freeze's inflexibility. His policy went into effect a couple of weeks before Labor Day, toward summer's end when gasoline prices always hit their seasonal peak. Home heating oil prices were then seasonably low. Refiners, stripped of their usual ability to adjust the relative prices of these products, in the fall of 1971 kept making gasoline long after they normally would have shifted to producing more heating oil. The unsurprising result was that when a harsh winter came in 1971–1972 to the Northeast, so did shortages of heating oil. Natural gas for heating was unfortunately also in short supply that winter. Many midwestern factories closed, as did some schools. Thermostats were lowered everywhere.

During the next decade, the U.S. government would struggle—with limited success—to keep U.S. energy prices low in the face of a rising tide of worldwide prices. Nixon's devaluation of the dollar ironically made this task considerably more difficult. Oil-producing nations lost purchasing power throughout the world as the value of the dollar fell because their oil prices were set in dollars. In September 1971, a month after Nixon's speech, at an OPEC meeting in Beirut, its member states increased oil prices by nearly 9 percent explicitly to compensate for the devaluation of the U.S. currency. And the value of the dollar continued to decline for several more years. By mid-1973, the dollar price of gold had risen to more than $90 an ounce; by the end of the decade, it exceeded $450. By the mid-1970s, the dollar had declined by more than 30 percent relative to the German mark and the Swiss franc and by more than 20 percent against other European currencies and the Japanese yen. Middle East oil producers, having embarked on ambitious long-range plans for spending their oil revenues, naturally reacted to offset this ongoing decline in their purchasing power by raising prices. To accomplish this goal they ultimately had to make sure that they—and not U.S. or British oil companies—had control over both the price and quantity of their oil output.

Thus, entirely unwittingly, Nixon's New Economic Policy exerted upward pressure on foreign oil prices and put a ceiling on domestic energy prices just as our nation was becoming more and more dependent on foreign oil. Regulators wrestled with the question of how to control prices and continue to encourage domestic production. In August 1973, they adopted a complicated two-tier pricing scheme, which allowed higher prices for "new" domestic oil than for "old" domestic oil. Refiners, wholesalers, and retailers all stood to profit if they could buy the old stuff, for old and new oil looked and worked the same. Imported oil had its own price based on worldwide market conditions. A small army of Internal Revenue Service (IRS) agents was deployed to enforce cumbersome rules that attempted to ensure a "fair" allocation of the cheap and more expensive oil. This is what passed for energy policy in the United States in the early 1970s.

Some version of the oil price controls instituted in August 1971 stayed in place until 1981, when they eventually expired—many years after other price controls had terminated. This policy was disastrous. Throughout that crucial period, the controls would keep U.S. prices generally below worldwide market prices, thereby inhibiting the downward pressure on demand that would otherwise have accompanied the dramatic worldwide price increases of the 1970s. Keeping the price of oil artificially low not only decreased incentives to conserve energy, but also diminished the likelihood of successfully developing and marketing alternative energy sources.

Unfortunately, Richard Nixon was far from the only U.S. politician to elevate short-term political expediency over sound economic and energy policies. The political struggle over whether to decontrol the price of oil would prove to be the most dominant and contentious issue of U.S. energy policy throughout the decade, inhibiting policymakers' ability to respond to a brand-new set of energy conditions. In August 1971, however, U.S. policymakers were oblivious to the fact that the long postwar era of abundant oil at low prices was ending.

As subsequent chapters detail, Richard Nixon's price controls proved surprisingly resilient. It would be September 1981 before price controls on oil were finally gone—more than a decade after they first came into place. Then they disappeared quietly when the authority to impose such controls simply expired. During their time, oil price controls decreased the effectiveness of conservation efforts and held down both domestic production of oil and domestic development of alternatives to oil. Perhaps even more important, they dominated the legislative efforts regarding energy policy and tied our political process in knots. Once Nixon had put them in place, finding a congressional majority to lift the controls became impossible.

Following his August 1971 speech, Richard Nixon returned his main focus to foreign-policy issues: ending the war in Vietnam, détente with the Soviet Union, and his forthcoming trip to China. Meanwhile, Britain's prime minister Edward Heath was making a major foreign-policy decision of his own. Despite some misgivings, his conservative government was in the process of fulfilling a promise made in 1968 by his predecessor, the Labour Party prime minister Harold Wilson, to withdraw all British troops from the Persian Gulf by the end of 1971.

Since early in the nineteenth century, Britain had played a key role in securing the peace and prosperity of the gulf region. The gulf, after all, was a very important pathway to India. And for much of this time, many observers had described the gulf states, including Iraq, as part of Britain's informal empire. Once oil was discovered in the Middle East, Britain had important commercial as well as political interests to protect. Following World War II, however, Britain decolonized, and by the late 1960s its own economic problems had led it to diminish significantly its military presence around the world. When Harold Wilson announced his plan to withdraw all British troops from the Middle East, the sheiks of the region were shocked. They begged the British not to leave and even offered to fund the troops' continuing presence.

It is difficult even now to know whether Britain's decision to leave the gulf was a mistaken short-sighted cost–benefit calculation or the region's growing nationalism had properly persuaded the British government that maintaining any military presence there was simply too risky to continue. In either event, when Britain abandoned the Persian Gulf in November 1971, the geopolitics of oil changed. A great power vacuum emerged east of the Suez, a vacuum that the Soviet Union, Iraq, Libya, and Iran were anxious to fill.

In the Middle East, a new future was under way. Richard Nixon, who had longstanding and cordial relations with the shah, favored Iran as the most important new regional power. Always seeking a power balance, he wanted a strong counterpoint to Iraq's ambitions, which were being promoted by the Soviet Union's ongoing shipments of arms. In 1971, these developments in the gulf region were unsettling but not alarming. Just two years later, in October 1973, the alarm bells would go off.

2 Losing Control over Oil

If our government and the public at large had been lulled into apathy during the 1950s and 1960s by plentiful oil at cheap and stable prices, the illusion that such conditions would last forever was shattered on October 20, 1973. It was on that day that the Arab oil-producing states declared a total embargo of oil shipments to the United States.

Soviet-supplied Egypt and Syria had launched surprise Yom Kippur attacks on Israel two weeks earlier. For the next ten days, Israel pleaded with the United States for arms and supplies. The Nixon administration finally responded with a shipment intended to arrive under the cover of darkness, but high winds at a fuel stop in the Azores delayed its arrival until daylight on October 17. That same day, with the airlift now public, Secretary of State Henry Kissinger boldly told a meeting of the Washington Special Action Group, the interagency body responsible for international crises that Kissinger chaired, "We don't expect an oil cut-off." On October 19, the president proceeded to request from Congress \$2.2 billion to pay for the arms and for additional aid to Israel.

The Arab states wasted no time in responding. They announced an embargo the next day, setting off the first great oil shock of the post–World War II era.

Americans would soon be lining up at gas stations to fill their tanks. Schools and businesses would close due to shortages of heating oil. The posted price of oil would rise from \$3 a barrel on October 16 to \$11.75—nearly a fourfold increase—by Christmas. Spot prices for Iranian oil would hit \$16 to \$17 a barrel that December. The take of the oil-producing nations would rise from about 90 cents a barrel in 1969 to more than \$10 a barrel in 1974. By 1980, the price of oil would reach \$39 a barrel. The greatest peacetime transfer of wealth the world had ever seen was under way, and this time the wealth was heading away from the United States, Europe, and Japan to the treasuries of the oil-producing nations, mostly in the Middle East.

For more than a decade, like an adolescent refusing to get out of bed, America's leaders had hit the snooze button every time the oil supply alarm went off: September 1960, Iran, Iraq, Kuwait, Saudi Arabia, and Venezuela form OPEC to protect themselves against declining oil prices—snooze; September 1969, Colonel Muammar al-Qaddafi seizes power in Libya in a military coup and within a year throws out U.S. and British forces, nationalizes foreign banks, and begins to take control of Libyan oil—snooze; March 1971, U.S. domestic oil production peaks and begins to decline—snooze; July 4, 1971, President Richard M. Nixon delivers the "first comprehensive energy message" to Congress, calling for a nuclear breeder reactor to be built by 1980, which he calls the nation's best hope for economical clean energy—snooze; October 1971, OPEC countries demand "equity participation" in the oil companies operating in the Middle East—snooze; December 1971, Britain withdraws all of its remaining troops from the Middle East—snooze.

With the actions of the Arab states in October 1973, however, the time had unmistakably come for this nation to wake up, wipe the sleep from its collective eyes, and tackle the energy issue.

The oil embargo was the culmination of a major shift in control over Middle East oil production away from U.S. and European oil companies, who for decades had controlled both output and prices, to the nations in whose lands the oil was located. By the end of 1972—at the latest—this shift of power over oil should have been apparent to policymakers in the oil-consuming nations, but complacency somehow persisted.

Western interests in Middle Eastern oil originated in the years following World War I in the international context of the British Empire. British political control allowed oil companies to obtain drilling rights to areas as large as Texas for as little as $30,000. In the late 1940s, the American and British governments came together to sign the Anglo-American Agreement on Petroleum, warning the newly established Middle Eastern governments that the United States and Great Britain would guarantee "all valid concession contracts and acquired [drilling] rights" with the full force of their governments.

In the early 1950s, Iran's prime minister Mohammed Mossadegh—who had been named *Time* magazine's man of the year in 1951—halted that nation's oil production in an effort to nationalize it and to exert greater control over world oil markets. Given the surfeit of worldwide oil supplies at the time, however, his gambit failed. In 1953, when it seemed Mossadegh might lose Iran to the Soviets, the United States and Great Britain backed up their words with actions when they engineered a coup led by the U.S. Central Intelligence Agency (CIA) to replace Mossadegh with Mohammed Reza Pahlevi—the new shah of Iran—a tyrannical but pro-Western leader who assured an inexpensive flow of oil to the United States and Europe.

Drilling for oil in the Middle East had begun in earnest during the 1950s when America's big five oil companies—Exxon, Mobil, Texaco, Gulf, and Standard Oil of California—joined together with British Petroleum and Shell to provide Middle East oil for rapidly growing postwar markets in Europe and Japan. For a long time, these so-called Seven Sisters controlled both production levels and prices and in the process earned enormous profits. While the United States was consuming the oil being produced at home, American and European oil companies were sending their Middle East oil to Europe and Japan. By the mid-1950s, Middle East oil was supplying nearly one-third of the energy consumed in western Europe. By 1970, refinery output in western Europe exceeded that of the United States, and Middle East oil supplied half of Europe's energy.

During the 1950s, the countries of the Middle East demanded a 50–50 split of oil revenues from the big oil companies. To help accommodate this request, the U.S. government shifted the increased costs away from big oil onto U.S. taxpayers by having the oil-producing countries' shares largely take the shape of income taxes rather than royalties. Under longstanding U.S. tax law, this shift allowed the countries' increased revenues to offset the oil companies' U.S. taxes dollar for dollar rather than paying increased royalties that simply would normally have been treated as the additional cost of doing business. So, for example, in 1950 the consortium of U.S. oil companies in Saudi Arabia (Aramco) paid $50 million in U.S. taxes and $66 million in royalties to Saudi Arabia; in 1951, when the Saudis increased their take to $100 million, U.S. corporate taxes were reduced to $6 million. The only conceivable justification for this policy—other than largesse to the big oil companies—was if it was considered an aspect of the U.S. effort to help Europe and Japan regain their industrial footing after World War II.

In any event, this arrangement, coupled with U.S. oil import quotas, was far more costly than anyone perceived at the time. U.S. domestic production was stimulated, and domestic reserves were squandered when oil from the Middle East was least expensive and most reliably available. Producing oil in Texas in the 1950s cost about $1 a barrel; in Saudi Arabia, a barrel of oil cost as little as five cents, ranging up to about ten cents in Iran and the other gulf states. In the decade and a half following the 1950 completion of the Trans-Arabian Pipeline, experts estimate that the five large U.S. oil companies took more wealth out of the Middle East than the British took out of their colonial empire in the entire nineteenth century.

In the 1960s, the big oil companies' assertiveness began butting heads with the oil-producing nations' newfound determination to gain control over their oil reserves. In the summer of 1960, faced with a temporary decline in the world's demand for oil and increased supplies by the Russians to Europe, Exxon unilaterally—without consulting any of the oil-producing states—reduced the posted price of its Persian Gulf oil. The revenues of the oil-producing nations were then based on posted prices, not on actual market prices, and in the subsequent year Exxon's move cost Saudi Arabia alone more than $30 million in lost revenues. A few months after Exxon acted, five nations—Iran, Iraq, Kuwait, Saudi Arabia, and Venezuela—joined together to form their own oil-production cartel: OPEC. Over time, 11 more oil-producing nations would join OPEC.

Despite this move, the oil companies, which were in their heyday, remained unfazed: "We don't recognize this so-called OPEC," said one of their leaders. And they didn't have to. Middle East oil was an important source of high profits for the oil companies and low costs for Western consumers, but even as late as 1969 the United States remained capable of satisfying its petroleum demand through domestic production. Had the producer countries refused to cooperate with the oil companies, the companies in the 1960s could have simply increased domestic production and cut the OPEC nations out of the oil market altogether. To be sure, this action would have hurt the companies' profitability and would not have been sustainable over the long run. In the short-to-medium term, however,

Western demand could continue to be met at relatively little additional cost to the average consumer, but the Middle Eastern governments would be deprived of their primary source of revenue. The companies adamantly refused to restore the 1960 cutback in posted prices.

During this period, the geopolitical context of Middle Eastern oil politics was changing. The tinderbox of the Middle East became central to the Cold War struggle between the United States and the Soviet Union. For the United States, relationships with the shah of Iran, the Kingdom of Saudi Arabia, and Israel were critical. The Soviets' major clients were Iraq, Syria, and Egypt. Each of the superpowers held an ace in the hole: its own ample reserves of oil. These reserves permitted both the United States and the Soviet Union not only to weather disruptions in the supply of Middle East oil that occurred occasionally during the 1950s and 1960s, but also to serve as an oil exporter when supplies elsewhere got tight.

In 1967, oil shortages hit Europe when the Suez Canal, which was then owned by the British and French, was nationalized and closed by Egypt's president Gamel Abdul Nasser. The oil companies made up this shortfall by increasing production of low-sulfur light crude in Libya, a country then ruled by the pro-Western but ineffectual King Idris. Libya had for some time been a minor oil-producing nation. In 1962, for example, its production totaled just more than 180,000 barrels a day. During the 1960s, however, Europe demanded more and more Libyan oil. It was attractive both because of its relatively low sulfur content, which reduced its environmental damage, and the cost and reliability advantages of shipping oil the short distance across the Mediterranean Sea rather than from the more distant Persian Gulf. By 1970, Libya's daily production had reached more than 3.3 million barrels, nearly 20 times its level less than a decade earlier. This output meant Libya was ranked as the fourth-largest producer among the OPEC nations, just behind Saudi Arabia, Iran, and Venezuela.

Proximity to Europe and low-sulfur content were not the only differences between Libyan and Persian Gulf oil production. Unlike in the gulf countries, Libya's oil concessions had not been negotiated with the Seven Sisters of big oil. Libya instead had granted most of its drilling rights to independent companies such as Getty, Occidental Petroleum, Continental, Marathon, and the Bunker Hunt Oil Company. In 1970, independents produced 55 percent of Libyan oil, compared to their overall average of just 15 percent in OPEC countries (which meant that the independents' presence was negligible in some of the more important oil countries, including Saudi Arabia and Iran). Crucially, some of these independent oil companies were entirely dependent on Libyan oil for their sales in Europe.

During the early 1960s, when Libyan production was just beginning to ramp up, there was a glut of oil on European markets, and because of U.S. oil import quotas the U.S. market was not an available alternative. This confluence of factors led to vigorous price competition that produced a decline in the price of Middle East oil throughout most of the 1960s. In Rotterdam, Netherlands, for example, the wholesale price of gasoline was seven cents a gallon in the spring of 1960 and five cents a gallon in the spring of 1970, notwithstanding overall price inflation during this period. The largest U.S. oil companies were somewhat insulated from this downward price spiral because of their ownership of refineries and retail gasoline stations and because of U.S. protectionism, but even they felt pressured by the competition.

The other oil-producing Middle East nations also felt the effects of the rising output and sales of Libyan oil throughout the 1960s. Despite the industrial world's ever-increasing appetite for oil, the only way for these countries to respond to the growth of Libyan production and its downward pressures on prices was to decrease their own output. But the idea of decreased output along with downward worldwide price pressures caused great consternation in the capitals of oil-producing nations, especially Iran and Saudi Arabia, both of which had embarked on ambitious domestic spending programs. Their only options were to try to grab a greater share of overall oil revenues and to drive output down to a level where price increases became inevitable.

But the situation with Libyan oil was about to get much more complicated. On September 1, 1969, a military coup overthrew Libya's King Idris, and the coup's 27-year-old leader, Muammar al-Qaddafi, took over. Qaddafi was an extreme Muslim ascetic and Arab nationalist whose anticommunism was unfortunately equaled by his hatred of the West. He soon successfully expelled all American and British troops from their large Libyan military air bases. Next he demanded a substantial increase in the price of Libyan oil—at a time when about 30 percent of Europe's oil imports were coming from his country. The executives of the major oil companies grossly underestimated Qaddafi's determination, political skill, and economic power, and they essentially ignored him.

In January 1970—paying no attention to the abundance of oil on world markets that throughout the 1960s had protected the oil companies from the oil-producing nations—Qaddafi demanded an increase of 40 cents a barrel in the posted price of Libyan oil. The companies refused. Then, rather than attacking one of the major oil companies who were calling the shots for the industry, Qaddafi moved in May 1970 to enforce his price increase by cutting the production of one of the smaller independent companies, Occidental Petroleum, by 300,000 barrels a day. The cut reduced Occidental's output by more than one-third. Unlike the majors, which had sources of oil spread throughout the Middle East and elsewhere, Occidental depended totally on Libyan oil to supply its refineries in Europe. It owned no oil from elsewhere to make up the shortfall. Qaddafi knew that.

Occidental's CEO, Armand Hammer, who had made his fortune trading with the Soviets, attempted to ease the pressure on his company and thwart Qaddafi by asking Exxon to make up Occidental's shortfall by selling to it at cost the oil it needed. On July 10, 1970, in a New York City meeting with Hammer, Kenneth Jamison, Exxon's chairman, refused.

Hammer scrambled unsuccessfully to obtain elsewhere the oil he needed, and he held out briefly against Qaddafi. In early September 1970, however, Hammer shocked the rest of the oil industry by agreeing to a 30 cents a barrel increase in the posted price of Libyan crude, cutting his company's profits in half. Within the next month or so, every oil company operating in Libya followed suit, except for Shell, which Qaddafi then kicked out of the country.

The industry's capitulation to Qaddafi's demands ended the 50–50 profit split between the companies and the oil-producing countries that had prevailed since the 1950s. Libya upped its share of profits from 50 to 58 percent. More important, power over Middle East oil had shifted unmistakably and irrevocably away from the companies to the oil-producing nations. Qaddafi, emboldened by his success, asked for and got another 50 cents a barrel increase in early 1971.

Following Qaddafi's lead, Abu Dhabi, Iran, Iraq, Kuwait, Qator, and Saudi Arabia in December 1970 formed a committee to seek higher prices for their own oil. Representatives of 15 companies, having also learned their lesson from Libya (albeit a bit too late), got together in New York on January 13, 1971, and agreed that if Qaddafi cut anyone's production, the others would replace the decrease at cost. The U.S. State Department, however, after discussions with the shah of Iran, who promised five more years of stable prices, encouraged the companies to negotiate prices separately with the gulf states and then with Libya. But the separate negotiations gave an advantage to the oil-producing nations, not the companies, and the shah did not keep his promise. Each time the companies agreed to a price increase with one country, a different country would ask for a higher price, producing an ongoing upward price spiral.

In February 1971, negotiations in Tehran with the gulf states led to a 40 cents a barrel increase in the price of oil. The following April an agreement with Libya reached in Tripoli provided for a somewhat greater increase. All the producing nations promised to maintain these prices for five years—a promise Henry Kissinger would later describe as holding "a world record in the scale and speed of its violation." Meeting in Beirut that fall, OPEC insisted on and received another price increase of about 9 percent, which it claimed to be compensation for the dollar devaluation of August 1971.

OPEC also more ominously demanded "equity participation" in the oil companies. This was the beginning of the end of the longstanding oil concessions and a turning point as the oil-producing nations established control over the oil in their lands. Negotiations over oil prices between the companies and the countries would soon become a relic of the past. Prices henceforth would be set by the Middle East countries, which over time obtained ownership and control of their oil reserves.

The structural factors that had made OPEC ineffective for the first decade of its existence had changed. In March 1971, Texas oil producers announced that they had reached peak oil production and that their output would begin to decline. By 1973, the United States was consuming 6.3 million barrels of oil per day more than it produced, Japan was consuming 5 million more than it produced, and Europe was consuming 13.1 million more than it produced. The Middle East was exporting 20.2 million barrels every day. Middle Eastern petroleum reserves were estimated to exceed 316 billion barrels, whereas those in every other region of the world had fallen to less than 50 billion. The Middle East governments were now in control. The oil companies' new role was primarily as their sales agents and managers.

As this dramatic change swept over the Middle Eastern oil market, the Nixon administration sought to address it largely through the foreign-policy framework of the Nixon Doctrine, the administration's general approach to relations with the "third world." The Nixon Doctrine was based on the premise that, unlike during the immediate post–World War II years, the United States was no longer capable of acting on its own to protect American interests in every region of the world. Instead, the United States would work to promote American interests through independent but like-minded allies, each ultimately responsible for its own defense but supported by the United States through military and economic aid.

In applying the Nixon Doctrine to the Middle East, the administration focused on establishing Saudi Arabia and Iran as the "two pillars" of anti-Soviet stability and free-flowing oil. In that effort, it endeavored to harmonize U.S. interests with those of Saudi Arabia and Iran. Specifically, we enabled the oil-producing nations to "invest in U.S. 'downstream' oil operations such as refineries or tanker construction" as a means of encouraging "economic interdependence among oil producing and consuming nations." The United States also tried to woo these producers through military aid. In 1972, extending a policy that had begun in the 1960s, Nixon granted the shah of Iran access to the most advanced U.S. weapons, including F-14 and F-15 aircraft. And in May 1973, Henry Kissinger approved the sale of F-4 Phantom fighter jets to Saudi Arabia to "significantly strengthen our relationship with Saudi Arabia at a time when it is beginning to emerge as a significant international actor because of its oil wealth."

The United States was working simultaneously to enhance coordination among oil-consuming nations to reduce the potential harm of any oil cutoff and to deter the OPEC governments from imposing one. As James Akins wrote in the April 1973 issue of *Foreign Affairs* in a prescient—but generally ignored—article entitled "The Oil Crisis: This Time the Wolf Is at the Door," because the United States would soon be confronting a permanent sellers' market, an oil embargo would harm the United States and the West in general far more than it would hurt the oil-producing countries "unless we could assume complete Western and Japanese solidarity, including a complete blocking of Arab bank accounts and an effective blocking of deliveries of essential supplies to the Arabs by the Communist countries—in other words, something close to a war embargo."

The primary vehicle for this attempt at coordination by the oil consumers was the Organization for Economic Cooperation and

Development (OECD), then a group of 23 of the world's most developed countries. At U.S. impetus, in March 1973 a working group of the OECD Oil Committee published a report outlining a number of areas for potential cooperation, including resource sharing in the event of an oil cutoff, common taxation and import policies for oil, common measures to reduce domestic petroleum consumption, and joint research and development (R&D) into alternative forms of energy. Although different forms of cooperation had more appeal to different countries—for instance, emergency sharing schemes were far less appealing to the Americans than to the Europeans, and coordinated negotiations with OPEC were less appealing to the Europeans—it was "generally agreed . . . [that the] oil supply problems of today and tomorrow are problems impacting upon each member country. . . . Their satisfactory resolution can best be achieved through cooperative action taken by the consumers."

The two threads of American strategy to secure gulf oil interests were tied together and codified in the summer of 1973 through the work of National Security Study Memorandum (NSSM) 174, a comprehensive U.S. government study titled *National Security and U.S. Energy Policy*. Although the report did not add many new strategies to those the United States had already been pursuing, it synthesized the administration's diverse policies into one guiding document and expanded on the details of each of those policies.

On the "two pillars strategy," NSSM 174 focused primarily on continued strengthening of the U.S. bilateral relationship with Saudi Arabia. (Because Iran had essentially been a U.S. client state since the 1953 coup, less attention was placed on strengthening rather than simply maintaining that relationship.) NSSM 174 reiterated that "the essence of [our] approach is to engage Saudi national security concerns and appeal to Saudi economic self-interests, thereby creating powerful vested interests in continued cooperation with the U.S." The specific policies described included "a greater willingness by the U.S. to meet additional Saudi requests for sophisticated [military] items"; a high-level investment mission to advise the Saudis on investing oil revenues in U.S. industries; and, accompanying the previous two U.S. gestures, "a general U.S.–Saudi understanding with regard to future rates of production to help meet U.S. and non-communist world needs." Regarding consumer sharing, the report reiterated the proposals for OECD cooperation and reported that the OECD had established an ad hoc working group to consider their implementation. That group was to make its report on October 15, 1973.

These U.S. policies ultimately failed, however, because they did not adequately take into account the rapidly developing connections between oil and another intractable Middle Eastern foreign-policy conundrum: the Arab–Israeli conflict. Throughout the early 1970s, Egyptian president Anwar Sadat staged an ultimately successful effort to persuade the gulf states to use oil as a political weapon in Egypt's conflict with Israel. In 1972, Sadat boldly diminished Egypt's ties with the Soviet Union. Both of his maneuvers were designed to pressure the United States into moderating its support for Israel. Sadat mistakenly reasoned that the United States would have less reason to support Israel if Egypt were no longer a Soviet client and would have a new reason not to support Israel if our now crucial Middle East oil supply was threatened. Sadat hoped the United States would pressure Israel into concessions in negotiations, but by 1973 he decided that war was the only option.

Early in 1973, Saudi Arabia's King Faisal ibn Abdul Aziz, whom Sadat had told about his audacious plan to attack Israel, began warning the Nixon administration that Saudi Arabia and other Arab nations would use the oil weapon if Washington failed to force Israel to return Arab land it had been occupying since the 1967 Arab–Israeli War. Throughout 1973, King Faisal—supposedly a great friend to the United States—threatened to halt oil shipments to America if the latter's policy failed to become more even-handed between Israel and the Arab nations. His oil minister, the soft-spoken, Harvard-educated Ahmed Zaki Yamani, told the same thing to U.S. cabinet members in April. Faisal himself also said as much to oil company executives in May and again publicly in September in interviews with both *Newsweek* and the BBC. But Richard Nixon and his energy policy advisors refused to take these threats seriously.

This failure resulted from three major shortcomings in the Americans' strategic vision. First, the Nixon administration continued to view the geopolitics of the Middle East predominantly through a Cold War lens, concluding in NSSM 174 that the Saudis would be unlikely to use oil as a weapon because they would prefer "to retain their ties to the West as a guarantee of their own security and independence." The administration failed to understand that by this time neither Egypt nor Saudi Arabia viewed the Arab–Israeli conflict as an element of the Cold War.

Second, the Nixon administration misperceived Middle East regional dynamics as occurring in two distinct spheres: a Mediterranean sphere and a Persian Gulf sphere. "Although the Persian Gulf Arab states are good Arabs and, therefore, enemies of Israel," NSSM 174 stated, "they do not have the same economic, military and political involvement in the Arab–Israeli dispute as the Eastern Mediterranean Arab States."

Finally, the Nixon administration's analysis of the Middle East focused on the interests and personalities of local elites, ignoring pressures on them from their societies. Even in moderate Arab countries such as Kuwait, the press lamented "the great American betrayal" of its Persian Gulf friends incited by "American Jews and their lobbyists" and called for retaliation through a "cut in the flow of oil." Arab leaders' options were limited. In an October 16 meeting with Henry Kissinger, the Saudi foreign minister said that even though the Saudis did not want conflict with the United States, "no Arab leader" could accept the U.S. support for Israel "or his regime would collapse." Kissinger failed to get the message, observing that "we are not used to seeing moderate Arabs" and that he had "expected more difficulty." Four days later the oil embargo began.

Although the prospect may seem far-fetched in retrospect, it was not obvious why the United States did not intervene militarily to protect its oil interests when the embargo struck. After all, our dependence on Middle Eastern oil had grown out of British imperialism in the Middle East, and, as noted earlier, in 1945 the British and U.S. governments had essentially guaranteed that military force would protect Middle Eastern oil interests. Indeed, on September 1, 1973, just as Libya's Muammer al-Qaddafi was nationalizing 51 percent of the oil company concessions that he had not already acquired, President Nixon boldly harkened back to the 1953 Anglo-American-sponsored coup in Iran: "Oil without a market, as Mr. Mossadegh learned many, many years ago, does not do a country much good."

In fact, the question of military intervention was very much on the table as Nixon's policy advisors considered the U.S. response to the oil embargo. As early as October 10, after the Arab–Israeli War began, but before the embargo was initiated, Defense Secretary James Schlesinger warned Henry Kissinger in a phone conversation that "we may be faced with the choice . . . between support of Israel [and support of] Saudi Arabia, and if interests in the Middle East are at risk the choice between occupation or watching them go down the drain."

In the months following imposition of the embargo, American discussion of military intervention in the Middle East became even more widespread. The report *Economic Interdependence and the Nation's Future*, originating at a seminar hosted by the United States Navy, argued that the United States "must be able to defend itself against those who would seek unilateral advantage from the system of international [economic] cooperation. In other words, the cooperative system works only if the U.S. is prepared to retaliate against those who violate its rules." At a January 7, 1974, staff meeting Henry Kissinger inquired, "If we wanted to intervene in the Persian Gulf today, what would be our capabilities? Is it possible to get an analysis of that question?" Kissinger's advisors responded, "Yes, we can do it," though they said it would "depend on the precise circumstances" of the intervention the United States wanted to launch.

Kissinger ultimately decided against military intervention. Given the widespread dissatisfaction with the Vietnam War, it was clear that the American people had no taste for another military conflict. And as the Watergate scandal came into full bloom, the Nixon administration had neither the credibility nor the will to try to persuade the public otherwise. In earlier times, Kissinger lamented at a staff meeting, "The idea that a Bedouin kingdom could hold up Western Europe would have been absolutely inconceivable. [The Westerners] would have landed, they would have divided up the oil fields, and they would have solved the problem." But realistically, he sighed, in the postimperialist world of 1973 and with a U.S. government in disarray, "We can't do it . . . that obviously we cannot do." In December 1974, Kissinger made this view public by telling *Newsweek*, "I don't think [protecting oil interests] would be a cause for military action."

The 1970s were not the 1950s. Vietnam had made Americans skeptical about the deployment of military power, and now we were trapped by small Middle East Arab nations' ability to try to wreak havoc on our lives simply by turning off the oil spigot.

To say our government was in disarray when the oil embargo hit in October 1973 is a gross understatement. That same weekend, in what became known as the Saturday Night Massacre, Richard Nixon insisted on firing Harvard law professor Archibald Cox, who had been appointed as an independent special prosecutor to investigate the June 1972 Watergate break-in of Democratic National Committee offices and its subsequent cover-up by the White House. Cox's transgression was his refusal to accept Nixon's proposal for only a limited review of taped White House conversations. Both Attorney General Eliot Richardson and his deputy William Ruckelshaus refused to fire Cox and resigned in protest. Robert Bork, who as solicitor general then became acting head of the Justice Department, ultimately fired Cox. Many members of Congress were incensed and introduced bills of impeachment against Nixon. A few weeks later, in a November 17 press conference, Nixon famously insisted, "I'm not a crook." His presidency was on its last legs.

Richard Nixon was preoccupied with trying to hold on to his office, and he thought—wrongly as it turned out—that lifting the oil embargo might be the kind of feat of presidential leadership that would save his presidency. So he became obsessed with the embargo.

The embargo's impact in terms of reduced oil imports was actually quite limited because the oil companies were able to shift nonembargoed, non-Arab oil to the United States while sending their embargoed Arab oil elsewhere. Glib pronouncements became commonplace. Mike Ameen, a vice president of Aramco (the consortium of U.S. companies operating in Saudi Arabia), quipped, "There will be more sex during the day and more blankets at night." Melvin Laird, then a White House counselor, advised, "I'd buy a sweater."

In November 1973, Nixon made a major energy policy address. He ordered public utilities to stop shifting from coal-based to oil-based electricity generation, called on homeowners to lower their thermostats, and urged a maximum speed limit of 50 miles per hour. Congress then finally approved the Alaska oil pipeline and provided $140 million for research into more fuel-efficient vehicles. The Interior Department announced an experimental program to extract oil from shale on federal lands in Colorado, Wyoming, and Utah.

Although oil generally remained plentiful, its worldwide price quadrupled, and the economic effects of the dramatic oil price rise shocked the U.S. economy. The stock market in November 1973 had its two worst consecutive days since October 1929. Domestic airlines eliminated 5 percent of their flights, canceling hundreds of departures. Costs for housing materials jumped 20 percent. Electricity rates in New York City went up 5.4 percent and in the suburbs by more than 25 percent. Consumers rushed to buy wood-burning stoves, electric saws, and long underwear. Students at Boston's New England School of Art used a plastic tent warmed by body heat to keep their nude models warm. Sears, Roebuck & Co. eliminated all Christmas lighting. (The light makers protested in a full page ad in the *New York Times*, but to no avail.) Oregon imposed strict conservation measures. General Motors (GM) closed more than a dozen plants for a week.

By December, the embargo was imposing a great psychological toll in the United States. Truckers, enraged by the combination of higher fuel prices and lower speed limits, blocked highways all along the east coast and in the northern Midwest. They also parked their eighteen-wheelers in front of gas pumps, blocking motorists from using them. One angry motorist threatened to kill a station attendant who refused to sell him gas one Saturday night. A gas station owner shot the pumps of a competitor who refused to follow the president's request to close. The winter in the Northeast was marked by diminished supplies of home heating oil. The Nixon administration ordered Americans to lower their thermostats 6 to 10 degrees, started rationing home heating oil, and threatened to ration gasoline.

John Love, a former Colorado governor—who in the summer of 1973, when named head of the newly created White House energy policy office, confessed, "I am no energy expert," and then proved it—was replaced that December by William E. Simon, deputy secretary of the Treasury. Nixon described Simon as his new "energy czar" with the power to allocate oil supplies. In his first few weeks, Simon published a new set of allocation rules that he later described as a "monumental blooper." The new rules would have forced refineries to cut gasoline production by one-fourth. After a very bad day of headlines and a large stock market decline, that order was replaced with one reducing gasoline production by only 5 percent. Simon's tenure was marked by many similar shortcomings. This kind of confusion and attendant shortages were doubtlessly due more to the government's efforts to control prices than to the actual disruptions in oil supplies resulting from the embargo, but in combination these efforts and events produced a dreadful situation.

By January 1974, many schools were shut for weeks at a time, and motorists were waiting in long lines to top off their gas tanks. When a driver in Pittsburgh filled his tank with only 11 cents worth of gas and tried to use a credit card to pay, the station attendant spat in his face.

Don Jacobson, who ran an Amoco station in Miami, said, "These people are like animals foraging for food. If you can't sell them gas, they'll threaten to beat you up, wreck your station, run over you with their car." By February, many states and cities had adopted every-other-day gas rationing in which people with even-number license plates could buy gas only on even-number days, those with odd-number license plates only on odd ones. Many states also required drivers to purchase at least half a tank. An eight-day truckers' strike that month disrupted transportation of goods and services; led to some factory and mine closures; resulted in scarce supplies of food, gasoline, and other products; and caused at least 100,000 workers some unemployment.

The December 1973 OPEC oil price increase from $5.12 to $11.65 a barrel raised the oil bill for the United States, Canada, western Europe, and Japan by $40 billion a year. Henry Kissinger continued to urge coordination and cooperation by the oil-consuming nations, but to little avail, even as he suggested that the higher oil prices might bring on a "worldwide depression," threatening economic growth and causing inflation. By early 1974, the American public had lost patience and hope. They were looking for villains and blamed the Nixon administration, the Arabs, and the oil companies for their troubles.

Senator Henry "Scoop" Jackson, who was planning a presidential run, decided to focus his attacks on "big oil." In February 1974, he brought executives from the seven largest oil companies before the Senate's Permanent Subcommittee on Investigations. In the same hearing room in the glare of the same kleig lights where the Watergate hearings had captured the attention of the American public the preceding summer, senators excoriated the executives for their inadequate supplies and their excessive profits. The oil executives soon became the nation's scapegoats when their companies reported record profits for the first quarter of 1974.

On March 16, 1974, following a U.S.-negotiated cease-fire between Israel and the Arabs, OPEC lifted the embargo. *Time* magazine reported a month later that Americans had returned to their "heavy feet," traveling more, routinely exceeding the 55-mile-per-hour speed limit, raising their thermostats, and pushing conservation to the recesses of their hearts and minds. Alan Greenspan, who would soon become chairman of the Council of Economic Advisers, along with other leading economists, predicted that the price of oil would fall $1 to $2 a barrel by the summer. One Nixon administration economist predicted a $4 drop by 1976. They all were wrong. Oil prices would not fall until the 1980s.

On August 9, 1974, Richard Nixon left office in disgrace, and Gerald Ford became the thirty-eighth president of the United States. While he was president and in fact until his death in 2006 at age 93, Ford was often ridiculed by the press as both clumsy and somewhat dim, although he was neither. Ford, a long-time Michigan congressman, was a graduate of the Yale Law School and had been an All-American football player at the University of Michigan, playing both center and linebacker. He was, however, only the fifth president not elected to office, and, uniquely, he had not been elected vice president either, having assumed that office in November 1973 after Nixon, invoking the Twenty-fifth Amendment for the first time, appointed him to replace Spiro Agnew, who had also resigned in disgrace.

When Gerald Ford was president, neither the public nor the press forgave him for pardoning Richard Nixon "for all offenses against the United States which he . . . committed or may have committed" while he was president, although in retrospect Ford surely had made the right decision. Nor did the public forgive Ford for the highest inflation rates in the nation's history. Ford lost himself a great deal of credibility when, as the most visible element of his battle against inflation, he urged people to wear "WIN" buttons (standing for "whip inflation now"), an idea that Alan Greenspan would subsequently label "unbelievably stupid." The inflation Gerald Ford confronted was, of course, largely rooted in higher worldwide energy prices.

In the fall of 1974, President Ford took his advisers to Camp David to fashion an energy policy that became the centerpiece of his State of the Union Address in January 1975. He began that speech to Congress by saying, "I've got bad news, and I don't expect much, if any, applause." He didn't get much. His energy proposals, he said, "will impose burdens on all of us with the aim of reducing our consumption of energy and increasing our production." The centerpiece of his program was the decontrol of oil prices as well as the price of new natural gas, which had been regulated since 1954, when the Supreme Court ruled that the Federal Power Commission (FPC) had the power to set prices for the nation's 40,000 gas producers. Ford had the power unilaterally to end price controls on oil (not natural gas), but not the political strength to do so. He wanted Congress to share the political heat for decontrol. Ford also asked for a $3 a barrel tariff on imported oil, more drilling on the outer continental shelf, greater use of nuclear power, and a windfall profits tax on oil companies, along with a number of conservation measures. But congressional Democrats, determined to keep energy prices low for American consumers, not only rejected virtually all of Ford's proposals and extended oil price controls at least until 1979, but also rolled prices back to their mid-1973 level.

The legislation that emerged from Congress did contain some significant conservation measures, most important the automobile fuel standards discussed in chapter 3, and Ford reluctantly signed the bill—a decision that kept oil prices subject to federal controls for another six years. In retrospect, Ford wondered whether he should have vetoed it. He said: "If I could [have acted] unilaterally, on my own, I probably would have decontrolled period, immediately. There would have been a little turmoil for a short period of time. But we would have opened the markets and the markets would have corrected things rather quickly. But that was not the real world. The real world was that I had to deal with Congress." William Simon, Nixon's former energy czar and Ford's Treasury secretary would later describe Ford's decision to sign the bill as "tragic," calling it "the worst error of the Ford administration." Ford had little success in persuading Congress to adopt his ideas for addressing energy policy. But he certainly understood what the problem was. He told a reporter, "The Saudi's have us by the balls."

Arthur Burns, still heading the Federal Reserve, presciently viewed the energy crisis as threatening a "permanent decline in our nation's economic and political power in a troubled world." Alan Greenspan described our nation as playing "Russian roulette" and "susceptible to blackmail." Frank Ikard, who headed the politically powerful American Petroleum Institute, was even more dramatic. "We will have to adopt a whole new way of life," he said. "The American love affair with the large automobile has come to an end." But, like announcements of Mark Twain's death, that claim was premature.

Frank Zarb, whom Ford appointed to head the Federal Energy Administration and who would be Ford's "energy czar" throughout his presidency, said in retrospect that everything the Ford administration did—what he called the "five pieces" of energy policy—"coal, natural gas, nuclear power, oil, and conservation"—were designed to reduce the nation's dependence on foreign oil. But with remarkably little success. Americans' love affair with large automobiles, SUVs, and small trucks would continue into the twenty-first century, as would our attitude that cheap energy is our birthright.

The new structure of policy challenges and constraints for the United States that emerged when we lost our longstanding easy access to cheap oil in the 1970s has remained in place to the present day, even if its prominence has waxed and waned. Our nation remains dependent on imported oil, much of it from the Middle East. Control of that oil remains in the hands of Middle Eastern governments, not Western oil companies. Those governments have used that oil to obtain wealth and to influence regional geopolitics not only in the United States, but also in Europe and Japan.

Moreover, the shock and trauma of the 1973 embargo itself—reinforced by a second energy crisis in the late 1970s—shaped the U.S. approach to energy policy over the coming decades, shaking confidence in the ability of the U.S. military and economic power to dominate global energy markets. Faith in oil as an engine for economic growth and prosperity and the certainty that American industriousness and technological prowess could solve any problem had been mainstays of our energy policy before 1973. Now all would be in doubt. And this doubt would be reflected in—and strengthened by—the failures of post-1973 foreign-policy initiatives, domestic-policy initiatives, and quests to find alternative energy sources, all in an effort to secure free-flowing, reasonably priced energy without sacrificing the country's other foreign-policy objectives in the Middle East.

This failure to develop any economically successful energy alternatives would become the dominant fact of U.S. energy policy for the remainder of the twentieth century and at least the first decade of the twenty-first. Telling the story of how and why that has happened is the task of subsequent chapters. But in order to comprehend that failure, we must first examine another dramatic change in U.S. public policy that took hold in the 1970s: the coming of age of the environmental movement in America.

3 The Environment Moves Front and Center

It has become commonplace—even though no one fully believes it—to date the emergence of the modern environmental movement from April 22, 1970, the first Earth Day. John Steele Gordon, writing in *American Heritage* magazine nearly twenty-five years later, described that day as "one of the most remarkable happenings in the history of democracy." "Fully 10 percent of the population of the country, twenty million people, demonstrated their support for redeeming the American environment," Gordon said. "They attended events in every state and nearly every city and county. American politics and public policy would never be the same again."

For such an auspicious occasion, Earth Day had rather modest origins. Wisconsin's Democratic senator Gaylord Nelson, searching, as he put it, "for some idea that would thrust the environment into the political mainstream," announced at a conference in Seattle in September 1969 that there would be a "nationwide grassroots demonstration on behalf of the environment." Nelson said that he got the idea for Earth Day from the anti–Vietnam War "teach-ins" then occurring at college campuses across the country, and he invited everyone to participate. A decade later, Nelson would describe that day as the time when "the environmental issue came of age in American political life."

For the first four months after Nelson's Seattle announcement, two of his staffers managed Earth Day logistics out of his Senate office, but in mid-January he recruited Denis Hayes, who had recently graduated from Stanford University, to serve as Earth Day's national coordinator. Hayes, who takes credit for christening the event "Earth Day," staffed his office mostly with college students. Together with Nelson, they embodied the tensions inherent in the movement. Nelson was a mainstream Democratic senator from the Midwest. Hayes was something of a radical who refused to meet with Richard Nixon's chief domestic policy aide John Ehrlichman, saying, "We will not appeal any more to the conscience of institutions because institutions have no consciences." "If we want them to do what's right," Hayes said, "we must make them do what's right." But Hayes did not want Earth Day to mimic the confrontational approach of leftist activists. He wanted it to "involve the whole society." Later, reflecting back, he elaborated, "We didn't want to alienate the middle class; we didn't want to lose the 'silent majority' just because of style issues."

Because the event was somehow scheduled to coincide with the anniversary of Lenin's birth, some skeptics remained suspicious of its real purposes and origins, but the celebration of Earth Day in 1970 exceeded all its organizers' aspirations as a demonstration of broad-based public environmental awareness and concerns. That day's events reflected a complex mixture of counterculture theater, youthful play, and mainstream concerns about environmental damage—not such a surprising concoction, given the times.

Nor was it surprising that the Earth Day celebration turned into a forum for broad challenges to the status quo. Maine's senator Edmund Muskie, an early front-runner for the 1972 Democratic presidential nomination, captured the complex cross-currents of the day when at the University of Pennsylvania he said: "Those who believe we are talking about the Grand Canyon and the Catskills, but not Harlem and Watts are wrong. And those who believe we must do something about the SST and the automobile, but not ABMs and the Vietnam War are wrong." At the other end of the political spectrum, Walter Hickel, Richard Nixon's secretary of the interior, who had been Alaska's governor and had long supported oil production there and elsewhere, was booed off the stage at the University of Alaska when he began cheerleading for immediate construction of the Alaska pipeline.

Private industry also participated in the Earth Day celebration. Electric companies were especially determined to be counted among those waving environmentalist flags. At the University of Illinois, students disrupted an Earth Day speech by a representative of that state's largest electric utility, Commonwealth Edison, by throwing black dust on one another and coughing uncontrollably. In New York, Consolidated Edison displayed an electric car outside its headquarters and lent New York City's mayor an electric bus to ride around in. But this display proved to be just symbolism; it failed to signal the beginning of electricity-powered transportation vehicles. No one seems to know what happened to those electric vehicles after the day ended.

Unlike the confrontations over civil rights and the Vietnam War, which also had gathered large masses of people into public squares in an effort to change public policy, Earth Day had an upbeat, all-inclusive quality to it. Earth Day supplied a positive counterpoint to the nation's conflicts over these issues; tear gas canisters and police batons were noticeably absent. Earth Day's gatherings were more like parades, less like demonstrations. The nation's media loved the spectacle, and the television and newspaper attention it garnered had a major effect in galvanizing the legislative, administrative, and judicial environmental pronouncements in the decade that followed. As *Time* emphasized, "water, air, and green space know no class or color distinctions." The day's tone was overwhelmingly optimistic. The widespread feeling was that if only our nation would pass intelligent laws and employ our technological prowess, our environmental problems would surely be vanquished.

Earth Day's success did not just spring up suddenly from nowhere. The preconditions for demonstrating broad public support for environmental protection had occurred during the prior decade. The 1962 publication of Rachel Carson's *Silent Spring*—which focused on environmental risks and health dangers emanating from the chemical industry and in particular the use of pesticides such as DDT—was especially inspirational. When Carson spoke eloquently of "the most modern and terrible" technologies turned "against the earth" and lamented the silencing of the "robins, catbirds, doves, jays, wrens, and scores of other bird voices," she captivated Americans. Carson died in 1964 after a struggle with breast cancer, but her legacy continued to inspire environmental activists long after her death. In her obituary, the *New York Times* described *Silent Spring* as hitting the chemical industry and the general public with the "devastating impact of a biblical plague of locusts."

Carol Yannacone of Patchougue, Long Island, was one person who translated Rachel Carson's message into action. In the spring of 1966, she learned of a fish kill in nearby Yaphank Lake resulting from the use of DDT to control mosquitoes. Victor Yannacone, her lawyer husband, filed suit against Suffolk County the next day. The suit was partially financed by the Rachel Carson Memorial Fund of the National Audubon Society, and although the judge denied the relief sought, the adverse publicity and public outcry the case generated convinced Suffolk County never again to use DDT. In this suit, Victor Yannacone established a pattern for the massive environmental litigation that would soon follow by introducing substantial amounts of scientific evidence and relying on scientific experts rather than simply on allegations of aesthetic or economic harm.

Victor Yannacone and a group of scientists then banded together to form the Environmental Defense Fund. The first lawsuits the group brought were against municipalities (largely in Michigan and Wisconsin) to ban the use of DDT. Many other suits, including a large number affecting energy production, would follow. Less than two decades later the Environmental Defense Fund would have a $3 million budget, a staff of 45, and a membership of 50,000. (In 2009 its budget was nearly $105 million, and its staff was composed of 350 scientists, economists, and lawyers.)

The National Environmental Policy Act of 1969 (NEPA), with its new requirement of environmental impact statements, created a cottage industry for lawyers and scientists. In the words of the national environmental correspondent for the *New York Times*, it had "unexpected revolutionary effects." The environmental impact statements on the Alaska pipeline alone required a stack of paper 10 feet high and cost $9 million to put together. By the middle of the decade, federal agencies were producing 1,000 such statements a year, a total of 7,500 between 1970 and 1976. About 20 states mimicked the federal law, and by the mid-1970s, California was producing 4,000 environmental impact statements annually.

Environmental organizations often challenged the adequacy and accuracy of environmental impact statements in court. According to the best estimate, 852 lawsuits were decided in the 1970s under NEPA alone. The Clean Air Act, also enacted in 1970, produced another 208 decided cases, and the Clean Water Act of 1972 accounted for another 313. If you count the environmental lawsuits that were settled before decision, approximately 3,500 cases were brought during the 1970s. Disasters and near-disasters have also played a special role in galvanizing the public to press for greater environmental protection.

On January 28, 1969, a week after Richard Nixon's inauguration, in the Santa Barbara Channel, alongside one of the most beautiful stretches of California's coastline, shadowed by rolling hills dotted with multimillion dollar estates, cattle ranches, and cedar forests, a gap between a twelve-inch drilling hole and a six-inch drilling bit in an offshore Union Oil platform opened up a much wider space between oil exploration and what most Americans came to see as the necessity to protect our environment. Through this gap, 21,000 gallons of oil a day began to leak into the ocean. Five days later the slick covered an area 20 miles long and 10 miles wide. More than 200,000 gallons of oil polluted the sea before the leak was plugged twelve days after it began. The beaches of the city of Santa Barbara and most of Ventura and Santa Barbara counties had been fouled.

Santa Barbara has long been the home to many wealthy, politically sophisticated people who are very much aware of the beauty of their natural surroundings. A new conservationist group, Get Oil Out (GOO), quickly gathered more than 100,000 signatures urging that offshore drilling be stopped. Walter Hickel tried to minimize the incident. So did oil company executives. One of them wondered aloud why there was so much fuss over a "few dead birds." This was the wrong fire to pour oil on. Most politicians knew that. Both the California and federal governments soon imposed moratoriums on new offshore drilling and sales of new oil leases. The federal moratorium also included the Gulf of Mexico. With few exceptions, these moratoriums remained in place until the OPEC oil embargo hit in 1973.

Nearly a year after the Santa Barbara spill, after prolonged controversy and much jockeying by legislators and litigators, a federal court in California approved the settlement of a $1.3 billion lawsuit against Union Oil. By then, despite public objections, some existing platforms had resumed drilling in California, and new drilling platforms had even been constructed in the federal tidelands of the Santa Barbara Channel. Nevertheless, more than a year after the spill most oil companies except Humble, Union, and Sun Oil had still not resumed drilling off Santa Barbara, and public efforts to halt offshore drilling had acquired a whole new momentum.

The spill inspired a great deal of litigation, and environmentalists obtained injunctions stopping or delaying further drilling off the California coast, in the Gulf of Mexico, in the Florida Everglades, and off the Maine Coast, eventually preventing the Interior Department from authorizing construction of the Alaska pipeline. The six-inch gap in Union Oil's drilling machinery and the damage it caused had inspired a wave of environmental uproar and activism.

And Santa Barbara's was not even the largest oil spill during this period. Two years earlier a tanker carrying 119,000 tons of crude oil had broken apart in the rough sea off Land's End, England. This Torrey Canyon spill, as it became known, dumped many times more oil—nearly 35 million gallons—onto England's coastline. Television audiences in the United States witnessed the primitive and unsuccessful efforts to limit the damage. Public outrage over this and the Union Oil spill became a national phenomenon with a lasting impact on energy policy throughout the 1970s and beyond. (Forty years later, in the spring of 2010, an explosion on a BP [formerly British Petroleum] well a mile deep in the Gulf of Mexico would dump far larger torrents of oil into that body of water—another major environmental disaster taken up in chapter 14.)

Six months after the Union Oil spill, late in the morning of June 22, 1969—less than a year before Earth Day—the Cuyahoga River burst into flames in Cleveland, Ohio. The five-story fire, which lasted only half an hour but destroyed wooden railroad bridges high above the river, was believed to have been ignited when sparks from a passing train hit an oil slick. Reporting on the incident and the general condition of the river, *Time* said: "Anyone who falls into the Cuyahoga does not drown, he decays." This fire also spurred environmentalists' action and helped stimulate the Clean Water Act of 1972.

Environmental protection, despite its roots in the leftist agendas and activism of the 1960s and in upper-class conservationism dating from Theodore Roosevelt's time, had become a major concern of the masses by the 1970s. When Richard Nixon signed NEPA on January 1, 1970, he said: "The 1970s absolutely must be the years when America pays its debts to the past by reclaiming the purity of its air, its water, and our living environment. It is literally now or never." His characterization—"now or never"—made it to the back of the next edition of *Silent Spring*.

In his 1970 State of the Union Address, Nixon reiterated his commitment to environmental issues. "The great question of the seventies is," he said, "shall we surrender to our surroundings, or shall we make peace with nature and begin to make reparations for the damage we have done to our air, to our land, to our water?" He went on to describe environmentalism as a "cause beyond party and beyond factions," adding that it "has become a common cause of all the people of our country." He said that clean air, clean water, and open spaces were "the birthright of every American," and he promised the most "comprehensive and costly program in this field in our nation's history." Soon thereafter Nixon sent Congress a plan to reorganize the federal bureaucracy for pollution control into a single new agency, and on December 2, 1970, the Environmental Protection Agency (EPA) opened its doors. But Nixon's enthusiasm for environmentalism was not to last. Less than a year later he described environmentalists as "going crazy" over water pollution legislation.

By then, environmental activists knew that the forthcoming "environmental decade" would demand from them more resources and more effective organization. As we have seen, they had already begun filing lawsuits and advocating legislative and regulatory changes to further their goals. Environmental organizations had expanded their memberships, and new ones had begun to form.

Perhaps the one contest that best illustrates why books have been written about the 1970s with titles such as *The Environmental Decade in Court* is the dispute over the Storm King Mountain power plant. In 1965, before Congress had passed any of its modern environmental protection laws, a group of individuals challenged a hydroelectric power facility that the FPC in 1963 had licensed Consolidated Edison to build alongside the Hudson River near Storm King Mountain, a lovely hill on the Hudson's west bank just south of Cornwall-on-Hudson. This landmark lawsuit established a significant precedent that allowed a group of citizens (in this case calling themselves the Scenic Hudson Preservation Conference) who could show no direct economic harm to challenge in court those activities that threatened environmental aesthetics. Scenic Hudson claimed that the proposed Con Ed plant would carve away part of Storm King Mountain, convert part of the nearby Black Rock Forest into a reservoir, and place unsightly electric transmission lines across the Hudson River.

But this is just one side of the story. William Tucker, who wrote a long screed in *Harper's* extremely critical of Scenic Hudson's challenge, pointed out that over time Con Ed modified its plans to put the transmission lines under water and agreed to relocate the facility so that the beauty of Storm King Mountain would be untouched. Tucker insisted: "It was obvious to anyone willing to look that the opposition was coming from the petty aristocrats of the Hudson Valley who were annoyed that their solitude was being invaded by the advancing society." To Tucker, Storm King represented little more than a troubling instance of elite NIMBY behavior.

Whatever one thinks of the Storm King plaintiffs, they were successful. Their lawsuits managed to delay the project until 1980, when Con Ed abandoned it.

Scenic Hudson used every tool at its disposal to prevent Con Ed from building the Storm King plant. It challenged every FPC decision in both federal and state courts and successfully petitioned the FPC itself to hear additional testimony and delay its decisions. Scenic Hudson also hired a public-relations firm to generate public aid for its cause, and it garnered support from the editorial page of the *New York Times* and ultimately from the government of New York City. It arranged for a favorable story in the widely read *Reader's Digest* and, using an up-market variation on civil rights sit-ins, enlisted fifty boats to engage in a "sail-in" in the river near Storm King. The challenges Scenic Hudson mustered in more than a decade of litigation included claims on behalf of nearby residents (even though the residents of the nearest town overwhelmingly supported Con Ed) and assertions about the plant's potential for more air pollution and fish kills.

Ultimately, of course, the peak electricity for New York City had to come from somewhere else, such as Indian Point nuclear facilities, 25 miles north of New York City, and expanded facilities in the city's Astoria area. The failure to bring the Storm King plant on line contributed to New York City's higher electricity prices for city residents in the 1970s. Notwithstanding the costs, however, Scenic Hudson's success inspired subsequent environmental litigation.

In 1970, two of Scenic Hudson's advocates, David Sive and Stephen Duggan, became the founders, along with a group of other lawyers, scientists, and law students, of the Natural Resources Defense Council. Sive has often been called the "father" of modern environmental law. The Defense Council today is one of America's most influential environmental organizations, boasting 1.2 million members, a staff of 300 (mostly lawyers and scientists), and a budget of nearly $90 million in 2008. During the 1970s, the Natural Resources Defense Council, the Environmental Defense Fund, and other environmental organizations turned litigation into an art form that delayed or halted numerous plans for new energy-production facilities.

Venerable conservation organizations such as the Audubon Society and the Sierra Club, which heretofore had focused principally on preserving national parks or protecting natural resources, also broadened their agendas and became more determined to affect public-policy decisions. Michael McCloskey, then the Sierra Club's executive director, observed, "Suddenly there was a whole new agenda that seemed to have nothing to do with the old issues, an agenda that focused almost exclusively on pollution and waste." He added, "We were severely disoriented suddenly to find out that all sorts of new personalities were emerging to lend something new, mainly people out of the youth rebellion of the 1960s who had all sorts of notions that just came out of nowhere."

In 1971 in San Francisco, a few lawyers founded the Sierra Club Legal Defense Fund (which in 1997 changed its name to Earthjustice). The Legal Defense Fund was largely independent of the Sierra Club organization. It had its own trustees, staff, and financing, but it undoubtedly benefited from its association with the nation's oldest conservation organization. Like so many of the environmental organizations born in the 1970s, it often resorted to litigation as a means to slow, discourage, or halt energy projects, but it also learned quickly to exert its muscle to influence legislation and administrative decisions via NEPA's requirement of environmental impact statements and other means. In forming the Legal Defense Fund, the Sierra Club acknowledged that its role had greatly expanded—from an organization focused on conservation to one also active in promoting the new environmentalism.

Of the 75 nonprofit environmental groups created in the United States before 1950, all but 7 were concerned with conservation rather than with the general quality of the environment. In the 1960s, the number of environmental groups grew from 119 to 221 and in the 1970s nearly doubled again to 380. Membership also exploded in the 1970s, with the National Wildlife Federation, for example, growing from 540,000 to 811,000 members and the National Audubon Society nearly quadrupling from 120,000 to 400,000 members. Not only did the 1970s see the creation of many more groups with broader agendas and greater memberships, but the full-time staffs of these environmental organizations grew even faster than their numbers—from 316 in 1961 to 668 in 1970 to 1,732 in 1980. These staffers, mostly lawyers and scientists, were able to litigate, testify at congressional and administrative hearings, and conduct original research on the issues they cared about. The *New York Times* coverage of environmental issues more than tripled from 285 column inches annually in the 1960s to more than 900 a year in the 1970s.

The idea that dispute resolution by courts is a sensible way to fashion energy and environmental policy is a peculiarly American indulgence, attributable, at least in part, to the difficulty of producing satisfactory outcomes through legislation or administrative agencies. Legislation, in particular, is far more responsive to short-term problems than to chronic conditions. And the long time horizons involved in environmental remediation as well as the inevitable scientific uncertainty and disputes often serve as a barrier to effective legislation. Moreover, given the crucial role that the courts played in the civil rights revolution of the 1950s and 1960s and would soon play in establishing women's rights, it is not surprising that environmentalists in the 1970s turned so aggressively to the courts to vindicate their claims. The courts not only became a key forum for debate over energy and environmental policy, but also then played a starring role in strengthening the claims of environmental advocates vis-à-vis those parties urging unfettered economic development. Notwithstanding the efforts of some commentators, such as William Tucker, to characterize environmental lawsuits as little more than an indulgence of the "leisure class" designed to maintain its comfortable status quo, the environmental movement that burst onto the American scene with such force in the late 1960s and 1970s enjoyed widespread public support.

That public support and critical leadership in the Congress supplied great momentum for new environmental laws through the 1970s. Nearly 20 important environmental statutes were enacted during that decade—most with large bipartisan majorities in support—covering, for example, clean air and water, toxic substances, endangered species, and the uses of federal lands and other assets. During the 1970s, the EPA also grew from an impotent infant agency into a powerful administrative hulk. Some of the early legislation was stunning in its scope and goals. The 1970 Clean Air Act, for example, demanded that the automobile industry produce a virtually pollution-free car within five years (a requirement that obviously was subsequently eased) and instructed EPA to establish pollution standards that would protect all Americans from adverse effects of air pollution regardless of costs or feasibility (an unrealistic mandate that also never took hold).

How these laws got through Congress has been much discussed elsewhere, so I do not detail that story here but instead highlight just a few important aspects. First, perhaps surprisingly, business interests provided much of the impetus for the early legislation, supporting both the first federal statute regulating air pollution (the Motor Vehicle Air Pollution Act of 1965) and the first federal statute regulating air pollution from factories and electric power plants (the Air Quality Act of 1967). Businesses regarded federal legislation as a way to secure national consistency and shift regulation away from more aggressive state and local governments to what they believed would be a more benign and responsive federal legislature. Second, business interests grossly underestimated the potential costs of early environmental laws, in particular NEPA. Third, new business interests, such as pollution-control equipment manufacturers, had much to gain from environmental legislation, and opposing businesses did not become sophisticated in lobbying against environmental legislation until the mid-1970s. Fourth, widespread and growing public support of environmental issues created policymaking and credit-claiming opportunities, especially for three 1972 presidential candidates: Democratic senators Edmund Muskie of Maine, "Scoop" Jackson of Washington, and Richard Nixon. Fifth, environmental and consumer protection organizations expanded and became much more effective Washington players.

However, the automobile industry, the oil industry, and coal interests had powerful allies in Congress. These allies included Congressman John Dingell of Michigan and Senators Russell Long of Louisiana and Jennings Randolph and Robert Byrd of West Virginia, all of whom would play critical roles in fashioning federal responses to environmental concerns. Senator Randolph, for example, headed the Senate Public Works Committee, the parent of Senator Muskie's Air and Water Pollution Subcommittee, and Randolph's amendments steered federal air pollution legislation in the direction of technological "solutions" to air pollution problems rather than toward the substitution of less-polluting fuels for coal—a path that federal legislation has continued to follow to this day. His West Virginia colleague Robert Byrd, who was the nation's longest-serving senator by the time of his death in office in 2010, began serving in the Senate's Democratic leadership in 1967 and soon thereafter became eastern coal interests' greatest ally.

A clash between the goals of economic growth and environmental concerns became inevitable. Such a clash is not surprising. American environmentalism, after all, had always been about limits, having originated in isolated acts of resistance to unbridled growth. Both John Muir, who founded the Sierra Club in 1892 and his ally/adversary Gifford Pinchot, the first head of the U.S. Forest Service, who coined the term *conservation ethic*, explicitly recognized the inherent tension between conservation and economic expansion. Conflicts between progrowth and environmental activists ran deep in the history of the American environmental movement. But they were now being seriously intensified. In 1968, for example, Paul Ehrlich's best-selling book *Population Bomb*, which predicted that every American would face water rationing by 1974 and food rationing by the end of the decade, spurred a "zero population growth" movement shrouded in pessimism and fueled by apocalyptic predictions.

In 1972, the Club of Rome published *The Limits of Growth,* based on models of population, consumption, and resource growth developed by four researchers at the Massachusetts Institute of Technology. This little book offered a very pessimistic picture of future prospects for economic growth. Its headline prediction was that economic expansion could not continue forever because of resource limitations, especially the availability of oil. The book garnered great media attention, stimulated much hand wringing, and provoked substantial blowback against economic development. It sold more than 30 million copies worldwide and was translated into 30 languages, making it an all-time environmental best-seller, although its central idea—that we should stop economic growth or at least slow it down until a more appropriate new technology appeared—was far from universally accepted.

Energy policy soon became the battleground for debating the appropriate relationship between limits on resource use and our nation's seemingly unlimited demand for energy. The oil embargo in the year following publication of *The Limits of Growth* underlined the precariousness of our energy situation and heightened public attention to this debate.

Significantly, the general idea of limits—indeed, of downsizing—resonated in the 1970s with an important subset of America's youth. In 1968, Stewart Brand, who played a leading role as one of Ken Kesey's "Merry Pranksters" traveling the country in Tom Wolfe's *Electric Kool-Aid Acid Test,* published the first edition of his enormously successful *Whole Earth Catalog.* This book listed, described, and provided access to thousands of "tools" for self-awareness and self-sufficiency—books, maps, outdoors gear, and tools—along with a variety of methods for generating one's own energy from solar, hydro, wind, or geothermal sources. It appealed especially to those young people in the counterculture who were withdrawing from urban settings to rural communal enterprises, but it was read and admired far more widely. Its 1972 edition sold 1.5 million copies and won the

National Book Award. The *Catalog*'s central idea of shedding the handcuffs of technological dependency by taking control of one's own surroundings appealed enormously to the generation coming of age in the 1970s—even to that vast majority who had no intention of retreating to rural settings. Small was beautiful, decentralization always desirable. In these quarters environmental enthusiasm was coming to feel much like religious fervor.

One important advocate for a smaller-scale, decentralized "soft energy path" was the physicist Amory Lovins. In an influential 1976 article in *Foreign Affairs* entitled "Energy Strategy: The Road Not Taken," Lovins argued for more efficient use of energy, especially in electricity production, and for much more reliance on solar, wind, biofuel, and geothermal electricity production. He repeated this argument, and it received further attention in two chapters of a widely read Rockefeller Foundation report entitled *The Unfinished Agenda*. Lovins has been advancing related arguments ever since. He was a cofounder of the Rocky Mountain Institute and has published 29 books, virtually all advocating much greater reliance on renewable energy sources.

William Tucker, in his *Harper's* article on the Storm King Mountain controversy, captured the skeptics' reaction to the goal of embarking on a soft energy path. Tucker acknowledged that the country faced "enormous and critical environmental problems" and recognized that there are "enormous numbers of people who are dedicated to trying to solve them," but he added: "My quarrel is with the political 'environmentalism' that offers no reasonable alternatives but proposes solutions which entail delaying or abandoning present, feasible, and proven technology and 'waiting for' solutions that are 'soft,' 'attractive,' and 'just on the horizon.'" "The leisure-class environmentalists," he said, "will be perfectly content to leave things the way they are, regardless of the economic consequences." Along with many of their political leaders, much of the public believed the solution to the nation's energy problems resided in some technological "silver bullet" waiting to be discovered just around the next corner.

Many of the battles over energy and environmental policy, such as those over resuming drilling for oil in the Santa Barbara Channel or building the Storm King Mountain hydroelectric facility, were fought at the local level, often in state courts. Everyone had come to understand that the NIMBY approach could be a powerful force. Less heed had been paid to the fact that the battles over the siting of energy production were being fought most effectively by well-educated and well-financed constituencies reluctant to accept trade-offs. Energy producers soon realized that it was easier to site facilities in poorer or rural neighborhoods. Nevertheless, disputes arose nearly everywhere.

Environmental organizations and their supporters generally embraced efforts to promote energy conservation but typically opposed efforts to stimulate domestic energy supplies. In California and the states bordering the Gulf of Mexico, the fights were over offshore drilling for oil. In the Northeast, these organizations successfully attacked proposals to build oil refineries. And they fought plans for nuclear electric plants virtually everywhere they were proposed. Coal always spewed too much sulfur into the air. Strip mining threatened to spoil pristine western landscapes. Natural gas, as the cleanest of fossil fuels, gained some new allies and became more desirable, but the FPC's insistence on maintaining artificially low prices inhibited its supply and distribution.

Given that their successes in stalling and stymieing new energy-production facilities were coupled with their disappointing failure to hasten the widespread adoption and use of alternative and renewable energy sources, environmental organizations soon realized the benefits of returning to their roots in energy conservation. Environmentalists knew that artificially low energy prices were a barrier to conservation and greater energy efficiency. Nevertheless, they resisted the raising of prices, which would create a conflict with their usual allies on the political left.

Nowhere was this conflict better illustrated than in Ralph Nader's testimony before the House Rules Committee in hearings on President Ford's proposal to decontrol oil prices in the summer of 1975. Nader, who had become famous as a consumer advocate through his attack on the Chevrolet Corvair in his 1965 best-seller *Unsafe at Any Speed*, had in the 1970s turned the attention of a number of the public-interest groups he controlled to environmental issues. Before Congress, he emphasized the importance of conservation: "Without a doubt," he said, "the top priority of Congress today should be saving energy." This, he said, was the "quickest new energy source we have" because "40 percent" was being wasted. But Nader also insisted, without explanation, that higher decontrolled energy prices would do little for conservation. This outcome, he disingenuously claimed, was "well known." He also asserted that tremendous inflation and unemployment would accompany oil price decontrol.

By this time, events in the Middle East had sharply raised the prices of oil and natural gas, notwithstanding Congress's efforts to maintain below-market rates in the United States. Nevertheless, the price controls remained in place, and President Ford and his successor, the Democrat Jimmy Carter, turned instead to exhortation, subsidies, and regulation as mechanisms for limiting energy consumption.

In 1975, President Ford signed the Energy Policy and Conservation Act, creating incentives and mandates for the conservation of energy in appliances, buildings, equipment, and, most contentiously, automobiles. We are familiar with the 55-miles-per-hour speed limit this bill introduced—rarely obeyed and infrequently enforced though it was. But by far the most important provisions of the bill established the fuel-economy standards for automobiles, the so-called Corporate Average Fuel Economy, or CAFE, rules.

Ford, who had long served in Congress as the representative of Grand Rapids, Michigan, about 160 miles from Detroit and home to an automobile manufacturer early in the twentieth century, urged a voluntary agreement on fuel-efficiency standards between the government and the auto industry. Ford called for an overall industrywide improvement of 40 percent in fuel efficiency, a number that GM, the industry's least fuel-efficient company, had suggested. He was fully prepared to loosen automobile pollution standards to achieve it. Congress, however, was determined to enact mandatory standards. The only questions were how tough the standards would be and whether President Ford would sign the legislation.

The oil industry cared little about this issue, but the auto industry lobbied Congress intensively. In 1975, the U.S. auto industry claimed to account directly or indirectly for nearly one-quarter of U.S. employment and to constitute 18 percent of U.S. gross domestic product (GDP). The big three U.S. automobile manufacturers—Chrysler, Ford, and GM—were facing stiff new competition from imports, especially from Japan, which made smaller, more fuel-efficient cars. And economic growth had stalled; the U.S. economy was in recession when the fuel-economy issue was being debated. At the end of 1974, one-fourth of U.S. auto workers faced at least temporary layoffs.

The U.S. automakers argued that no mandatory standards were necessary, claiming that the market was already demanding more fuel-efficient vehicles. They also contended creatively that fuel standards might backfire, asserting that if such standards raised the price of new cars, the American public would simply hold on to its old gas-guzzling, polluting vehicles. GM's representative urged Congress to respect the wishes of American drivers, stating: "The small car cannot adequately perform all of the functions required by automobile owners." He never said what those functions were. Ford Motor Company's chief economist advanced a clever class-based fairness argument against fuel standards, claiming that "small cars are bought mainly by people who are younger, more highly educated, and have smaller families." "Why," he asked, should we favor "their preferences over other people's?" GM's chairman, Thomas Murphy, delivered a speech to the Greater Detroit Chamber of Commerce suggesting that if decisions on the size of the car were taken out of the marketplace that might make the Bicentennial Era "the twilight of American freedom." The auto workers' unions, which were very worried about competition from abroad, largely agreed with the arguments of the big three automakers.

The auto industry and its unions had a key ally in John Dingell, the chairman of the Energy and Power Subcommittee of the House Commerce Committee, who represented Michigan's Fifteenth District near Detroit. Dingell managed to water down the penalties for failing to meet the fuel-efficiency standards: $50 per mile per gallon below the standard for each car sold. After much tugging and pulling, the standards ultimately enacted by Congress were not very different from what the industry had urged: 18 miles per gallon by 1978, gradually increasing to 27.5 miles per gallon by 1990 (taking into account some administrative delays in the 1980s). Political journalist Elizabeth Drew of *The New Yorker* described the bill as "an automobile fuel-efficiency proposal that was, in effect, a product of the Ford Motor Company." The CAFE standards nevertheless allowed Congress and President Ford, who ultimately signed them into law, to herald change and claim tough action even though their real success was far from certain. Drew quoted a "man on Capital Hill" as observing that "one of the most important things involved in this is not energy policy, but the public's perception of Congress as an effective instrument."

Even today, debates rage over CAFE's effectiveness. Its requirements certainly did nothing to reduce the number of miles driven. Nevertheless, many experts believe that CAFE was the most significant aspect of the 1975 energy legislation and generally agree that the CAFE standards actually did come to have some bite when oil prices tumbled downward in the mid-1980s. Some analysts, including the influential energy analyst Daniel Yergin, tout CAFE as the key instance of U.S., European, and Japanese conservation efforts in response to the oil embargo and price increases of the 1970s. Yergin estimates that CAFE reduced U.S. oil consumption by about "2 million barrels per day from what it otherwise would have been—just about equivalent to the 2 million barrels per day of additional oil production provided by Alaska." Other analysts, however, claim that "careful studies of the effects" of the CAFE standards "can identify no significant influence on American manufacturers' production prior to 1981." These contrary claims can largely be reconciled if automobiles' greater fuel economy during this period resulted from the increased oil prices that preceded the enactment of CAFE.

Given the lag in the way automobile production responds to changes in gasoline prices and the fact that the fuel economy of the fleets of U.S. automakers was already extremely close to that required by CAFE, it is difficult to believe that the CAFE requirements had much effect in the 1970s. Changes in the U.S. auto-manufacturing industry in process before CAFE was enacted were at least as important. And the public's love affair with large cars continued. Chrysler's Lee Iacocca said, "[A]lmost as soon as the oil started to flow again [the American public] had turned and streamed back toward larger cars." In a mid-1970s poll typical of Americans' attitudes toward energy conservation, more than twice as many Americans said that Detroit should keep making smaller cars than said they would buy a smaller car themselves. The American public remained unwilling to change its lifestyle in the interest of energy conservation.

An exception to CAFE for light trucks—supposedly enacted to help farmers—turned into a large loophole that exempted SUVs from the CAFE requirements altogether. GM was able to convert one "light truck," a large military vehicle called the "Hummer," into a family car without running afoul of CAFE. The "light truck" loophole stimulated production and purchases of gas-guzzling SUVs well into the first decade of the twenty-first century. SUVs rose from 1.8 percent of the market in 1975 to 6.3 percent by 1987, then to 16.1 percent by 1997 and 25 percent by 2002.

Moreover, because the CAFE requirements were based on the *average* fuel economy of a manufacturer's fleet, they helped the Japanese automobile manufacturers, who easily met them. In response to Americans' ongoing taste for larger cars, Japanese manufacturers soon began to make larger, less-fuel-efficient vehicles such as the Lexus and Infiniti. Japanese imports climbed from 800,000 cars in 1973 to 2.5 million in 1980, when they accounted for 20 percent of U.S. sales.

The enactment of CAFE in 1975, whatever its effect in encouraging gasoline conservation from the mid-1970s through 2005, may nevertheless still prove important. In 2008, Congress enacted legislation strengthening the CAFE requirements (this development is discussed in chapter 12).

Like Gerald Ford, President Jimmy Carter also attempted to bring conservation back to the forefront of national energy policy. In 1977, he described "reducing demand through conservation" as the "cornerstone" of his energy policy. "Our first goal," he said, is "conservation." "Conservation," he added, echoing Ralph Nader's non sequitur, "is the quickest, cheapest, most practical source of energy." As subsequent chapters discuss, Carter described energy policy as "the greatest domestic challenge our nation will face in our lifetime" and larded his policy proposals with moral fervor. "It is a matter of patriotism and commitment," he said. But Congress bought little of his conservation agenda.

The contradictions of our energy policy in the 1970s were now apparent: Congress would endeavor to keep oil and gas prices low to benefit energy consumers, whereas presidents and environmental organizations would exhort citizens to use less. But why would a homeowner or business make large capital investments in energy-saving windows or insulation, for example, when natural gas and heating oil were cheap? And artificially low prices for oil and gas also hampered the environmentalists' quest for a "soft" energy path. They made competition with those fuels much more difficult for energy produced from the sun, wind, or other nonfossil sources.

Nevertheless, by the early 1980s conservation efforts could claim some major achievements. Electricity consumption had flattened, and gasoline consumption was actually declining. But the avaricious energy-consuming nature of the American economy remained fundamentally unchanged. Families continued to move to large suburban and exurban houses if they could afford to do so. Americans' love for large automobiles and their disdain for mass transit remained preeminent. A report card on the 1970s efforts at conservation might give the results a passing grade, but they fell very far short of either environmentalists' or our political leaders' aspirations.

Indeed, the environmentalists' greatest successes came not on the demand side of the energy crisis, but rather on the supply side. Here, the environmental legislation of the 1970s, along with the combination of newly created environmental organizations and new advocacy techniques adopted by old ones, had an undeniably major impact. Litigation to enforce new legislative requirements, especially for environmental impact statements, made placing new sources of energy supply in service much more expensive and difficult. Environmental activists had mastered techniques that at a minimum served to delay energy projects and make them more costly, but that in many instances also succeeded in killing projects altogether. They "succeeded" in hindering or halting offshore oil drilling, new oil refineries, production and gasification of coal, importation of liquid natural gas, and electricity plants of all sorts. But perhaps the greatest success of the environmental movement was in derailing what many considered the great hope for achieving the ever-elusive goal of energy independence: nuclear power.

4 No More Nuclear

At four in the morning of March 28, 1979, unit 2 of the Three Mile Island Nuclear Generating Station (TMI-2) shut down. The power plant, located in the middle of the Susquehanna River, stood 10 miles from Harrisburg, Pennsylvania, the state capital. A valve that controlled the flow of water that cooled the reactor had failed to close, allowing a large amount of water to disappear from the reactor's cooling system. The turbines that carried the heat away closed down, and the nuclear core began to overheat. More than 100 alarms went off in the control room, yet the operators still did not know that the pressure relief valve was stuck open. They erroneously allowed thousands of gallons of radioactive coolant to be pumped into an adjoining building. Taking a step that was mistaken but consistent with their training, TMI-2's operators shut off an emergency cooling system that otherwise would have worked automatically. Although the reactor itself shut off, it was damaged beyond repair, and its uranium fuel continued to generate enough heat to turn much of the remaining coolant water into steam. When the steam was sent into holding tanks, the loud noise awakened and frightened nearby residents.

Although temperatures in some areas of the unit exceeded 5,000°F, the thermometers in the core never registered higher than 750°F. This made the operators complacent, so they fiddled about during the next couple of hours while the fuel rods started to disintegrate, producing more heat and beginning to melt the nuclear core. (Scientists would learn many years later that at least 45 percent of the core had melted.) By about seven that morning the core was being cooled, and the damage was contained, although no one was confident of that. By eight that night, the operators had restarted the coolant pumps, reestablished the water flow, and stabilized the conditions in the reactor's core.

At just before eight on the morning of the malfunction, Pennsylvania's governor Richard Thornburgh had been notified of the accident during a breakfast with state legislators. It would be ten days before he would be able to tell the residents of Pennsylvania that the crisis had passed or to urge those who had left the area to "come home again." During that interval, confusion reigned. On the day the accident happened, the utility that owned and operated the plant, Metropolitan Edison Company, assured state and federal officials and the public that it had everything under control and that all of its safety equipment worked properly. Both claims were false. The company failed to mention that it had found above-normal radiation levels in the area surrounding Three Mile Island.

The Nuclear Regulatory Commission (NRC) behaved somewhat better but not right away. The day after the accident Joseph Hendrie, NRC's chairman, told a congressional committee that TMI-2 had been "nowhere near" a meltdown, something that he could not have known at the time and subsequently proved untrue. Harold Denton, the director of Nuclear Reactor Regulation, and his supporting cast of about 100 NRC staffers did not arrive at Three Mile Island until Friday afternoon, March 30, two and a half days after the incident. Earlier that morning NRC's executive management team in Washington had recommended evacuation of all residents within five miles of the plant. This suggestion stemmed from radiation readings taken by a helicopter crew that happened to be flying directly above the plant's exhaust stack at the very moment that TMI-2's operators, asking approval from no one, opened a valve allowing steam containing radioactive material to escape into the air. Less than an hour later, before any evacuation order was issued, but after a local radio station had announced one was on its way, the NRC team rescinded its recommendation.

But panic had already infected the local residents. Rumors of ground and water radiation spread, prompting many grocery stores nearby to post signs insisting "we don't sell Pennsylvania milk." Emergency sirens had also blasted—for no apparent reason. And shortly after noon on Friday, Governor Thornburgh recommended closing all schools within five miles of the plant, and he urged that pregnant women and preschool children leave the area. That night Walter Cronkite, the nation's most credible journalist, began the *CBS Evening News* with the statement, "[We] are faced with the remote, but very real, possibility of a nuclear meltdown at the Three Mile Island atomic power plant."

The next day, Saturday, March 31, televisions warned that the NRC was concerned about a potential explosion resulting from a large bubble of hydrogen gas in the TMI-2 containment building. Given the lack of oxygen, an explosion was never a real threat, but panic once again took hold in the nearby area. On Sunday, in a rather successful effort to convince the public that the plant and the surrounding area were safe, President Carter and Governor Thornburgh toured the plant together with television cameras trailing behind them. Nevertheless, more than 40,000 people had hit the road by the end of the weekend.

Many of those who stayed behind would not have been reassured when they went to the Harrisburg movie theater playing the thriller *The China Syndrome*, which had opened nationwide two weeks earlier. The movie's title came from the idea that if a nuclear power plant's core had a meltdown in the United States, it would be so hot that it would burn through the earth all the way to China. The movie portrayed a nuclear accident at the "Ventana power plant" near Los Angeles, along with a government and utility company conspiracy to conceal the causes, damages, and risks from an unsuspecting public. The actress Jane Fonda, a notorious Vietnam War protester, was nominated for an Academy Award for her role as a television reporter, who, with her cameraman, played by a young Michael Douglas, fights to uncover the facts. They were aided by Jack Lemmon, who also was nominated for an Oscar for his portrayal of a courageous plant operator seeking to reveal the truth. In an eerie coincidence, at one point a character remarks that a nuclear plant accident would make "an area the size of Pennsylvania uninhabitable." The movie, which proved extremely successful, contributed to public skepticism of both nuclear safety and the credibility of its utility and government promoters.

Time's April 9 cover story "Nuclear Nightmare" was typical of the press coverage of Three Mile Island. Radio stations made jokes, one calling the weather "partly cloudy with a 40 percent chance of survival." Another joked: "What's the five-day forecast for Harrisburg? Two days." Students at nearby Dickinson College wore T-shirts saying "Kiss Me, I'm Radiated." A fearful public did not find any of this very funny. It would take 15 years to complete the cleanup of the TMI-2 site and the processing and removal of 700,000 gallons of contaminated coolant and large quantities of solid waste.

The Three Mile Island accident and the panic it engendered marked the end of widespread hopes and expectations of the 1960s and early

1970s that nuclear power would soon supply the major share of the nation's electricity. Many thought that both oil- and coal-based electricity would soon become obsolete. Nuclear power's opponents had enjoyed some major victories before Three Mile Island, but they had also suffered important defeats. And, to be sure, the costs of nuclear power plants had skyrocketed. But the events of March 28, 1979, and the days that followed signaled the end of the nuclear power era, such as it was, in the United States. For the rest of the twentieth century and the first decade of the twenty-first, no new nuclear plants would begin construction in this country.

In little more than a decade, nuclear power had morphed from the certain answer to our nation's electric generation needs into a pariah industry. Environmental organizations deserve much of the credit—or blame—for this transformation, but administrative agencies, courts, Congress, the companies who built the plants, and the utilities that operated them must share it.

As early as 1903, less than a decade after radioactivity was discovered, a British scientist suggested that the radioactive atom might provide the world with an "inexhaustible" source of power. In 1914, H. G. Wells suggested in his book *The World Set Free: A Story of Mankind* that nuclear energy would be discovered, used as a weapon, and then become the primary source of energy. And even before World War II, physicists knew that uranium fission could produce enormous energy, enough to make a nuclear bomb or provide nuclear power. The bomb came first.

Shortly after the war, in 1946, Congress passed the Atomic Energy Act, creating the powerful Atomic Energy Commission (AEC) with regulatory power over all nuclear activities. This legislation also created the Joint House–Senate Committee for Atomic Energy (JCAE) to guide congressional policy on nuclear energy and to oversee the AEC. The push for nuclear power began in earnest soon thereafter. But even the most enthusiastic proponents at the AEC realized that it would be at least the mid-1960s before nuclear power plants could supply any important quantity of our nation's electricity.

Favorable political circumstances for nuclear power in this Cold War era could not overcome unfavorable economic conditions. Oil was cheap; coal inexpensive. Both provided more certain and cheaper power than nuclear energy. And, as we shall see, the cost advantage of fossil-fueled electric power plants survived even the oil price explosion of the 1970s because the costs of nuclear power plants ballooned far beyond what anyone had anticipated.

In the early 1960s, the AEC was predicting substantial cost savings from nuclear power, and a nuclear nirvana myth soon took hold, promising that electricity from nuclear would be so widespread and so cheap that electric companies wouldn't even bother to meter residential use. By then, the federal government had already spent billions of dollars aiding the design and commercialization of light water nuclear reactors. Electric utilities across the country had responded by ordering unprecedented numbers of nuclear plants.

By the time Richard Nixon gave his "Project Independence" speech in November 1973 in response to the oil embargo, a number of nuclear plants were already in operation. New Jersey's Oyster Creek plant, for example, which had been ordered a decade earlier, had been producing electricity since 1969. Significantly, that plant had enjoyed no direct government subsidies. Because General Electric (GE) had sold the plant to Jersey Central Power and Light on a fixed price basis at a substantial loss, nuclear power had been the cheapest option available to the utility at that time.

Prior to 1965, a total of 20 nuclear power plants were ordered in the United States. In the following decade, from 1966 to 1975, utilities placed orders for 204 more nuclear plants. In 1972, the AEC forecast that by the year 2000 the United States would have a thousand nuclear power plants in operation. This projection implied that the AEC would issue new construction permits and operating licenses virtually every week for the next 28 years. In the peak years of 1973 and 1974, 99 orders were placed. This period became known as "the great bandwagon market." A total of 53 different electric utilities and 12 architect-engineering firms were involved in nuclear power projects. Manufacturers—most notably GE and Westinghouse—sold more nuclear electricity-generating facilities in the first half of the 1970s than the manufacturers of oil and coal power plants combined. Only another 15 plants, however, were ordered in the second half of the decade, and none after 1979. Many of the nearly 250 total sales were ultimately canceled. Fewer than half ever came online.

Despite all the planning, design, sales, and regulatory activity going on in the early 1970s, even by then harbingers of rough sledding ahead could be seen. First, there was no standardization of plant designs. Each of the four chief competitors, GE, Westinghouse, Babcock and Wilson, and Combustion Engineering, sold different products. GE's boiling water reactor was significantly different from the various pressurized water reactors sold by the other three. In an unfortunate effort to achieve economies of scale, all the companies sold ever-larger reactors without redesigning the plants from scratch. Engineering changes, production delays, and cost overruns became commonplace.

Once the electric companies got caught up in the frenzy to obtain nuclear power, the manufacturers were able to shift the risks of escalating costs to the buyers, who in turn tried to collect them from their customers through higher electricity rates. The utilities sometimes succeeded in recouping costs through higher rates, sometimes not. Meanwhile, the costs of construction were rapidly escalating. Construction costs per kilowatt rose at a rate of 18 percent per year—170 percent in six years (in constant dollars) for completed reactors that had construction permits issued between 1967 and 1973. And this figure ignores interest expenses during construction. By the mid-1980s, there was no region of the country where it was cheaper to build a nuclear plant than a coal-fired facility—assuming a nuclear plant could be built there at all. In the coal-rich regions of the country, where the costs of transporting that bulky substance were low, electricity from nuclear power grew to be twice as expensive as electric power from coal. How could a utility justify charging its customers twice the price?

Time is always money, but in the high-interest-rate years of the 1970s delays between ordering a plant and bringing it on line were extremely costly. Such delays alone led many cash-strapped utilities to cancel orders and absorb their sunk costs. In this context, the opponents of nuclear power, who were increasing in numbers, well understood that postponing a nuclear plant might lead to its abandonment. This knowledge produced a multipronged attack on the nuclear industry and its proponents—one that eventually implicated all three branches of government.

In the executive branch, the opponents' target was the AEC. The Atomic Energy Act of 1946 had given that agency ownership of all atomic materials, facilities, and information left over from the Manhattan Project, which had developed nuclear reactors as part of the World War II effort to produce a nuclear weapon. Congress also gave the AEC control over the distribution of technical nuclear information and of all nuclear patents. This founding legislation created within the AEC the General Advisory Committee, composed of distinguished physicists, to advise the agency on scientific and technical issues. Within a year after its creation, the AEC was an outspoken booster of nuclear energy—at a time when there was no economic demand for this alternative to fossil fuels. Shortly thereafter, the AEC began providing economic subsidies in an effort to build a nuclear power industry. From the outset, then, the AEC had a cozy relationship with the industry it was charged with regulating.

From its earliest days, the agency was aware of the barriers to nuclear power—especially concerns about safety. Its Reactor Safeguard Committee (which was so cautious that it earned the sobriquet "Committee for Reactor Prevention") was headed by the renowned physicist Edward Teller. It focused on the potential for accidental release of radioactive materials into the surrounding area and the potential for nuclear meltdown. In 1947, the AEC concluded that a demonstration project of nuclear power was eight to ten years away and that the prospect of any significant supply of nuclear energy was twenty years down the road. As it turned out, that prediction was right on target.

In the early 1950s, the agency contracted with both GE and Westinghouse to develop and manufacture nuclear power reactors. In 1954, Congress, at the behest of the Eisenhower administration and the urging of the JCAE, passed a new Atomic Energy Act, which established private patent rights in nuclear power production and authorized the AEC to license privately owned nuclear reactors. The potential for overwhelming liability claims in case of an accident, however, was severely dampening the private producers' enthusiasm, so in 1957 Congress limited private liability and provided public insurance for nuclear accidents. The AEC also sought to drum up public enthusiasm for nuclear power through its "Atoms for Peace" Program, complete with a Disney film called *Our Friend the Atom* and intended to inculcate positive attitudes toward nuclear power in the nation's schoolchildren.

But Congress would not forever remain the agency's protector. Beginning in the late 1960s, concerned with the AEC's potential conflicts as both booster and regulator of nuclear energy, Congress enacted a number of environmental statutes stripping the AEC of its exclusive jurisdiction over nuclear energy matters. To take just one example, the Water Quality Improvement Act of 1970 transferred to the EPA crucial regulatory power over the types of cooling systems to be used in nuclear power plants, and it gave the EPA the responsibility for regulating thermal discharges from nuclear plants. The AEC was increasingly being viewed more as a cheerleader for nuclear power than as a neutral regulator. And its results could also be called into question. In 1968, after the AEC had spent $1.2 billion in subsidies to develop reactor technology, twice the amount spent privately, very few plants were operational. The largest of these plants could produce only half the power of the smallest nuclear plant then being constructed, one-sixth the power of the largest.

In 1971, Richard Nixon tried to stem declining public confidence in the AEC by appointing James Schlesinger (later his defense secretary) to head the agency. Schlesinger—a smart, somber, buttoned-down, supremely self-confident, gray-haired, pipe-smoking, long-winded former college professor—insisted that the agency would no longer "[foster] and protect the nuclear industry" but would instead "perform as a referee serving the public interest." Two years later Schlesinger was succeeded by Dixie Lee Ray, a colorful marine biologist, who traveled around in a chauffeured limo with her miniature poodle and large Scottish deerhound and who sought to restore public confidence by reorganizing the AEC staff to place greater emphasis on the AEC's safety research programs. She even allowed Ralph Nader access to all AEC reports and ordered Consolidated Edison to protect fish in the Hudson River. These acts did not create friends in the nuclear industry, but even these kinds of changes fell far short of satisfying the agency's critics, including both opponents of nuclear power, who continued to regard the AEC as blessing unsafe reactors, and members of the nuclear industry, who insisted that the AEC's new regulations and procedures were delaying power plants for years and making them much more expensive to build.

In 1974, Congress, with President Nixon's and the AEC commissioners' own support, eventually decided to reconfigure the regime. The Energy Reorganization Act of 1974, a post–oil embargo law aimed primarily at national energy self-sufficiency, terminated the AEC and transferred its power to two new agencies: the Energy Research and Development Administration and the NRC. The former was charged with supporting a broad program of energy R&D, giving "no unwarranted priority" to any particular energy technology. The latter was designed as a bipartisan commission made up of three operating units, the office of Nuclear Reactor Regulation, charged with responsibility for licensing plants; the office of Nuclear Material Safety and Safeguards, charged with safeguarding reactors' nuclear materials; and the office of Nuclear Regulatory Research, charged with supporting research requested by the other two offices. No executive agency had the mission of being a nuclear booster any longer.

Elimination of the AEC was not the only blow to proponents of nuclear power. A few years later they would also lose much of their sway with Congress. As I have noted, the same 1946 legislation that had created the AEC also created the JCAE to oversee the AEC. The nine members each from the House and the Senate were generally close to the nuclear industry and strong supporters of nuclear power. Many were chosen for their experience with issues of national security. The Atomic Energy Act gave the JCAE all authority over nuclear energy legislation and supervisory power over the AEC, and Congress also gave it the full investigative powers of regular congressional committees. It soon became powerful and prestigious, closely guarding its monopoly over atomic energy.

In 1954, the JCAE had guided legislation through Congress, encouraging the development and use of nuclear power for peaceful uses, authorizing private ownership of nuclear reactors licensed by the AEC, and making clear that the AEC was responsible for developing, promoting, and regulating nuclear power. By the late 1970s, however, several powerful longtime JCAE members had departed Congress, and nuclear power had become very controversial. Many recently elected members of Congress were anxious to exert real influence over nuclear electricity issues. Moreover, a series of congressional reforms earlier in the 1970s had diminished the sway of many committees and their chairmen, opening up the legislative process to new players and dispersing previously concentrated committee and staff power.

As part of legislation limiting the number of committees on which an individual member could serve and restricting the number of chairmanships any senator could hold, Congress in 1977 abolished the JCAE. This change precipitated a dramatic increase in congressional oversight of nuclear power. The JCAE's responsibilities were transferred formally to 10 standing committees, 5 in the House, 5 in the Senate. After that, congressional oversight resided with more than two dozen subcommittees. Several of these committees and subcommittees were far less supportive of nuclear power than the JCAE. For example, the House Interior Committee and its

Subcommittee on Energy and the Environment, both chaired by Arizona Democrat Morris Udall, a staunch environmentalist, were very critical of both the nuclear power industry and the NRC. In the Senate, West Virginia Democrat Jennings Randolph, who chaired the Committee on Environment and Public Works, not surprisingly often put the interests of the coal industry ahead of those of the nuclear industry. The opponents of nuclear power did not let up, and by the late 1970s there were many new congressional forums where they could attack.

As with other environmental and energy issues, opponents also turned to the judicial branch to slow and in some cases stop nuclear power. The key development was the July 1971 decision of the D.C. Circuit Court of Appeals in *Calvert Cliffs Coordinating Committee v. Atomic Energy Commission*. In this case, a group of local residents, who together with the Sierra Club and the National Wildlife Federation constituted the Calvert Cliffs Coordinating Committee, attacked the AEC's regulatory regime for licensing nuclear plants. The plaintiffs claimed that both the AEC's general rules and their specific application in connection with a proposal for a construction permit for a nuclear plant at Calvert Cliffs, Maryland, violated NEPA.

In a highly critical opinion, Judge Skelly Wright labeled the AEC's process "ludicrous," adding that "the commission's crabbed interpretation of NEPA makes a mockery of the Act." He said, "Environmental protection is as much a part of their responsibility as is protection and promotion of the industries they regulate." Judge Wright interpreted NEPA as requiring the AEC to consider environmental consequences at every stage of the licensing process, whether anyone was opposing a license or not. The court also held that the environmental impact statements required by NEPA must detail not only the environmental consequences of the actions approved by the AEC, but also alternatives that might lessen any adverse environmental impact. The AEC's job was not to behave "like an umpire and resolve adversary contentions at the hearing stage." Instead, the agency "must itself take the initiative of considering environmental values at every distinctive and comprehensive stage of the process." Judge Wright's opinion also made clear that if a court regarded as inadequate an AEC environmental impact determination or statement—or that of any other agency, for that matter—it should reject the agency's determination and halt any licensing or construction.

Soon after the court's decision, James Schlesinger, still head of the AEC, announced that the agency would not urge the Supreme Court to reverse the judgment. Instead, the AEC hired environmental specialists to help prepare cost–benefit analyses of proposed facilities that would take into account environmental concerns, and it began to issue environmental impact statements in connection with both nuclear plant construction permits and operating licenses.

Under its new procedures, the AEC delayed utilities' ability to engage in land clearing or other activities preparatory to construction until environmental concerns were resolved. The new rules also resulted in the redesign of water-intake systems, reactor-cooling systems, and radioactive waste–recovery systems as well as the rerouting of transmission lines at about half-a-dozen nuclear plants. The agency did not issue any new nuclear plant construction permits or operating licenses for 17 months following the *Calvert Cliffs* decision. Notwithstanding its efforts to comply with the court's requirements, the AEC and its successor, the NRC, lost several more lawsuits involving nuclear licenses in the next few years.

In 1978, the Supreme Court reversed a 1976 D.C. Circuit decision imposing extensive additional procedural requirements on the NRC licensing process in a case involving the Vermont Yankee nuclear power plant. The Supreme Court criticized the D.C. Circuit's "activism" and labeled its opinion "Kafkaesque." The *Vermont Yankee Nuclear Power Corp. v. Natural Resources Defense Council* decision upholding the NRC's procedures was welcomed by both the commission and the nuclear power industry, and it served to dampen somewhat the lower courts' willingness to second-guess the new agency, but winning this particular battle provided small solace in light of previous defeats and the soon-to-follow accident at Three Mile Island.

Responding to environmental concerns at the federal level was just one reason for the delays in bringing on line the large number of nuclear plants that were ordered in the great bandwagon market of the late 1960s and early 1970s. Because electricity rates to consumers were generally set when public utilities requested new rate authority from their state's public-utility commissions, the states offered additional forums for challenging nuclear plant construction and operations. If a utility managed to convince its commission to raise rates, that decision could be appealed in state court and often was.

The saga of Pacific Gas and Electric's (PG&E) two plants at Diablo Canyon provides a good illustration of how these larger struggles between antinuclear interests both outside and inside the government, on the one hand, and the utilities and their own government boosters, on the other, played themselves out. Today, each of those two plants, standing on 750 acres of magnificent land adjacent to the Pacific Ocean at Avila Beach, 12 miles south of San Luis Obisbo, California, supplies about 1,100 megawatts, enough electricity for more than 2 million central California homes. But getting to that output was hardly easy.

PG&E had originally purchased the land in 1966, but owing to a dispute with the previous owner over the terms of sale it did not get full control of the site or a construction permit from the AEC until two years later. The company's plans called for completion of the first plant in 1975 and the second a year later at a cost of $110 million. It had no idea how hopelessly off-base these projections would prove to be.

First, construction was slowed and costs increased by labor strikes and design flaws. By then, the first inklings of public opposition to the plants had begun to appear. Protestors began a series of sit-ins and blockades in an effort to keep company officials and construction workers from entering the property. The estimated price of completion soon began to rise, first to $450 million and then to $600 million.

The AEC originally dismissed the objections of opposition groups, and by the summer of 1973, when things were going well for PG&E, the *Los Angeles Times* featured Diablo Canyon in its "Trip of the Week" column, recommending that its readers take special guided bus tours to the construction site where they could enjoy magnificent views over San Luis Bay and watch the swaying crane 150 feet above ground. The first domed plant housing nuclear reactors was then mostly finished, and construction of the second plant was well under way.

But this was California, after all, the land of earthquakes. Environmental groups began raising questions about the plants' ability to withstand a quake on two faults off the nearby coast, pointing out that no geological surveys had been undertaken before the AEC had granted PG&E its license. The agency responded that it had no geologists on staff back in 1966. A mild tremor off the coast a year later underlined the concern. The San Luis Obispo Mothers for Peace and other groups sought a moratorium on further construction. They didn't get it, but two years later, as construction moved ahead, the new NRC agreed that the matter of seismic activity had to be resolved before an operating license could be issued.

By now, nine years had passed since the utility's purchase of the land, and the estimated cost of the project had risen to nearly $1 billion. And yet on it went: state Senate hearings, Coastline Commission hearings, Energy Resources and Conservation Committee hearings, and an antinuclear California ballot initiative, followed by a legislative end run enacting a watered-down version of the initiative and the initiative's defeat at the ballot box. Every branch of government, state and federal, as well as numerous pressure groups and protestors engaged in a battle royale that stretched on year after year.

In what at the time might have appeared to some as the last word, the scientists of the NRC's Advisory Committee on Reactor Safeguards unanimously declared in the summer of 1978 that the Diablo Canyon plants were safe. But the legislative hearings and court cases continued, as did the protests and the arrests, by now in the hundreds, among them 40 members of Mothers of Peace.

And then in March 1979 the cooling system of the reactor at Three Mile Island failed. By that time, PG&E had completed its construction (at a cost of $1.4 billion, it turned out), so activists turned their attention to preventing the issuance of an operating license. A concert featuring the comedienne Lily Tomlin and the singers Peter, Paul, and Mary, Graham Nash, Holly Near, Joan Baez, and Jackson Browne raised $90,000 for the opposition. Peter Yarrow (of Peter, Paul, and Mary) described the antinuclear movement as the third grassroots movement (after the civil rights and anti–Vietnam War protests) to "alter government policy which could injure us." Ralph Nader, the only person who appeared in a suit and tie, reminded the crowd of Three Mile Island and described the gathering "as a vanguard in a major citizen movement, which is the essence of patriotism . . . to defend our land, our country, and the rights of America to inherit a non-radioactive America." Nader fatuously predicted that by 2025 California "could run its entire economy on solar energy with a little geothermal thrown in." A protest at the plant a month later drew an estimated 25,000. Jerry Brown, the governor of California, told the cheering crowd that he would oppose licensing the plants.

Following another offshore earthquake (4.6 on the Richter scale), Brown, Mothers for Peace, and other antinuclear organizations requested further hearings on seismic activity by the NRC's appeals board, and the NRC then held six days of public testimony. Meanwhile, the California Office of Emergency Services declared the NRC's evacuation plans for residents within 10 miles of nuclear plants "inadequate," and it issued new requirements for an evacuation plan for everyone within 35 miles of any nuclear plant. When the NRC held another 10 days of hearings on this issue in May 1981, 3,000 antinuclear protestors showed up complaining that the 70-seat hearing room was too small. An exasperated PG&E executive told a U.S. House Interior Subcommittee hearing that the "nuclear plant at Diablo is the most thoroughly studied nuclear power plant in the United States, in the history of regulation and obviously in the history of the world." "The nuclear licensing process," he added, "fiddles while oil burns."

When in September 1981 the NRC appeals panel ruled that the Diablo Canyon plants were safe, there were more protests on the land and sea, ending in the arrests of more than a thousand people. Governor Brown and the antinuclear groups filed suit in federal court asking for more hearings. It took another two years for the case to make its way to the D.C. Circuit Court of Appeals, which, without ruling on the suit's merits, enjoined the NRC from giving PG&E a license to begin loading fuel into the reactor. The lawyer for the Los Angeles Center for Law in the Public Interest, which represented Mothers for Peace, described its members as ecstatic about the court's ruling. But the Mothers did not stay ecstatic for long. Four days later the court lifted its injunction, and within three hours PG&E began loading uranium fuel rods into the reactor.

On November 12, 1984, the plant provided its first electricity to customers. From land purchase to power generation, the struggle had lasted 18 years, and the plant had come on line a decade late. PG&E's interest costs alone were estimated to exceed $20,000 an hour, and by the time both plants were up and running, Diablo Canyon's total costs exceeded $4 billion, roughly 40 times the utility's original estimate.

Other utilities faced somewhat similar obstacles. Safety concerns have never been greater than for nuclear power, and environmental organizations and other opponents were able to use both state and federal regulatory, legislative, and judicial venues to seek and obtain delays and sometimes to achieve abandonment of projects. There were nuclear protests in Shoreham, New York; Rocky Flats, Colorado; Black Fox, Oklahoma;

and Seabrook, New Hampshire—to name just a few. In Seabrook, about 40 miles north of Boston, a consortium of 10 utilities had come together and obtained construction permits in 1971 for two nuclear reactors, then estimated to cost a total of $900 million. After many delays, including the arrest of more than 1,400 protesters in April 1977, construction was completed in 1986. In 1990, nine years behind schedule, Seabrook's unit 1 started providing electricity to New England customers. The reactor ultimately cost nearly $6 billion. Unit 2 was scrapped, and Seabrook's majority owner, the Public Service Company of New Hampshire, filed for bankruptcy—then the fourth-largest bankruptcy ever in the United States. (Nearly two decades later, in his 2008 presidential campaign, John McCain suggested that it might be time to build a second reactor at Seabrook. No one has yet risen to that bait.)

The federal government further complicated and ultimately inhibited the nation's move to nuclear power by advancing a technology quite different from the water-cooled nuclear plants being developed commercially. This expensive diversion was the so-called breeder reactor.

As described earlier, the commercial nuclear industry was based entirely on water-cooled systems of one sort or another using uranium-235 as a fuel. In the meanwhile, the federal government was throwing many of its nuclear power resources into a different technology, using a cheaper and more abundant form of uranium. First the AEC, then Richard Nixon and Gerald Ford bet on the so-called fast breeder reactor to become the holy grail of energy policy. A breeder reactor produces—or "breeds"—fuel (in this case plutonium) faster than it consumes it. Rather than being cooled by water, a breeder reactor is cooled by sodium-based liquid metal. This technology offered the promise of a cheap and inexhaustible supply of energy—energy policy's version of a perpetual motion machine.

In June 1971, Richard Nixon described the breeder reactor as "our best hope today for meeting the nation's growing demand for economical cheap energy," and six months later, James Schlesinger, the AEC chairman, announced a project to build the first U.S. fast breeder reactor. Westinghouse was selected to build the plant near the Clinch River in Oak Ridge, Tennessee, and it was to be operated by the Tennessee Valley Authority and Commonwealth Edison of Chicago, the company most experienced with nuclear power. The Clinch River "demonstration" plant was scheduled to begin providing 350 to 400 megawatts of electricity by 1980.

The Scientists' Institute for Public Information, chaired by the environmentalist Barry Commoner, challenged the project in court. Within a month, this group had the support of a group of 30 scientists, including two Nobel Prize winners. Opposition to the breeder reactor was even greater than that against commercial water-cooled plants because its plutonium is a lethal and essential ingredient in atomic bombs and is estimated to be potentially damaging to human health for up to 200,000 years. People were greatly concerned that the government's safeguards were not sufficient to be confident that small amounts of plutonium would not wind up in the wrong hands. Theodore Taylor, a physicist formerly with the AEC, told Congress that the plutonium was inadequately guarded and that he was worried that a small group of terrorists could steal a small amount and make a nuclear bomb. A practical plan to store the waste safely was regarded as an essential but missing element of any effort to produce a breeder reactor.

There were further court decisions and volumes of environmental impact statements concerning the breeder reactor. Congress debated whether to cancel the R&D funding for breeder technology that by 1975 had already cost taxpayers more than $10 billion, making it the largest cost overrun in the nation's history. By then, 62 congressional and executive branch committees and agencies were working on breeder reactors and nuclear technologies.

Little more than a month after taking office, President Carter proposed slashing the breeder's budget and suggested that he might cancel the Clinch River project. In April 1977, in what the president described as a "major change" in U.S. policy, he announced that we would not use plutonium to fuel nuclear reactors. A few weeks later, in a letter to Congress by James Schlesinger, a ubiquitous presence in the formation of energy policy in the 1970s, acting now in his capacity as Carter's chief energy advisor, described the Clinch River demonstration project as "highly expensive" and "superfluous," even though he himself had announced the project's creation while head of the AEC five years earlier under Nixon. Clinch River eventually died a quiet death. The plant never produced a single watt of electricity.

In the wake of Three Mile Island—facing profound public skepticism, experienced and well-organized opposition from citizen and environmental organizations, as well as massive government regulation and escalating costs—the move to nuclear power in the United States entered a hibernation from which it is yet to emerge. Our most recent nuclear power plant began generating electricity in 1996, more than two decades after its construction commenced in 1973. Today, we get about 20 percent of our nation's electricity from 104 nuclear power plants, far less than was expected and hoped for when the 1970s began. In 2010, those plants accounted for nearly three-quarters of the electricity produced from sources that emit no greenhouse gases.

But even today the environmental and safety concerns about nuclear plants have not been put fully to rest. After several decades, we continue to debate how and where to dispose of nuclear waste. In 2010, three decades and more than $10 billion after the idea took hold, President Obama called for ending funding for a permanent nuclear waste repository at Yucca Mountain, about 100 miles northwest of Las Vegas, but a three-judge panel of the NRC's Atomic Safety and Licensing Board said that a 1982 law required the administration to pursue this avenue. To date, instead of being locked away in a secure mountain enclosure, the highly radioactive fuel rods have since the mid-1980s been "temporarily" stored in steel and concrete containers located at or near the power plants that have used them.

Moreover, the nuclear power plants built a generation ago are not all aging well. A discovery in May 2009 of a leak of about 100,000 gallons of coolant water from an underground pipe at the 40-year-old Indian Point 2 nuclear plant in Buchanan, New York, raised a new set of concerns about public health and safety just as many of the plants that came on line in the 1970s are applying to renew their operating licenses. Water leaks have also occurred in at least three plants in Illinois and one in New Mexico. Reprising the old conflicts, the NRC said these issues were trivial, whereas the chairman of the House Subcommittee on Energy and the Environment, Democrat Edward Markey of Massachusetts, claimed the leak "may demonstrate a systemic failure of the licensee and the commission."

So the beat goes on. Nearly a century after H.G. Wells's prediction of a "world set free" by abundant and inexpensive nuclear energy, that vision remains unrealized.

The halt in developing nuclear power struck a serious blow at our nation's effort to achieve energy independence with a carbon-free fuel (not that anyone was paying any attention to carbon emissions then). Such a halt did not happen everywhere, though. Both France and Japan, for example, built large parks of nuclear reactors beginning in the 1970s, and nuclear power now provides nearly 80 percent of France's electricity and about 33 percent of Japan's. In the United States, however, the path toward nuclear power hit a dead end in 1979. Facing the political and economic limitations on our ability to rely safely on this "fuel of the future," we turned back to our past—toward coal, the dirty, bulky, inconvenient, but abundant compound that a century earlier had fueled the Industrial Revolution.

5 The Changing Face of Coal

In a country still reacting to the 1973 Arab oil embargo, many policymakers regarded coal (along with nuclear power) as the nation's energy savior. In contrast to price controls on oil, Richard Nixon released all controls on the price of coal in March 1974. The United States was the "Saudi Arabia of coal," and coal had become, in Gerald Ford's words, our "ace in the hole." We had more coal reserves than any other nation in the world—enough to supply power for 300 to 500 years, depending on the estimate. Coal, supposedly aided by developing technology that would make it cleaner, more efficient, and able to fuel automobiles, was going to free America from OPEC's oil cartel. Coal companies, like starving men at a banquet, rushed to capitalize, and they were soon enjoying sales and earnings increases like never before. Congress in 1974 allowed the EPA to suspend emissions regulations to encourage power plants to use more coal and, along with our presidents, urged oil-burning power plants to switch over to coal-burning furnaces.

When coal replaced wood 200 years ago, it had fueled the Industrial Revolution. After World War II, plentiful and cheap coal had powered a second energy innovation that ushered in a new era of economic growth. When the oil producers of the Middle East flexed their muscles, coal was no longer viewed as a fuel of the past—despite its being dirty, difficult, and dangerous to mine, costly to transport, harsh on the environment when burned, and riddled with a history of labor unrest.

In December 1977, negotiations between the United Mine Workers (UMW) union and the mine owners' Bituminous Coal Operators Association reached an impasse over whether UMW locals should have the right to strike whenever a majority of miners at the site agreed, regardless of whether there was a national contract in place. The miners believed the right to strike was fundamental to local monitoring and maintenance of safety conditions in the mines as well as a key to rehabilitating their own corrupt union by returning decision making to the workers themselves. The Operators Association saw the local right to strike as union authorization of increasingly common wildcat strikes, a phenomenon that significantly undermined workforce stability. Neither party would budge, and a strike seemed inevitable.

Most miners stopped working 48 hours in advance of a midnight deadline; coal companies did not schedule Monday production; and in anticipation, electric utilities, steel mills, and other major industrial users had already stockpiled enough coal to last up to 90 days. *Time* speculated that the strike would initiate "a long stoppage, perhaps twice as long as the 32-day walkout in 1974." As expected, when the contract expired on December 6, a total of 165,000 miners walked off the job.

Despite the central role of strikes in the history of our country, now early in the twenty-first century, with shrinking car giants and weakened unions, we have grown unaccustomed to the idea of the big strike in which organized workers confront their management by withholding the vital resource of their labor, thus demonstrating their centrality to the smooth functioning of society. In the late 1970s, however, such a strike was still possible.

As the coal miners' strike wore on through the winter months of 1977–1978, the drastic cutback in coal production led to an energy squeeze felt most dramatically in coal-producing states across America's heartland. Although the unionized workforce included only half of U.S. miners and was concentrated in the eastern states, attempts to substitute coal from western, nonunion states remained difficult then due to both transportation logistics and threats of violence by striking miners, who sometimes held up coal deliveries at gunpoint. By February 1978, the prestrike coal reserves built up by electric companies were disappearing. The city of Columbus decided to switch off its street lights, and Pittsburgh department stores dimmed their lights during the day, let display windows go dark at night, and scheduled a 30 percent cut in business hours. Across Pennsylvania, Ohio, and Indiana, government officials told residents to give up dishwashers, hairdryers, and other nonessential household appliances to conserve energy. Power companies urged an end to evening sports events and all outdoor lighting and proposed widespread cutbacks in retail business schedules. If these voluntary conservation measures failed, power companies suggested that mandatory factory shutdowns might be the next step, leading in turn to a devastating spiral of layoffs and an economic slowdown. Even the stability of the automobile industry came within the strikers' reach because Chrysler and GM depended on Ohio coal-fired electricity plants' remaining open and operational. Unlike the 1973 oil embargo, this coal crisis was inflicted from within our own borders, and although this fact helped to reduce its psychological toll, potential adverse economic consequences loomed large.

Leading the UMW was Arnold Miller, a self-described hillbilly and "proud of it." A "soft-spoken, lantern-jawed, curly-haired, second generation miner," Miller knew what life was like for members of the UMW: he had worked in the West Virginia coal mines beginning at age 16, from 1938 until 1970, when arthritis and black lung disease forced him into early retirement. Leaving the mines allowed Miller to enter union politics, and although he had led his local UMW chapter and had served as president of the Black Lung Association, he was rather inexperienced as a leader when his name came up as a potential challenger for UMW president in 1972.

He had decided to run for the leadership of a troubled institution. The UMW had been built into one of the nation's most powerful unions by a charismatic leader, John L. Lewis, who had served as its president for a remarkably long span, from 1920 to 1960. Lewis had won miners a place among the highest-paid industrial workers in the United States and had obtained generous health benefits and pension plans for them. From 1960 on, however, the union had been led by Tony Boyle, and it had been plagued by massive corruption and mismanagement and become best known for Boyle's sordid activities, which included embezzlement and ultimately murder. Indeed, after leaving office in February 1978, Boyle was found guilty on charges that he ordered the murder of a rival for UMW president, Joseph "Jock" Yablonski, and his wife and daughter. Yablonski, a popular union official, had challenged Boyle for the presidency in 1969 and threatened to expose serious corruption within the union's finances and retirement funds. Boyle told two colleagues, "Yablonski ought to be killed or done away with," and although Boyle was victorious in his reelection, he paid three men $20,000 from the UMW treasury to enter Yablonski's home in the middle of the night on December 30, 1969, and shoot his family as they slept. The killers, who were certainly not professional criminals, left a trail of beer cans and whiskey bottles for investigators to discover along the road to Yablonski's house.

When it came time for the 1972 election, the rank and file reasserted their belief in the UMW as an organization run by and for miners: Miller, who was determined to turn the UMW into a democratic organization, won with 70,373 votes to Boyle's 56,334. The *New York Times* declared, "Nothing like it had ever happened in the labor movement before."

Arnold Miller's election signaled a return of some legitimacy to the UMW but not the end of labor unrest. Unauthorized wildcat strikes became common in the 1970s. By the time of the big strike, these smaller work stoppages had cost the industry more than 2.3 million workdays compared to only 117,000 days in the 1960s. And contract renewal negotiations had led to a month-long strike in November 1974. Much of the miners' unrest stemmed from the fact that underground mining was (and remains) one of the most dangerous occupations in the United States. One miner explained what it was like to live with the knowledge that his job might be lethal: "When I kiss my wife goodbye every day, she doesn't know if I'm going to get back home that night." Health and retirement benefits as well as payments to widows were issues of continual concern. Even the right to strike locally, a central issue in the 1977 contract dispute, was related to concerns about safety. Although wildcat strikes were sometimes caused by small grievances, most of them were safety related, such as over the availability of emergency-evacuation equipment, the mines' structural integrity, and whether foremen were adequately responsible for the miners' safety. The miners' insistence that a local right to strike was the only way to protect themselves from unsafe working conditions underscored the inadequacy of congressional regulation of mine safety.

Despite organizational disarray, under Miller's leadership the UMW continued its legacy of promoting greater health and safety regulation of the coal-mining industry. Congress is often moved by anecdote, and it often legislates in response to accidents that rise to national attention. The first major mine safety regulation of the postwar era, the Mine Safety Act of 1966, was passed as a long-delayed response to an explosion that had killed 220 miners in 1952. The act required annual inspections for mines employing 15 or more workers. These requirements were strengthened with the passage of the Mine Safety and Health Act of 1969, legislation motivated by a 1968 explosion that killed 68 in Farmington, West Virginia. Adding to this legacy, Miller helped pass the more stringent Mine Safety and Health Act of 1977, a law that added safety and ventilation requirements and mandated four complete annual inspections of underground mines.

In emphasizing underground safety, however, this legislation didn't reach accidents such as the 1972 flood of Buffalo Creek, West Virginia. After three days of heavy rain, early one March morning a dam created by the residue from washing coal, called a "slag heap," gave way. A wall of water and sludge 20 to 30 feet high demolished Buffalo Creek Hollow, a 12-mile valley with a dozen coal mines and 16 mining towns. The flood carried homes, trees, cars, and even a wood frame church downstream. People, too—Drema Hopson's two daughters were swept out of her arms and drowned. Of those affected, 125 died, and 4,000 were left homeless. Pittston Company, the owner of the subsidiary Buffalo Mining Co., called the disaster "an act of God," but the company itself was subsequently found liable for much of the flood's damage. Reports released by the U.S. Geological Survey disclosed that in 1967 West Virginia state officials had been told that the dam at Buffalo Creek, although stable, could be breached by water. The miners insisted that they had known the dam was dangerous for years.

Despite the very real dangers of coal mining that to this day remain—29 workers were killed in an April 2010 explosion at the Upper Big Branch Mine in Montcoal, West Virginia, owned by Massey Energy Company, an operator that had been cited for numerous safety violations—the 1970s legislation undoubtedly increased mine safety. In a 1974 speech, Miller remembered miners from the past who had "labored their lives away in the bowels of the earth and reaped as their reward a back bent like a stunted tree and lungs that did not work because they were full of coal."

The increased regulations also increased the price of coal production. After the 1969 act, one coal company estimated that production had dropped by 15 to 25 percent, and costs had increased up to 30 percent. Within five years, 500 marginal eastern mines closed due to higher operating expenses, and by 1974 productivity per worker had dropped from a high of 16.5 tons a day to 12 tons.

But an even more consequential shift in the production of coal was taking place far west of Appalachia in a region of the country where unionized labor had never taken hold. The so-called Powder River basin of southwest Montana and northeast Wyoming covers 20,000 square miles, an area larger than New Hampshire and Vermont combined. The Powder River, which in the decades surrounding the Civil War often washed away the blood of soldiers, settlers, and Cheyenne and Sioux Indians, is only one of many rivers that run through this expansive, wind-swept area. Underneath the thin topsoil there lay hundreds of billions of tons of coal—40 percent or more of the total U.S. supply. In the early 1970s, its extraction was just beginning.

Much of this coal was owned by the federal government, and the leasing policies for removing it would remain controversial for decades. The Northern Cheyenne, the Crow, and the railroads had the rights to most of the rest. This complex ownership produced numerous stops and starts in mining. The Crow tribe came to view its coal rights as providing a path from poverty to prosperity. In 1972, it granted Westmoreland Resources the rights to mine for a royalty of 17.5 cents for each ton of coal. Westmoreland simultaneously entered into contracts to sell 77 million tons of coal to four midwestern electric utilities for the next 30 years. After the 1973 oil crisis, however, the Crows renegotiated their royalty up to a range from 25 to 40 cents a ton, and when the new contract was finally signed in December 1974, they also got an advance payment in excess of $1 million. By 1978, Westmoreland was mining and selling 5 million tons a year. The Crow royalties amounted to about $200 annually for each member of the tribe—far from the prosperity they had hoped for.

The strip mining of coal made many Montana and Wyoming ranchers miserable—and not just over what they regarded as the desecration of their land to supply electricity to people they cared little, if anything, about in faraway places such as Dubuque, Iowa, and Minneapolis, Minnesota. The additional workers (and their families) accompanying the 1970s mining boom produced new demands for housing, schools, medical facilities, and police and fire protection. As many as half-a-million people moved into this sparsely populated area.

The town hit probably hardest by the influx of miners was Gillette, Wyoming, the town closest to the Powder River basin. Between 1970 and 1977, its population doubled from about 7,000 to 14,000 inhabitants, nearly 40 percent of whom lived in mobile homes. Little towns such as Forsyth, Montana, grew from 2,000 to 2,700 people in two years, also with many of the newcomers living in trailers. Bar fights between ranchers and miners soon became common. Bumper stickers appeared on cars with such sentiments as "Let the Bastards Freeze in the Dark."

But not everyone was unhappy. The young Forsyth mayor Gene Tuma pointed to businesses such as the new feed store and the new Ford dealership and said, "It's an expanding situation. The economic impact is really great." As one resident observed, "It's back to the olden days; and the ranchers having taken over Indian lands, now face the same forces the Indians fought—the railroads, the mining interests, and the Federal Government."

The railroads not only owned some of the coal, but were also building many new lines to transport the coal to the utility companies who bought it. They constructed numerous new spur lines to the mines to pick up the coal, and the Burlington Northern Railroad in the Powder River basin built the longest continuous stretch of new railroad line in the United States since the early 1930s.

Western coal had long been mined in Wyoming and Montana, but production had slowed after the railroads replaced their steam locomotives. The heat content of western coal is lower than that of coal mined in the eastern United States, but it has about one-seventh the sulfur content of eastern coal. To some extent, the 1970 Clean Air Act had made western coal more desirable. But the spike in oil prices, coupled with the new emphasis on energy independence, provided the catalyst for the boom in western coal production.

Western coal, which lies near the surface, is cheaper to mine than eastern coal but more expensive to transport the long distances to where it is burned to produce electricity. Because mining the old-fashioned way was no longer as profitable—especially given the advent of Nixon's price controls in 1971—coal moguls turned their attention west, to what would become the future of mining. Instead of going underground, they would "strip mine" deposits along the surface. This process allowed them to replace workers with increasing mechanization. Although strip mining had been around since the 1930s, by 1974 it accounted for about half of all U.S. coal production. And although it had become common in Appalachia by the early 1970s, its importance for the future was mainly in the West, where large amounts of coal could be found in seams close to the lands' surface. As *Time* reported, "All the major companies have lately bought or leased rights to hundreds of millions of tons of coal that lie close under the plains of the Dakotas and Montana, the semi desert of New Mexico, the basins of Colorado and Wyoming." By the late 1980s, Wyoming had become the nation's largest coal-producing state—a status it retains today.

The extraction of western coal soon unsurprisingly drew fire from environmental organizations. The Sierra Club and the National Wildlife Federation filed a lawsuit to halt strip mining in the Powder River basin, alleging that NEPA demanded an environmental impact statement assessing the environmental consequences of strip mining on the region as a whole rather than on just one area in a project-by-project basis. The federal district court disagreed, but the D.C. Circuit Court of Appeals issued an injunction in early 1975 stopping the mining. In an opinion by Judge J. Skelly Wright, joined by Chief Judge David I. Bazelon, this court held that NEPA required a statement evaluating the combined environmental impact of five separate projects. In his opinion, Judge Wright brooded about the impact of coal mining on a part of the country "best known for its abundant wildlife and fish and for its beautiful scenery—a region isolated from urban America, sparsely populated, and virtually unindustrialized." In January 1976, the Supreme Court lifted this injunction and in June issued a decision rejecting the environmental organizations' claim that NEPA required an environmental impact statement assessing the combined effect of major strip mines on the region.

But what happened to the strip mines once the coal was gone? As companies left mountainsides scarred and torn, the residents—and environmentalists—were outraged. Even in 1971, before surface mining had truly taken off, it had affected 1.8 million acres in Appalachia, and the U.S. Department of Interior had estimated that it would cost at least $250 million to repair damage already caused. Not just aesthetics were at stake; more than 10,000 miles of once-clear streams were "contaminated by acids, sediments and metals draining from exposed coal beds," and debris from mines threatened residents with monster landslides. In the early 1970s, proposals for legislation on both the state and federal levels were brewing, but it took until 1977 for Congress to pass (and the president to sign) the Surface Mining Control and Reclamation Act (SMCRA). Congress had passed two earlier versions in 1974 and 1975, but President Ford had vetoed both, claiming that the legislation would restrict the nation's energy supply by increasing the costs of coal production, raise inflation, and harm the coal industry when the country most needed it to thrive. In 1977, President Carter signed the bill, fulfilling a promise he had made in 1976 while campaigning in Appalachia.

SMCRA had a dual purpose—regulation of active mines and the reclamation of abandoned mines. Like much other environmental legislation, this law relied on the states to formulate regulatory programs and seek approval from the federal government. Until states had federally approved programs in place, the federal government had authority to oversee basic SMCRA compliance. (When Ronald Reagan was elected president in November 1980, Carter's Office of Surface Mining scrambled to approve 15 state program applications, fearing that under Reagan's leadership lax state programs would be allowed to proceed. That fear was not unfounded; even in the 1990s, several states enacted provisions suggesting that the state's program be "no more stringent" than the baseline federal program, an interpretation that frustrated SMCRA's intent to put a floor, not a ceiling, on this environmental regulation.)

The act's purpose was frustrated on many levels. SMCRA required most strip-mining operations to restore the "approximate original contour" of the land when possible, and a key tenet of the legislation required companies to purchase reclamation bonds to ensure that the land would be rehabilitated once mining was complete. But the bonds didn't cover key expenditures to protect environmental safety, such as the costs of separating acid-producing material from other debris, and rarely provided sufficient funds to cover the true cost of rehabilitation given the inflation during the years of mining. Federal oversight of state regulatory activities also lessened over time—by 1996, the Office of Surface Mining required federal intervention only when state decisions posed a threat to the environment or public health. Although the law required minimal duties to rehabilitate abandoned strip mines, it did impose substantial new costs on the industry—and on the federal government. By 2005, the federal government had paid more than $5.5 billion to reclaim old mines and more than $1 billion to support state SMCRA programs. Like oil, coal was an energy source whose price to consumers understated its true costs.

Commercial mining of western coal not only brought the issue of land reclamation into sharp relief, but also dramatically changed the role of mine workers in the United States. With dynamite, bulldozers, giant shovels, and gargantuan dragline equipment, coal companies could peel back the earth and dig out the coal beneath—a venture requiring only a few (and rarely unionized) workers whose productivity was high. Per worker day, strip mining was more than twice as productive as underground mining, and it was much safer: black lung disease and underground accidents were all but eliminated by surface mining. The effect on the UMW was profound.

By mid-February 1978, UMW's big strike still had not ended, and with reserves of coal dwindling across the Midwest, President Carter brought both sides to the White House to press for a settlement. He threatened further intervention if the parties could not come to an agreement, including invoking his power to compel the workers to return to the mines. Negotiations broke down again in early March, when the UMW rank and file voted down the proposed contract by a two-to-one margin. Carter felt he had no choice but to trigger the national emergency provision of the Taft–Hartley labor law. A federal district court judge granted a temporary injunction ordering the miners to return to work on March 9. In a televised speech, Carter explained, "My responsibility . . . is to protect the health and safety of the American public." "The law will be enforced," he added. But the miners ignored the injunction. One miner explained, "Taft–Hartley is O.K. for wartime. But it's not going to get me back down there except at gunpoint. Even then, no gun can make me work faster than I want to, and I can work awful slow when I put my mind to it."

Despite his promise to enforce the law, Carter recognized the difficulties of making recalcitrant miners go back to work, and his administration made little attempt to put the court's order into effect. When the government went back to court on March 19 and asked that the injunction be made permanent, Judge Aubrey Robinson Jr. questioned the existence of a national emergency, particularly given the government's lackluster attempts to reopen the mines. The Carter administration took no further court or administrative action to end the strike. Across America, two-thirds of people polled thought that Carter had performed poorly during the strike, doing too little too late.

The ones who suffered most acutely from the strike were the miners themselves. Without wages or health insurance (which lapsed when the miners went on strike) or pension checks once the union's funds dried up in February, the miners were hurting. Resourcefulness abounded—miners in the hills of Appalachia hunted for food, and their families canned fruit and vegetables; miners also stockpiled coal just as the utilities had. But mining families faced serious hardship as entire rural communities became unemployed, lacking money for food and hospital bills, let alone Christmas presents for their children. Jerold Hamrick, a thirty-four-year-old miner and father of four explained, "The miners know they're hurting, but they've suffered this so long they're gonna stick it out." As months passed, however, miner sentiment grew less unified while financial burdens piled up. Nonunion western mines increased their production to fill gaps caused by the strike in the east, and warmer weather both lessened demand and facilitated transportation of coal to areas in need. By the time the third contract proposal went out for a membership vote, it appeared that the contract had at least a 50–50 chance of passing. As one miner quipped, "Principles are nice, but you can't buy food with them."

When the miners voted on the third proposal on March 19, 1978—Good Friday—about 57 percent of the rank and file voted in favor. The new contract offered a wage increase of 37 percent over three years, up to $11.40 per hour, but it also included some important benefit concessions on the part of the UMW. The miners would pay an annual maximum contribution of $200 for medical care (previously free of charge), and retired miners would see a $50 pension increase, rising up to $275 per month (instead of the $500 per month benefit demanded by the UMW). The strikers also failed to earn a guarantee on the one issue that had spawned the strike in the first place: the local right to strike. Coal operators, who had lost 2.5 million man-days of work in 1977, wanted contract provisions to punish or fire wildcat strikers, but these provisions failed to make it into the final agreement, where the question of punishment was left to an arbitration board. By the time a new contract was ratified by the UMW membership, coal miners had been out of work for an agonizing 111 days.

In many ways, the 1977–1978 strike was the UMW's last great stand. To the extent that there had been any doubt that mining underground the old-fashioned way was less safe and less profitable than strip mining, the strike erased it. Replacing workers with mechanization, despite the large up-front capital costs, seemed like a winning strategy to the coal owners and operators. And the environmentalists were hardly more friendly to eastern than to western coal.

Furthermore, the strike had come at a bad time for the industry, just when President Carter was calling for doubling coal production and encouraging electric utilities to switch from oil and natural gas to coal. It called into question the reliability of the coal industry, especially in the eastern United States, and demonstrated that labor unrest—even without any increased reliance on coal as an energy source—might teeter the nation over into a national crisis.

In 1970, when Congress passed the Clean Air Act, sulfur dioxide emissions had reached an all-time high. Coal-fired power plants were responsible: their sulfur dioxide output had doubled every decade between 1940 and 1970. In 1971, President Nixon proposed a "sulfur tax" to curb one of "the most damaging air pollutants"—one he said that was "linked to increased incidence of diseases such as bronchitis and lung cancer." This tax would have been the first pollution tax of its kind; although Nixon argued—convincingly and well ahead of his time—that such a tax would establish "the principle that the costs of pollution should be included in the price of the product," few outside of his administration supported the measure. Coal companies claimed it was impossible to curb emissions at a reasonable price, and environmentalists shortsightedly criticized the level of the tax, arguing that the companies would pay because it would be cheaper than investing in cleaner technologies. They also complained that the tax would be passed on to consumers, which, of course, would have dampened demand for the polluting product. A sulfur tax was never adopted. This would be far from the last time that coal and other energy-producing interests and objectors from environmental organizations would defeat proposals to tax emissions.

Ironically, despite higher emissions, the air quality in cities was improving because of one simple trend: taller smokestacks. The height of smokestacks was a heavily contested issue during this time, one negotiated between coal companies, governments, and environmental groups both in and out of court. Although higher smokestacks meant that pollution was less concentrated in industrial areas, the pollution spread to remote areas. Throughout the 1970s, scientific evidence began to link "acid rain"—which was killing fish, disrupting ecosystems, and causing trees to brown in places such as the Adirondack Mountains—to sulfur dioxide emissions, sometimes emissions very far away. This link was an important disappointment of the 1970 Clean Air Act: even though it had regulated sulfur dioxide emissions, air pollution from acid rain increased after its passage, and taller smokestacks only exacerbated this problem. Despite convincing research to the contrary, in 1980 the head of the National Coal Association characterized acid rain as "a campaign of misleading publicity."

The cost of reducing emissions was surprisingly cheaper than anyone imagined—in fact, 20 times cheaper than some industry estimates—because of the availability of low-sulfur coal in the western states and improving technology. Plants could switch to low-sulfur coal, or they could purchase expensive "scrubbers," devices that sprayed rising smoke with a liquid or powder that chemically bonded to the sulfur and removed it from the exhaust. Because of their political sway, eastern coal interests played a decisive role in shaping amendments in 1977 to the Clean Air Act, which mandated the use of scrubbers. If that legislation had simply prescribed emissions standards, it would have favored lower-sulfur, non-unionized, western coal and imposed greater costs on eastern coal, which had long benefitted from not being required to reflect in its prices the environmental harm it was causing. In what Bruce Ackerman and William Hassler describe in their book *Clean Coal, Dirty Air* as a "bizarre coalition between clean air and dirty coal forces" in "the service of regional protectionism," high-sulfur, eastern coal producers and the UMW joined together with some environmental organizations such as the Sierra Club to insist on mandatory universal "scrubbing"—an expensive process to remove sulfur from coal. The environmental groups viewed mandating scrubbers as a way also to improve environmental quality in the western United States. This coalition decided to "exploit congressional ignorance to serve their own, mutually incompatible purposes," and, as a result, Congress produced a "hopelessly incoherent mix of statutory language and legislative history." "Clean air symbols," Ackerman and Hassler say, "were being manipulated in plain view on behalf of the coal mining industry." Electricity consumers paid the price for this odd enviro–business alliance.

In 1978, environmentalists and pro-coal business interests joined together again in an unlikely coalition to encourage the nation to shift from oil-based to coal-fired electricity. This so-called Coal Policy Project, which one wag described as "the industrial lions and the ecological lambs," had come together when leaders of Dow Chemical Corporation and the Sierra Club sought common ground on a range of issues, including how coal should be mined and how electricity sources and prices should be regulated. Eastern coal once again benefitted when the coalition urged that the bulk of the nation's future coal should come from the Appalachian states, Illinois, and Indiana and from deep underground mining. As chapter 13 describes, a somewhat similar coalition joined forces three decades later to promote a cap-and-trade policy in legislation to address the risks of climate change from greenhouse gas emissions.

As chapter 12 details, it wasn't until 1990, during the presidency of George H. W. Bush, that Congress specifically addressed acid rain, requiring power plants to reduce sulfur dioxide emissions by 50 percent by 2010. Congress took a different approach to reduce coal-fired electric plants' emissions of sulfur dioxide and the acid rain they produce: it created a market-based trading system, called "cap and trade," which allows polluters to choose how to reduce pollution and, through the trading system, to decide where to reduce pollution. Although efforts to reduce sulfur dioxide emissions have been largely successful, some environmentalists call for more stringent regulation, citing acid rain as a continuing problem. In some areas of the United States, rainfall can be up to 10 times more acidic than usual, wreaking havoc on the ecosystems where it falls.

But the problems with coal emissions do not end there. Sulfur dioxide, although invisible, deflects light—and it is one of the main reasons that average visibility in the eastern United States is only 14 miles. Burning coal also produces one-quarter of the nation's nitrogen oxide, an important contributor to smog, and coal emissions have led to high mercury levels in fish, causing brain damage to fetuses and children. One study, cited by Barbara Freese in her book *Coal: A Human Story*, estimates that power plant emissions kill more than 30,000 people each year and are responsible for "tens of thousands of hospitalizations, hundreds of thousands of asthma attacks, and millions of lost work days yearly." Freese does the math, arguing that these results, if true, prove that coal emissions should be a top-priority public-health concern: "It would mean that coal kills nearly as many people per year as traffic accidents . . . and causes more deaths than homicide . . . or AIDS."

In the 1970s, however, when coal was poised to become the linchpin for the future of an energy-independent America, these environmental concerns were trumped by the ever-present and seductive promise of a technological fix. The hope was that technology would filter out pollutants, making coal a cleaner-burning fuel and would enable a commercially viable coal-to-gas-and-liquid process, a "synfuels" process—so that even cars and airplanes might soon run on fuel products made from our abundant domestic supply of coal. In 1961, President Eisenhower had established the Office of Coal Research (OCR) in the Department of the Interior, and when the oil embargo put a squeeze on energy supplies, more attention than ever before was put into the development of these synthetic fuels. President Nixon's Project Independence began pouring money into the OCR—$43 million in 1973, which rose to $261 million in 1975. Before the oil embargo, the OCR's synfuels program had concentrated on coal gasification; after the embargo, making liquid fuel became its priority. When Jimmy Carter assumed the presidency in 1977, he made synfuels a key element of his new energy program for the nation.

Although coal liquefaction processes date back to the early twentieth century, the history of synfuel projects surprisingly runs through Nazi Germany. In the 1910s and 1920s, German chemists developed two processes to hydrogenate coal into oil. Later, in 1935, when the United States and European powers threatened an oil embargo in response to Mussolini's invasion of Ethiopia, Hitler decided to use this technology to render Germany independent of foreign oil. At the height of German production efforts, coal liquids accounted for about 90 percent of Germany's aviation and motor fuel. In the early 1940s, American officials began to take note. Feeling threatened, Congress (supported by key officials in the War Department) began a program to construct demonstration plants to produce American-made synthetic fuels. The U.S. Bureau of Mines went so far as to establish a mission that followed Allied armies in 1945, searching synfuel plants for any information that might be helpful to burgeoning American efforts. After the war, when the U.S. Operation Paperclip brought more than 600 German scientists to the United States, seven had been involved in Germany's synthetic fuel industry. From the mid-1940s to the mid-1950s, a total of six demonstration plants were built in the United States to experiment with synthetic fuels. After Nixon's enhanced investment in the OCR following the oil embargo, energy companies also partnered with the government to fund a synthetic fuel pilot project in Catlettsburg, Kentucky. The Nixon administration insisted that one-third of the costs of all new plants come from private sources. That project later failed due to mismanagement.

In his 1975 State of the Union speech, Gerald Ford called for enhancing the development of commercially viable synfuels technology, and he established the Synfuels Emergency Task Force to achieve his goals. That group concluded that the combination of costs and risks was too high to expect any commercial development without substantial government subsidies and price guarantees.

Despite the government's numerous failures to develop commercially viable synfuels from coal, Jimmy Carter was also determined to substitute fuels from coal for oil and natural gas—a goal that he intended to achieve by increasing domestic coal production (by 1985, to more than a billion tons per year—an increase of two-thirds) and developing synfuels. In 1979, when oil prices spiked again and public concern about energy policy once again took center stage, Carter announced plans for a government corporation to help bankroll production of synthetic fuels with the goal of producing an output of 2 million barrels per day by 1990. At one point, Carter estimated that this project could cost as much as $120 billion. The president ultimately proposed an $88 billion plan (which he wanted to be funded by a windfall profits tax on oil) to fund synthetic fuels from coal and other domestic sources. After months of debate, in May 1980 Congress approved Carter's plan to fund construction of a network of synfuel plants (without any funding from a windfall profits tax). This bill created the Synthetic Fuels Corporation, which, with $17 billion in start-up capital, was expected to act as a government-funded investment bank. Its purpose was to provide enough financial sweeteners necessary to stimulate industry to develop and improve synfuel technology. Funded by the new government entity, construction began on a commercial, public-private partnership facility intended to produce 50,000 barrels of synthetic fuel a day. Coal companies were on board with the project, and car companies were too. In 1980, they predicted coal-powered cars would be on the market by the 1990s. Another energy prediction gone wrong.

When oil prices leveled off and fell during the period 1980 to 1985, private partners pulled out of the Synthetic Fuels Corporation, which continued to face both technical and economic problems. After Ronald Reagan became president, he reorganized the corporation into his Clean Coal program, which was designed to rely on industry involvement as opposed to government leadership. It soon became clear that regardless of the price of oil, synfuel would likely never be economically competitive. Much of the political support for synfuel in Congress has come from an ill-fated ongoing effort to maintain and improve the economic viability of eastern coal. The real benefit of synfuel was supposed to be its potential to address the national security objective of energy independence, but on July 31, 1985, Congress cut off funding to the Synthetic Fuels Corporation, sounding the closing bell for Jimmy Carter's vision of a synfuel-powered nation. Even though the idea of synthetic liquid fuel from coal reached back in time to a technology developed more than a half-century earlier, in the end production simply proved too costly. Notwithstanding all the money poured into R&D, it soon became clear that coal would not shape our nation's ability to decrease our reliance on oil to fuel our transportation.

Coal operators are the first to say that the coal business has long cycled between boom and bust. By the late 1970s, coal prices had fallen from their mid-1970s highs. The combination of increased supplies from the western states and declining demand due to conservation efforts was taking a toll on coal companies' profits.

Despite its environmental costs, however, coal remains a hugely important ingredient in our current energy mix, especially for generating electricity. Indeed, coal currently supplies half of all the electricity in the United States (hydroelectric dams, natural gas, and nuclear energy fill in nearly all of the rest). The world of coal mining, meanwhile, has changed drastically since the 1970s. Three-quarters of the coal mines operating in 1976 have closed, and only about 72,000 miners remain working—less than half the number that participated in the 1977 strike. The UMW, once a powerful (and sometimes powerfully corrupt) union, now represents only 20,000 workers—and those workers produce less than one-fifth of the nation's coal.

The idea of synthetic fuel derived from coal has not yet disappeared, despite its failures. In 2007, the company Rentech claimed to be able to convert coal to a transportation fuel as long as it could be sold at a price of at least $45 a barrel. With oil hovering at higher than $100 a barrel at that time, the nation's energy supply once again seemed at the mercy of OPEC, and dollars were flowing in enormous quantities to petro-producing countries (most of which cannot be counted as friends). Absent a sequestration plan for carbon during the manufacture of its synfuel, however, Rentech's project would double the carbon dioxide emissions of automobiles and trucks, making it an unlikely solution to our transportation fuel concerns given worries about the risks of climate change from fossil fuel emissions, especially carbon dioxide.

If we are serious about moving to a world of carbon limits, coal will have to make up a far smaller percentage of our electricity-generating system. As subsequent chapters describe, carbon sequestration—a technique that might reduce emissions by burying carbon dioxide underground—is expensive and uncertain, and the problem of where to store the carbon that would be removed from the atmosphere is hard to solve. The term *clean coal* remains largely an oxymoron. The shortcomings of coal that led to its replacement by oil remain today. Technology has not changed that fundamental fact. Jimmy Carter's synfuel vision proved untenable, but he was spot on in his analysis of the need for a sound energy policy. More than 30 years ago President Carter urged Americans to take heed, warning, "The energy crisis has not yet overwhelmed us, but it will if we do not act quickly. It is a problem we will not solve in the next few years, and it is likely to get progressively worse through the rest of this century." Indeed, it has. Natural gas, however, offers a major exception.

6 Natural Gas and the Ability to Price

It was sunny, but a freezing and blustery 28°F that felt like 13°F with the wind chill on January 20, 1977, as Jimmy Carter, having just been sworn in as president, walked from the capitol toward the White House, holding hands with his wife, Rosalyn, and his daughter, Amy. The bone-chilling cold was not surprising that January afternoon; the winter had been far colder than usual. Three days earlier the temperature in New York's Central Park had been the lowest—one degree below zero—since weather records began being kept, shortly after the Civil War. And New York City was balmy compared to Cincinnati, Ohio, which also set a new record that day at minus 24°F. Similar records were being set all over the eastern half of the country.

The day following his inauguration President Carter urged the country to turn its thermostats down to 65 degrees, three degrees lower than Richard Nixon had asked for in the midst of the Middle East oil embargo. A week later James Schlesinger, now serving as the White House energy czar, reported to the House Subcommittee on Energy and Power that it had been 20 percent colder than normal nationally in November and December and that in the Southeast temperatures had been between 40 and 50 percent colder than normal. With the wind chill that January day, parts of Minnesota felt like 100 degrees below zero. The Ohio River had frozen for the first time in 30 years, trapping dozens of barges carrying fuel oil, road salt, and other products.

More than 200,000 workers had been laid off at least temporarily by then because their employers were unable to get natural gas to heat their plants. A similar number of schoolchildren had also been sent home. Governors of eight states in the Midwest and Northeast declared states of emergency. Natural-gas supplies were shrinking. The National Fuel Gas Company, which served western New York, northeastern Pennsylvania, and eastern Ohio, asked its school and business customers to close for at least a few days and requested that its residential customers lower their thermostats to 55 degrees. All four U.S. auto companies closed their plants, mostly in Michigan and Ohio, adding another 56,000 workers to the ranks of the unemployed. Hardships also spread to the Southeast. In Birmingham, Alabama, for example, United States Steel furloughed more than 4,000 of its 10,000 workers. GE, which used natural gas not only for heat but also to seal the light bulbs it manufactured, laid off 6,000 workers in lamp factories alone.

In the winter of 1977 and again the following winter, a genuine crisis in natural-gas supply caused enormous hardship and substantial economic losses for this nation, especially in states east of the Mississippi. The Carter administration later estimated that 1.8 million workers were at least temporarily unemployed that winter due to the shortages.

Not everyone was bothered, however. In the gas-producing states of Texas, Louisiana, and Oklahoma, natural gas was plentiful. These three states alone were consuming more than 40 percent of the nation's supply. In Texas, bumper stickers appeared saying, "Turn up the gas and freeze a Yankee" and "Let the Yankee bastards freeze."

People tended to blame the weather for the natural-gas shortages that were causing such hardship, but record-breaking cold was only partly at fault. Indeed, earlier in the decade during the less extreme winters of 1972 through 1975, many interstate gas pipelines could not obtain enough gas to fulfill their contractual obligations. For the year from April 1973 through March 1974 alone, gas curtailments involved as much lost energy as the nation experienced from the oil embargo in the first quarter of 1974, even though the 1973–1974 winter was not unusually cold. Because fuel oil frequently serves as a substitute for natural gas, these gas shortages increased the demand for oil, including imported oil, at the worst possible time. The culprit—though not everyone recognized it as such—was artificially low prices.

Natural gas was first used for street lighting, but it soon became a popular fuel for heating and cooking. By the early 1970s, it supplied nearly one-third of the total energy used in the United States. In 1972, about two-fifths of the natural gas consumed in the United States was used by industry, one-fifth for generating electricity. Residential consumers used about one-quarter of the natural gas produced, with the remainder used in commercial establishments. Because of priorities for delivery established by the federal government, residential customers were generally not threatened by the shortages, although throughout the Northeast and Midwest natural-gas distributors refused to add new customers.

Natural gas has been called the "perfect fuel" or the "energy prince" because although it is composed of hydrocarbons, it burns more cleanly than oil or coal. It is relatively free of carbon monoxide and sulfur dioxide. Substituting gas for high-sulfur coal or fuel oil, for instance, would eliminate the problem of acid rain. (Natural gas, however, is composed principally of a greenhouse gas that sometimes does leak into the atmosphere; when burned, natural gas emits carbon dioxide at about half the levels of coal.) Natural gas has long been abundant domestically, and virtually all of our consumption of it has been from domestic sources and Canada. Unlike nuclear power, this fuel does not pose a threat of a truly disastrous accident or catastrophe if it were to fall into malevolent hands or any insurmountable problem of waste disposal.

The biggest shortcomings of natural gas are that its energy content is relatively low by volume, and, as a gas, it is difficult to transport. To solve the latter problem, an extensive network of high-pressure pipelines, ranging in diameter from a little more than a foot to three and a half feet, have been constructed across the country. By 1970, the nation's pipeline grid totaled 250,000 miles. These pipelines transport gas principally from producing states, such as Texas or Louisiana, to locations all over the country where the gas is consumed. When the gas reaches the distribution companies in various cities, it enters either a secondary network of local pipelines or a storage facility.

About one-third of the nation's natural gas has historically been a byproduct of drilling for oil. There has never been a workforce comparable to the UMW for extracting natural gas. The political conflicts—and there have been many—have largely been between producers and consumers. Unsurprisingly, the former want high prices for their product, whereas the latter want low prices. Faced with shortages, however, many consumers would rather pay a premium than endure the hardships of inadequate supply. Of course, there are many more consumers than producers, and they are more widely dispersed. Pipeline companies are often caught in the middle: in general, they have preferred lower prices, which increase demand and allow them to transport more gas, but when shortages emerge and they are unable to fulfill their commitments, they, like consumers, would prefer higher prices and adequate supplies.

Pipelines are usually laid underground by only one supplier, making the system for delivering natural gas to homes, schools, and businesses a natural monopoly because consumers can't turn to alternative suppliers when the price rises. Thus, as with electricity in general, price regulation of natural gas was long regarded as an obvious response to its means of delivery. Other fuels for heating, such as oil, coal, or even wood, can be delivered to a homeowner in trucks by many competing suppliers. To be sure, one can convert a furnace to burn oil or coal instead of natural gas, but such conversions are expensive, and they impose other burdens on consumers: both oil and coal must be stored at the home or business until used, oil typically in underground tanks, coal in dirty soot-emitting bins. Natural gas, in contrast, flows from the pipeline into the home directly from the local distribution system and burns cleanly. Thus, state and local governments have long regulated the prices that can be charged by distribution companies.

The larger, national networks of pipelines that travel long distances to the smaller "gathering lines" that transport the gas from its wellheads (historically mostly in the Southwest and the Gulf Coast) to local distributors around the country also tend to be monopolistic. No one would consider building a competing pipeline. Although possible and now frequently done, converting natural gas into a liquid form for transportation has been both expensive and technologically challenging, so there was no practical alternative to the pipelines.

Pipeline construction is capital intensive but creates significant economies of scale. Owners of these networks assume the risks of a supplier's cutting off the flow of gas into a given pipeline and of a local distribution company's refusing to purchase the gas at the other end. As a result, companies wanted to control both ends of the supply chain. In the 1930s, shortly after the first pipelines had been built, 10 companies controlled 86 percent of the interstate transmission of natural gas. Just four companies—Standard Oil of New Jersey, Columbia Gas and Electric, Cities Service, and Electric Bond and Share—controlled about two-thirds of that. This concentrated control of natural gas supply led Congress to enact the Public Utility Holding Company Act of 1935, which ultimately resulted in the breakup of many of these companies but gave little or no price relief to consumers.

This antitrust legislation was soon followed by the Natural Gas Act of 1938. Because the pipelines often crossed state lines, regulation by the states was not practical, so this legislation granted the FPC regulatory control over interstate natural-gas pipelines. Thereafter, building or expanding a pipeline required permission from the FPC, the agency that also controlled the price of natural gas transported across state lines. For a long time, interstate prices were regulated, but intrastate prices were not. Intrastate prices depended only on market forces of supply and demand. The 1938 legislation remained in force nearly four decades later in January 1977, when Jimmy Carter came into office, but there had been some serious bumps along the way.

Setting prices for the transmission of natural gas through the pipelines had been relatively straightforward. The FPC allowed pipeline companies to recover the costs they incurred to move the gas through the pipeline, plus a return on their invested capital. But the issue of the FPC's right to regulate the price of the gas as it entered the pipeline "at the wellhead" remained unresolved. If the FPC was to limit the price to the consumer—and Congress wanted it to do so—setting rates for transportation was insufficient; the price of the gas itself had to be regulated. However, as a federal rather than a state agency, the FPC confronted a jurisdictional question because much of the gas came from independent local producers, who themselves did no interstate business.

The FPC ultimately claimed that it had no authority to set wellhead prices, but in 1954 the Supreme Court disagreed. The Court accepted an argument of the State of Wisconsin that under the 1938 law the state's consumers of natural gas deserved federal protection from the "excessive" prices that producers might extract at the wellhead. Although the Court required the FPC to determine "just and reasonable" prices, it did not tell the FPC how to set them, nor was the FPC itself confident about how to do so. The agency limited its task by interpreting the Court's decision as applying only to gas that crossed state lines, not gas that remained in the state where it was produced. This interpretation led to very different price regimes for interstate and intrastate gas.

To set rates, the FPC tried to use a "costs plus rate of return on investment" methodology similar to the one it had employed for the transmission of gas through pipelines, but given the large numbers of producers selling gas to the pipelines, the information determining the producers' costs and invested capital overwhelmed the agency. The FPC, facing nearly 3,000 applications for price review, decided in 1961 to set price ceilings based on average historical production costs in five separate regions. This "area rate" method proved workable, if imprecise and inelegant. The biggest problem was that because firms had begun drilling in the lowest-cost areas, this system gave them no incentive to expand into higher-cost areas. The prices they could charge would be inadequate to cover their higher costs.

From time to time, Congress tried to eliminate natural-gas price regulation, but before Jimmy Carter took office, it had never succeeded. In 1950, President Harry Truman vetoed legislation, sponsored by the millionaire oil and gas producer Senator Robert Kerr of Oklahoma and narrowly passed by both houses of Congress, to limit the FPC's jurisdiction. And in a truly bizarre turn of events, six years later President Dwight Eisenhower vetoed a price-deregulation bill that he had originally championed. Just as the Senate was about to pass this legislation, South Dakota Republican senator Francis Case, who also favored deregulation, revealed that a gas industry lobbyist had left him an envelope containing twenty-five $100 bills. An indignant Senator Case returned the money with great fanfare. (The Superior Oil Company, which had financed this payoff, was subsequently fined $10,000 for failing to register its lobbyists.) The Democratic Senate majority leader, Lyndon Johnson of Texas, accused Case of trying to derail the bill by exposing the payment, and Johnson pushed the legislation through the Senate anyway. President Eisenhower, who in his diary described "a great stench around the passing" of this legislation, then vetoed it.

By the late 1960s, the effort to keep prices low for consumers was taking a toll on the domestic supply of natural gas, and a shortage of natural gas began to emerge in interstate markets. However, in the unregulated intrastate markets, such as Texas or Oklahoma, gas was plentiful, albeit more expensive despite its lower transportation costs.

Interstate and intrastate prices had previously not diverged much, even though the producers' profits varied substantially. In 1968, for example, when the average price of interstate natural gas was 17 cents per thousand cubic feet, the intrastate price was about 23 cents. By the mid-1970s, however, the prices of gas that stayed in a single state were twice to four times the prices of gas that crossed state lines. Fortunately for buyers in the Northeast and Midwest during the bitter winter of 1976–1977, despite the price differential, more than half of total U.S. gas production was still finding its way into interstate markets because distributors and industrial users had entered into long-term contracts with the pipeline companies.

Because much domestic natural gas was then a by-product of oil drilling, when domestic oil production peaked, so did the production of natural gas. Interstate prices were so low that it was uneconomical to search for and retrieve natural gas alone. Fields that contained much more gas than oil were unattractive to drillers. In 1973, domestic natural-gas production peaked at 22.6 trillion cubic feet, falling 6 percent in 1974 and another 7.5 percent in 1975. Remarking on the decline of domestic natural-gas reserves, Ray Sauve of the FPC observed in 1976 that "in the past eight years, proved reserves in the lower 48 states have shrunk 42 percent. During that time we have consumed twice as much gas as we have found."

As interstate gas shortages grew more serious, local gas companies refused to add new residential customers, so builders shifted to oil (whose prices exploded following the embargo) or to expensive electric heat. One of the most absurd products of American home-building history—the "all-electric home"— entered the market and, despite the extremely high relative cost of electric heat, became a realtor's clarion call.

Both producers and consumers became more and more dissatisfied with the system of regulation. Producers, as they had all along, despised the depressed prices. Consumers, who previously had enjoyed artificially low prices, now faced shortages and found that low prices were of no avail when there was no gas to be had. Here is how the OECD described the situation: "United States policies . . . have been dominated by myriad regulations and various levels of regulatory agencies, sometimes in competition, which decreased system flexibility, delayed projects, and restrained price increases, thereby deterring both conservation and exploration."

In his April 1973 energy message, Richard Nixon urged Congress to decontrol natural-gas prices when existing contracts expired. Congress didn't bite, nor did it heed Gerald Ford's request to deregulate. Congress, confronting an oil shock that was draining voters' pockets, regarded the idea of decontrolling natural-gas prices as, in effect, allowing a foreign cartel to further dictate rising prices for our nation's second most important fuel—one produced domestically. The great irony, of course, is that the shortages that resulted from artificially low natural-gas prices required us to use and import more oil than we otherwise would have.

In 1976, when intrastate gas prices were averaging three to four times the interstate prices, the FPC finally took action on its own. The commission moved to a three-tier pricing system for interstate gas. "Old gas," which the FPC defined as gas in production by the end of 1972, would sell nationally at a price of $0.52 per million cubic feet; gas brought into production in 1973 and 1974 would sell for $1.01; and gas brought into production after January 1, 1975, for $1.42—with all prices increasing four cents annually for inflation. A few months later, the commission discovered that it had underestimated the increased costs of these changes to consumers by $500 million, and it lowered the price of 1973–1974 gas to $0.93 per million cubic feet and expanded its definition of "old gas."

The federal government assigned priorities to different users as it had with oil under Nixon's price controls. Residential users came first, industries using gas for feedstocks second, industries with gas boilers third, and electric utilities last. Gerald Ford continued to urge Congress to decontrol prices, but to no avail. This was the situation when Jimmy Carter entered the White House in that frozen winter of 1977.

In a stunning exercise of legislative prowess that many observers took as evidence that the new president and his fellow Democrats in Congress would be able to work together smoothly and with alacrity, on February 2, less than two weeks after his inauguration, Jimmy Carter signed into law the Emergency Natural Gas Act of 1977. That law, enacted only six days after Carter requested it and with vast margins in both the House and Senate, gave the president unprecedented powers over natural gas, including temporary authority to order transfers of gas from any interstate or intrastate pipeline to anywhere he directed. It also allowed interstate buyers to purchase gas in intrastate markets at unregulated prices, as approved by the president.

Despite its overwhelming support inside the Washington beltway, this law did not make everyone happy. Governor Dolph Briscoe of Texas complained to the Senate Committee on Interstate and Foreign Commerce: "To allow gas-starved interstate pipelines to suddenly enter our intrastate market and purchase this precious commodity amounts to nothing less than allowing consumers of other states, who for years have reaped the benefits of low-cost interstate gas, to now reap the benefits of an extensive drilling program, which has been financed entirely by the consumers of the State of Texas." In any event, everyone knew this law to be a quick fix, not a long-term solution. The 1977 natural-gas legislation terminated at the end of June that year in a warm summer month, long before shortages of natural gas had actually ended.

A few hours after signing the Emergency Natural Gas Act, Carter donned a beige cardigan sweater, light blue shirt, and red tie and sat down in an armchair in front of a blazing fireplace in the White House Library, just under the well-known Gilbert Stuart painting of George Washington that appears on the dollar bill, to address the nation. He talked about a catalog of issues, but he began with energy, congratulating Congress for its "quick action" on the Emergency Natural Gas Act and promising a comprehensive set of energy proposals by the end of April. He also announced that he would seek to replace the many different federal agencies that had responsibility for energy policy with a new cabinet-level energy department. He once again urged Americans to lower their thermostats to 65 during the day and 55 at night to "save half the current shortage of natural gas."

On the front page of the *Los Angeles Times* the following day, just below a picture of Carter preparing to speak, a headline announced, "Drillers Say Natural Gas Abounds—Deep Down." The article reported that, according to a group of wildcat drillers, the nation had "hundreds of trillions of cubic feet of undiscovered gas reserves . . . enough to fuel the entire country for 30 years at least." All that was needed, they said, was the technology to drill and the willingness to pay for more expensive gas.

But paying more for natural gas was not easy. In a bargaining fiasco reminiscent of Richard Nixon's refusal to accept the shah of Iran's offer to sell the U.S. enormous quantities of oil at $1 a barrel, James Schlesinger, negotiating on behalf of the Carter administration, failed to reach an agreement with Mexico's national oil company Pemex over the price of the 2 to 4 billion cubic feet a day of natural gas that Mexico wanted to export to the United States. Mexico wanted a price equal to that of fuel oil; Schlesinger insisted on the cheaper price of residual oil. When all was said and done much later in September 1979, after oil prices had once again spiked, the two countries agreed on a price. In the intervening years of the impasse, however, Mexico built a pipeline and gas-distribution system that enabled Pemex to sell virtually all of its natural gas domestically. The 1979 agreement provided for the United States to buy no more than 300 million cubic feet of gas a day—a ceiling far below the 2 billion cubic feet floor in place when the negotiations had begun—and gave Mexico the unilateral right to reduce its exports below even that level.

Having been trained as an engineer, Jimmy Carter viewed fashioning energy policy as essentially an engineering problem—one that simply required redirecting some resources and revising some longstanding policies and that could be solved if the president were willing to tackle the issues whole. And, like others who have held that office, Jimmy Carter regarded the president as the only elected official in a position to take into account and balance all the competing concerns of producers, consumers, and environmental advocates. He also viewed conserving energy, which he made the centerpiece of his efforts, as a moral imperative. He would wage a righteous crusade against waste. However, he greatly underestimated the political difficulties he would confront.

President Carter handed the responsibility for developing a comprehensive energy plan to James Schlesinger, who assembled a task force of experts to operate in secret to develop the National Energy Plan. The plan that emerged from this group and that Carter embraced was as ambitious and complex as any legislative proposal a president has ever sent to Congress. Three months after his inauguration—when his approval rating hovered about 75 percent and with a Congress controlled by his own party—Carter sent Congress a comprehensive energy plan containing 113 specific legislative proposals, including, for example, new taxes on gas-guzzling automobiles, taxes on utilities that burned oil or natural gas instead of coal, an oil price equalization tax, and a standby gasoline tax intended to create a floor on gasoline prices. He also proposed expanding dramatically the federal government's regulatory power over energy producers, suppliers, and consumers and offered new incentives for conservation. Although his proposals were extraordinarily complex, his overall policy goal was clear: to transition away from "cheap and abundant energy used wastefully and without regard to international and environmental imperatives to an era of more expansive energy with concomitant regard for efficiency, conservation, international and environmental concerns," as James Schlesinger put it.

The president addressed the nation from the Oval Office 89 days after taking office, describing his quest for a new energy policy as "the moral equivalent of war." Two days later, speaking before a joint session of Congress, he unveiled his National Energy Plan in all its detail and requested that Congress enact it promptly. It was not, however, until eighteen months later, in October 1978, on the last day of the congressional session, that the House and Senate reached an agreement on energy legislation and sent it to the president for his signature.

Like Ford, Carter had wanted to decontrol both oil and natural-gas prices, but he knew that even a Democratic president could not get the Democratic Congress to go along. A decontrol proposal would have been welcomed by powerful members of Congress from the oil- and gas-producing states, but widespread worries among other congressional leaders about the potential costs of taking responsibility for raising consumer prices eliminated any chance for its enactment. In addition, both Carter himself and Senator Henry "Scoop" Jackson (D.–Wash.), who chaired the Senate Energy Committee, were suspicious of what they called the "oil cartel." They were determined that any increase in oil prices be accompanied by a "windfall profits" tax on the oil companies. Given these cross-currents, Democratic representatives of oil-producing states and oil-consuming states were at loggerheads even though they wanted the first Democratic president since 1968 to succeed. By the late 1970s, the pressing need for more sensible national energy policies had long since become obvious, but the American public continued to regard cheap gasoline, inexpensive electricity, and low heating prices as an entitlement.

Carter's plan advocated an extremely cumbersome and complex formula devised by Schlesinger to increase oil and natural-gas prices, but the president did not propose deregulating the prices of either fuel, despite the fact that during his campaign he had urged deregulation of natural-gas prices. Rather than proposing complete deregulation of either oil or gas prices, which would have promoted his goals of encouraging production and discouraging consumption, Carter suggested a complex system of controls that would allow prices to rise gradually, especially for newly discovered oil and gas. Oil, for example, would be subject to a complex three-tier system of regulations. First-tier oil—from wells in production before 1975—would indefinitely be subject to a ceiling price of $5.25 a barrel; second-tier oil—from wells in production after 1975—would have a maximum price of $11.28 a barrel; and only oil from small "stripper" wells, which produce 10 or fewer barrels a day, would be permitted to sell at the worldwide market price. In order to foreclose a frenzy by wholesalers and retailers to acquire the cheapest oil, a so-called crude oil equalization tax would be imposed on oil priced below world levels, with all of its proceeds returned to consumers through income tax credits.

Carter's new pricing scheme would continue federal regulation of prices of interstate gas from existing wells and regulate the prices of both interstate and intrastate gas from new wells. The general idea was to allow natural-gas prices to rise to match the energy equivalent prices of oil. Carter told Congress that this equalization would provide producers an "adequate incentive for exploration, without taking the federal lid off prices altogether." Carter himself and many of his advisers knew, however, his plan with its byzantine price-control rules was incomprehensible to the public and to most members of Congress.

After announcing his plan, Carter did everything he could think of to sell it to the country. Officials of his administration not only appeared on the usual talk shows, such as *Face the Nation* and *Meet the Press*, but they also showed up in such popular venues as *The Tonight Show* with Johnny Carson, the *Merv Griffin Show*, and the *Dinah Shore Hour*. Lobbyists opposing the plan remained far away from public view, prowling the halls of Capitol Hill.

In the House of Representatives, Speaker Tip O'Neill (D–Mass.) formed a special committee to consider the legislation, and in August 1977, before Congress's summer recess, Carter's plan passed relatively intact. In the Senate, however, the legislation hit a wall. Democratic majority leader Robert Byrd of West Virginia, a man who always held the prerogatives of each individual senator in extremely high regard, separated Carter's plan into five separate bills, which he then referred to six separate committees for consideration. This approach fulfilled lobbyists' dreams. In October 1977, after 73 roll-call votes on five separate bills, the Senate passed a bill bearing only a little resemblance to what Carter had asked for. Connecticut Democrat Abraham Ribicoff correctly labeled Carter's plan a shambles. It would take another year for the House and Senate to reconcile their differences. The Senate voted more than 100 times on issues of natural-gas pricing alone.

Natural-gas prices proved to be the most controversial issue of the legislation. The Senate voted to decontrol prices altogether, whereas the House version of the bill retained controls but adopted a complex scheme that allowed prices to rise gradually over time. For months, the House–Senate conferees struggled over compromises that satisfied no one.

Consumer advocates resisted prices going up too fast; industry representatives complained that prices were scheduled to rise too slowly.

In March 1978, the nine Senate conferees announced a plan to eliminate federal price controls on new natural gas by 1985, but Michigan Democratic congressman John Dingell, who had been fighting against decontrol of natural-gas prices for 30 years, said that if the Senate conferees intended the compromise as a "take it or leave it" proposition, he "would be compelled to leave it." In a subsequent two-hour conference session open to the public, described by one reporter as like the Mad Hatter's Tea Party, the House offered its compromise response. After much rancorous talk, the conferees then adjourned for the Easter recess without setting any new meeting date.

Beginning on April 11, President Carter brought the parties to the White House for a series of meetings intended to break the logjam. Ten days later—a year and a day after Carter had presented his proposal to Congress—at a marathon 13-hour conference session that did not end until 3:30 a.m., the House and Senate negotiators, working closely with Schlesinger, who was now energy secretary, reached a tentative agreement on natural-gas prices. Under that agreement, both interstate and intrastate prices would be regulated by the federal government until January 1, 1985, when prices of new natural gas, most intrastate gas, and gas from new wells more than 5,000 feet deep in old reservoirs would be decontrolled—at least until July 1 of that year. Beginning then, for the next two years either the president or Congress would be authorized to reimpose price controls once for no more than an 18-month period. During the 1978–1985 period, natural-gas prices were to be established by the legislation, as were priorities for allocating gas in case of shortages.

The House Energy Subcommittee estimated that this plan would provide natural-gas producers (and cost consumers) about $23 billion more by 1985 than the House bill, $9 billion more than regulations then in effect. The Senate bill would have increased producers' revenues by $46 billion more than the existing regulations during the same period.

For the next several months, many members of the House and some Senators worked to derail this compromise, which, of course, had satisfied no one. Proponents insisted that the legislation would increase production; opponents said it would slow exploration and development. Proponents claimed it would help the dollar; opponents said it would weaken the dollar. Proponents believed it would curb inflation, opponents that it would increase it. Proponents said it would cut oil imports, opponents that it would increase them. In late September, when a compromise bill was about to be voted on, Robert Byrd described natural-gas pricing as "the most complex and divisive issue the Senate has faced in this 21-month session." A month earlier Jimmy Carter had described the legislation as "perhaps as complex as any that has ever faced the Congress."

The legislation's complexity was amply demonstrated at one conference committee meeting, when staffers attempted to explain to the conferees a priority for the allocation of natural gas to "essential agricultural uses," including the production of animal feed. Senator Howard Metzenbaum, a Democrat from Ohio never known as the brightest light in the Senate, objected, insisting that dog and cat food production should not supersede other uses. When the staff explained that the provision related to beef cattle and other essential agricultural purposes, Metzenbaum insisted on language explicitly excluding dog and cat food. Senators then suggested exceptions for sheepdogs, hunting dogs, and Alaskan sled dogs before rejecting Metzenbaum's amendment. Everyone agreed that this legislation contained many cats and dogs. One of President Carter's closest advisers, Robert Strauss, at a White House meeting just before Labor Day, confessed, "It no longer makes a difference whether the bill is a C-minus or an A-plus. Certainly it is better than a zero, and it must pass."

The six separate pieces of energy legislation that Carter signed into law on November 10, 1978, which included the Natural Gas Policy Act, were a far cry from the comprehensive energy strategy he had proposed a year and a half earlier. Many of his proposals, including virtually all of his proposed taxes to stimulate conservation and the production of alternative fuels, failed to make it. Carter got no oil decontrol, no crude oil equalization tax, and no stand-by gasoline tax (proposed to come into effect if lower consumption goals were not met). As James Schlesinger well understood, Congress was simply unwilling to impose immediate costs on the public for long-term benefits. As Schlesinger put it, "[The] basic constituency for this problem is in the future."

With respect to natural gas, the new law intended to permit prices of newly discovered gas to rise about 10 percent a year until 1985, when price controls were scheduled to end. The pricing formula Congress ultimately enacted was exceptionally complex, however, putting the early burdens of higher prices on industrial users, then shifting costs to residential consumers. Even though this law was ending price controls over time, in the interim Congress for the first time extended such controls to gas produced and used within the same state.

The Associated Press described the Natural Gas Act of 1978 as a "beastly tangle," "half intended to remove federal price controls over natural gas and half intended to continue and enlarge them." During the period from the end of 1978 until the scheduled decontrol in 1985, at least 26 different categories of gas were subject to various pricing regimes, which reflected a mixture of deregulation goals, distributional considerations, and favors to special interests. Prices ranged from 25 cents to several dollars per thousand cubic feet. Residential customers were assured lower prices than industrial or commercial users, but this distinction created enormous legal and economic complexities. For example, because the legislation immediately exempted from price controls any gas produced from wells of 15,000 feet or deeper but retained controls on gas from shallower wells until 1985, an expensive "deep-gas" boom occurred during 1978–1982, when for the same costs much more gas could have been produced from cheaper, shallower wells. The American Gas Association estimated that the 1978 law would increase household gas bills by a little more than 8 percent annually by 1985. In fact, residential gas bills rose more than twice as fast—an average of 19 percent a year from 1979 through 1983. But with its scheduled 1985 deregulation of natural-gas prices, the 1978 act did reverse a price control policy that had been in place since 1938.

Jimmy Carter's efforts to deal with oil and gas price controls, along with his quest to fashion more sensible and comprehensive energy policy, and the failure of these efforts were critical elements in both defining and hobbling his presidency—along with a terrible economy, brought about in large part by extraordinarily high interest rates set by the Federal Reserve in 1978 to halt inflation and the ongoing decline in the value of the dollar. One can easily imagine a diabolical smile crossing the face of Richard Nixon as he watched Carter—who had run for office as the non-Nixon (based on little more than the promise that he "would never lie to the American people") and had entered office as the moral counterpoint to Nixon's dishonest, immoral, and illegal actions—struggle and then flounder. Nixon's 1971 devaluation of the dollar wreaked havoc with Carter's presidency, and his oil price controls not only hamstrung energy policy but produced a conundrum for Carter from which he could find no politically beneficial and policywise escape.

In 1980, Jimmy Carter would lose the presidency to Ronald Reagan. As chapter 8 describes, Carter's frustrations and pessimism were overwhelmed by the sunny, optimistic Reagan. In addition to helping secure Nixon's reelection in 1972, the New Economic Policy that Nixon had instituted in 1971 paid the Republican Party a second political dividend nearly a decade later—although it should not be given anywhere near full credit for this success. By the time price decontrol actually occurred in January 1985, oil prices had declined and natural-gas prices were relatively stable. The rise in prices since 1978 had induced a burst of gas exploration and development, producing an excess supply that was expected to keep natural-gas prices from rising for at least the rest of the 1980s. In the mid-1980s, the U.S. Geological Survey estimated that U.S. natural-gas resources were 600 trillion cubic feet, more than two and a half times the amount that had been estimated by a similar 1975 study. The natural-gas shortages of the winters of 1976–1977 and 1977–1978 had spurred a fundamental change in our nation's policy with respect to pricing that fuel, and in a few years shortages were replaced with surpluses.

The enactment of the 1978 Natural Gas Policy Act provided the main legislative drama starring natural gas during this period, but other important events occurred as well. For example, in 1977 Congress approved President Carter's proposal for a pipeline to transport natural gas from Alaska's North Slope to the lower 48. (The Energy Department now estimates that such a pipeline should actually be transporting natural gas by 2020.)

Transporting natural gas by pipeline is not the only option. It has long been technologically possible to liquefy natural gas at very low temperatures, minus 259°F. The gas can then be transported in liquid form on tankers, removed, regasified, and sent into a pipeline or a storage facility.

In response to the oil crisis in 1973, some countries in western Europe and especially Japan, faced with shortages of domestic natural gas, turned to imports of liquefied natural gas (LNG) to replace or supplement the newly expensive and now uncertain oil imports that they had been relying on. Since then, Japan has generally accounted for at least one-half of all LNG imports. In the United States, some companies in the 1970s also turned to LNG imports as their likely salvation. They had to overcome considerable reluctance to do so, however. Use of LNG in the United States had come to a halt after an LNG storage tank in Cleveland in 1944 had spilled liquid gas into a sewer system and the gas had then produced a series of explosions and fires that killed 130 people and left an additional 225 injured. No commercial LNG facilities were constructed thereafter until the 1960s.

In 1974, Boston's Distrigas Company became the nation's first importer of LNG, importing it from Algeria. (Five years earlier Phillips Petroleum had begun the first U.S. LNG operation in Alaska, exporting gas to Japan.) In March 1978, an LNG port opened in the Chesapeake Bay also to import gas from Algeria. Columbia Gas and Consolidated Gas Company distributed that gas to their 7 million customers in the mid-Atlantic states and Pennsylvania. Pacific Lighting Companies, which supplied natural gas to southern California, also announced their plans to build jointly with PG&E an LNG terminal at Point Conception off the Santa Barbara Coast to import LNG from Alaska and Indonesia.

In sharp contrast to his vocal opposition to the Diablo Canyon nuclear plant not far away, California governor Jerry Brown strongly supported this LNG facility. Brown this time brushed aside fears of tanker leaks or gas explosions and successfully urged the legislature to support the project. A longtime commentator for the *Sacramento Bee* speculated many years later that Governor Brown's support had been influenced by the fact that an oil-trading corporation owned by family members had close ties to Pertamina, the Indonesian company that would be exporting the LNG to the Santa Barbara site. U.S. banks, which had loaned Pertamina billions of dollars, also supported the LNG facility as potentially producing a revenue stream that the Indonesian company could use to repay its loans.

Fearful of earthquake faults beneath the site, Californians living near the proposed facility did not warmly embrace the prospect of an LNG terminal. Using familiar tactics, in April 1978 about 300 people jammed into a hearing of the California Coastal Commission to protest the idea. Environmental groups complained that such a faculty would ruin the "pristine coastline." They also raised the specter of LNG explosions that would cause coastal fires and expressed concerns for the effects that such a terminal might have on the area's fish and wildlife. They delayed the project long enough that natural-gas surpluses ultimately rendered it moot.

Later that year the Department of Energy turned down two other applications for large LNG facilities, signaling the Carter administration's opposition to substantially increasing imports. Carter was leery. Although he acknowledged LNG as "an important supply option," he expressed concern about the potential of a future embargo. Because much of the known natural-gas reserves were then located in the Middle East and the Soviet Union, Carter was fearful that a cartel similar to OPEC would form to control natural-gas output and prices, as had happened with oil. And as indicated in the previous chapter, Jimmy Carter vastly preferred the idea of producing LNG from domestic coal.

During the 1970s, some energy policy analysts urged that transportation vehicles should use natural gas as a fuel instead of gasoline or diesel. Natural gas burns cleaner than either diesel or gasoline and has historically been less expensive. The goal was to shift natural-gas use from electricity production to the transportation sector and to use alternative fuels, such as wind or solar power, for generating electricity.

Vehicles using compressed or LNG have been around for a long time; they first appeared in the United States in the 1930s. But they have never enjoyed widespread use in this country. They are generally more expensive than comparable diesel- or gasoline-fueled vehicles (by about $4,000 for a car and $35,000 for a bus), fueling stations are few and far between, and they tend to have a shorter range and require much more space for fuel than their diesel or gasoline counterparts. (Federal tax credits have been available, however, to offset much of this differential cost.)

To date, natural-gas vehicles have been used primarily for fleets of vehicles, such as metropolitan or school buses, taxi cabs, or airport shuttles, where there are economies of scale for fueling costs. Even now, however, they compose only a miniscule share of the transportation sector. Of the more than 250 million vehicles registered in the United States in 2006, only 116,131 were compressed natural-gas vehicles and fewer than 3,000 were LNG vehicles. Only one natural-gas-powered has been commercially available in the United States, the Honda Civic GX. It costs about $6,000 more than its gasoline-powered sibling, the Honda Civic EX, and it has never been a very successful product, not even in California. A home refueling system—the "Fuelmaker Phill"—is available that connects directly into a natural-gas outlet for home heating, but this fuel system costs between $3,500 and $4,500, plus the fee for installation. (Beginning in 2005, federal tax credits have been available to offset 30 percent or up to $1,000 of these costs, whichever is less.) Despite some prominent advocates, it does not seem likely that natural-gas vehicles will replace substantial numbers of gasoline or diesel transportation vehicles any time soon.

Difficult as the 1978 Natural Gas Policy Act was to enact and despite its many shortcomings, it has to be counted as one of the most important energy policy successes of the 1970s. By phasing in the deregulation of natural-gas prices, it eliminated the distortions between the intrastate and interstate market that had previously existed and permitted prices to rise high enough to stimulate sufficient domestic exploration and production to head off the kinds of shortages that had previously occurred. Deregulation also unsurprisingly brought with it more volatility in natural-gas prices, however, and natural gas played an important role in the famous accounting fraud scandal and downfall in 2001 of Enron Corporation, a Houston-based energy company, which had originated from a merger of two reasonably successful natural-gas companies in the 1980s before diversifying into other energy-related gambits, especially the trading of commodities and derivatives connected to natural gas and electricity. (Enron's stock price hit $90 a share before it became worthless.) Notwithstanding such setbacks, natural gas emerged from the 1970s as a more robust fuel source than ever before, especially in the parts of the country distant from where it is produced.

In recent years, largely due to technological advances, extremely large deposits of natural gas have been found in the United States and Canada. New drilling technologies have emerged that have revealed vast new domestic reserves of natural gas that can be removed from rock shale formations at reasonable cost. For example, removing natural gas from the large Haynesville shale formation in Texas and Louisiana in 2010 cost about $3 per million British thermal units (BTUs), about 60 percent of what such removal cost in the 1990s. Estimates of recoverable domestic natural gas have recently skyrocketed, especially when one includes all estimates of potential reserves. The Energy Information Administration has estimated that proved reserves, which are a small fraction of potential reserves, are at their highest level since 1977, and a consensus among analysts has emerged that domestic reserves, along with those in Canada, are adequate to supply both countries for many decades, if not a century.

Nevertheless, some shortcomings still remain. Although cleaner than coal or oil, natural gas is still a source of greenhouse gases, and supply disruptions continue to be of some concern. Other environmental risks are present, including the possibility that an excavation or pipeline failure might contaminate groundwater, even though the safety record for natural gas from shale has been good so far. On the other hand, in 2010 a PG&E natural gas pipeline exploded near San Francisco, destroying nearly 40 homes and killing a number of residents. So contests over safety and environmental regulations are certain to emerge. Natural gas may not be the panacea that many had hoped. But given its environmental advantages in the production of electricity over other fossil fuels, especially coal, and its safety advantages, particularly over nuclear power, natural gas is now much better positioned than many of its competitors. Other alternatives to coal and oil, preferred by environmentalists over natural gas, have not fared nearly so well.

7 The Quest for Alternatives and to Conserve

May 3, 1978, was a Wednesday and pouring rain, but to Jimmy Carter it was Sun Day. The bare-headed president refused the offer of an umbrella as he laid out his vision of a sun-powered nation at the Solar Energy Research Institute atop South Table Mountain in Golden, Colorado—a town that is sunny about 330 days a year. "We know it works," Carter said, adding: "Nobody can embargo sunlight. No cartel controls the sun. Its energy will not run out. It will not pollute our air or poison our waters. It is free from stench and smog. The sun's power needs only to be collected, stored and used." Carter promised to spend $100 million to install solar equipment in federal buildings, and he ordered his skeptical energy secretary James Schlesinger to develop an overall strategy for accelerating the use of solar technologies.

Carter's speech was just one of many Sun Day events. The day opened with a sunrise celebration at the top of Cadillac Mountain in Maine, where the sun first hits the United States, and ended nearly 5,000 miles west on an Hawaiian beach at sunset. In between, thousands of people joined in activities to express their support for greater reliance on solar energy. Despite the participants' enthusiasm, however, this event failed to capture the public's attention as Earth Day had, and it is long since forgotten.

The day was the brainchild of Denis Hayes, author of *Rays of Hope: The Transition to a Post-Petroleum World* and, as chapter 3 has described, the national coordinator of the first Earth Day celebration eight years earlier. A year after Sun Day, Jimmy Carter would appoint him to lead the National Solar Energy Research Institute, giving Hayes a leading role in the development of solar power. When Hayes announced plans for Sun Day six months before Carter's speech, he had said that the Carter administration wanted solar energy and competing energy sources to be put on an equal financial footing.

Barry Commoner, a professor of environmental science at Washington University and one of the nation's best-known environmentalists, along with many other supporters of solar power, regarded its decentralization and diffusion as giving solar energy its great appeal. His vision included solar cells on every rooftop, solar cells powering every streetlight. Commoner regarded it as an *advantage* that "unlike conventional energy sources, there is no economy of scale in the production of solar energy." Along with other advocates of solar power, he rejected the idea of solar energy substituting for oil or coal at electric power plants. He insisted that central power stations are a needlessly expensive way to use sunlight. "Sunlight is widely distributed," he said, "and so should be the means of capturing it." Tracy Kidder, writing in the *Atlantic Monthly,* described Commoner's attitude as "almost a religious faith." The most avid supporters of solar energy in the late 1970s were missionaries for the "small is beautiful" movement.

Jimmy Carter tried to keep the faith. On June 20, 1979, he climbed to the roof of the White House to show the press gathered there the solar panels he had ordered to be installed (at a cost of $4,000) to heat water for the White House mess and other areas of the White House's West Wing. Standing ramrod straight and holding his hands out in front of him as if he were giving a blessing, Carter said, "I hope we can set an example for the rest of the nation." President Carter was not only adding solar energy to the house where he lived, but intending to extend it throughout the country. That day he announced a "new solar strategy," which included tax incentives, government loans, and federal R&D to help reach his goal of getting 20 percent of the nation's energy from solar power and other renewable sources by the year 2000. The Solar Lobby—the leading national coalition of environmentalists, farmers, antinuclear activists, unions, industry organizations, and celebrities (such as Robert Redford)—was even more optimistic: it claimed that the nation could obtain one-quarter of its energy from solar power by the year 2000.

One of Denis Hayes's fellow Sun Day coordinators, Richard Munson, complained about "vested interests want[ing] to turn toward nuclear and coal power" and described the Sun Day coordinating committee's vision as "two revolutions." "The first," he said, "is a solar revolution to replace oil, coal, and nuclear power with the sun and to end the energy crisis." "The other," he added, "is a social revolution to begin creating institutions that are smaller and more humane, instead of bigger and more impersonal." He wanted to change how we get our power—and also how we live our lives.

Munson may not have realized it, but these two ideas conflicted. The goal to replace oil, coal, and nuclear power with solar power residence by residence and business by business is hopelessly utopian. It is difficult to imagine a more unlikely endeavor than the delivery of solar-powered electricity as the occasion to remove big government and big business from people's lives. Powering the nation from the sun—putting solar energy "on an equal financial footing with the competing energy sources"—demanded that federal and state governments invest heavily in R&D and provide substantial financial incentives for people to convert to solar. And it required that large electric utilities either embrace the program or be forced into it. But Munson, along with Commoner and others, insisted that solar energy is "best used when solar collectors are on individual houses or small groups of houses." "It gives you energy self-sufficiency," Munson said. "You have some independence."

Even with the rise in the prices of oil and natural gas that had spurred the effort to find new energy sources, however, solar electricity during this period cost as much as 100 times more than electricity generated from traditional sources. So it is not surprising that the country fell very short of Carter's target. In 1978, a group of scientists led by Henry Ehrenreich, a Harvard physicist, released a federally funded study predicting that solar cells and solar batteries would likely produce no more than 1 to 3 percent of the nation's electricity by the year 2000. Even that estimate proved quite optimistic; in 2000, solar energy accounted for less than one percent of the nation's electricity.

The president's effort to convert a large share of the nation's power to solar energy ran into two major roadblocks. The first was economic. It is both expensive and risky for a homeowner or business to commit individually to solar power. Purchasing and installing solar cells and installing the infrastructure to produce heat and send electricity to power appliances require a substantial up-front commitment of funds. In 1978, for example, a combined solar hot-water and space-heating system sold in New Jersey for about $18,000; it cost an additional $2,500 to replace a conventional system with a solar hot-water heating system. (The median home price in New Jersey then was about $60,000.) If all went well, the cells might provide power for a couple of decades. But people are leery of paying for 20 years of power for their homes in advance. Many middle-class families would need to borrow to finance a solar investment, and lenders do not rush to finance purchases of solar cells. What if the cells corrode or malfunction? What if a neighbor's tree grows to diminish your sunlight? What if it rains nearly every day for a month? What if you move—could you recoup this investment?

The second obstacle in Carter's way was technological. Sunshine is not distributed evenly across the United States, and even in sunny climes sunlight is always intermittent. One must store sunlight during the day to power a house at night. In the late 1970s, storing power was dicey. An average house would require 24 large, costly, short-lived, and potentially explosive lead-calcium batteries.

The biggest technological problem was the difficulty in producing photovoltaic devices that could convert sunlight into electricity cheaply enough to compete with coal, natural gas, or oil. Less expensive production required reducing the cost of the silicon that went into the solar cells, lengthening the cells' useful life, and producing more electricity from each cell. Making solar cells economically competitive required hundredfold cost reductions.

Jimmy Carter did what he could to address both of these barriers. Direct federal expenditures on solar energy R&D increased from less than $1 million in 1973 to $150 million in 1980. Carter also managed to convince Congress to enact the Photovoltaic Energy Research and Development Act of 1978, which set a goal of 2 million peak kilowatts from solar in the coming decade and authorized $1.5 billion in expenditures over 10 years. And the Department of Housing and Urban Development spent $5.2 million on individual $400 grants to subsidize nearly 20 percent of the cost of solar hot-water heaters for more than 10,000 families in eleven states.

President Carter also signed the Energy Tax Act of 1978, which began a program of providing substantial tax credits for solar, wind, and other renewable energy expenditures. Under that law, residential solar and wind equipment expenditures were eligible for an income tax reduction equal to 30 percent of the first $2,000 spent and 20 percent of the next $8,000.

Legislation in 1980 increased this tax credit to 40 percent of the first $10,000. These tax credits were modeled after a similar law in California. Several states joined in with tax credits of their own. New York, for example, offered a 65 percent tax credit. By 1982, there were about 22,000 solar units in that state. The federal tax credits remained in effect until they were terminated in 1986.

Legislation in 1978 also encouraged the production and use of solar and wind power by electric utility companies. The Powerplant and Industrial Fuel Use Act of 1978 restricted the use of oil and natural gas by utilities, and the Public Utility Regulatory Policy Act of 1978 required utilities to purchase power from small generators or cogenerators at an "avoided cost" price that was favorable to renewable energy sources.

Notwithstanding these expansive new levels of federal and state support, it became clear within a few years that the goals for solar energy set by Jimmy Carter and Congress in the late 1970s were unrealistic. Indeed, the entire solar effort was plagued by overstatements both of technical feasibility and economic value. The idea that solar power could replace the power produced from oil was discredited.

In May 1980, a confidential memo from Energy Secretary Charles Duncan, who had replaced James Schlesinger, leaked. Duncan's memo called for reductions in proposed federal spending on solar programs and an increase in funding for nuclear and fossil energy. The Carter administration continued to insist publicly that it was not backing off its commitment to solar energy. But not even Jimmy Carter always put the government's money where his solar ambitions demanded. The 1978 legislation that he requested allocated $19 billion for his synfuels program compared to $1 billion for solar energy.

In fact, power generated from water has for a long time been our most successful clean, alternative energy source. Our nation's first hydroelectric power plant was built in 1882 by the Appleton Edison Light Company on the Fox River in Appleton, Wisconsin. By the early 1900s, hydroelectric power accounted for more than 40 percent of our nation's electricity and still supplied about one-third in the 1940s. Hydropower had long been strongly supported by the federal government, which fully subsidized construction of most of our nation's dams and operated many of our hydroelectric plants. The Depression, along with droughts and floods in the West, resulted, for example, in the construction of the Grand Coulee

Dam on the Columbia River and the Hoover Dam on the Colorado as well as the establishment of the Tennessee Valley Authority. In the famous 1960 presidential campaign debate with Richard Nixon broadcast on both radio and television, John Kennedy described hydropower production as "the hallmark of an industrial society" and complained that we were not producing more.

But the advocates of solar power were not enthusiastic about hydropower, and the rise of the environmental movement hastened the end of large hydropower projects. Environmentalists opposed water-development projects out of concerns both for safety and the destruction of natural habitats through the construction of a dam. The last authorization of a large dam construction project occurred in the late 1960s, and Congress enacted a dam inspection act in 1972.

Then several accidents in the 1970s exacerbated safety concerns. The largest failure was the 1976 complete collapse of the recently completed Teton Dam in the Snake River plain of Idaho. That dam's failure killed 11 people and thousands of cattle. Dale Howard, a college professor who was visiting Teton when it collapsed, told *Time* that the scene was surreal, saying that "as the water hit some of the farm fields, you could see an eerie cloud of dust and mist rising up three to five miles away." The political cartoonist Pat Oliphant depicted this disaster by showing a beefy bureaucrat from the Interior Department's Bureau of Reclamation sitting at his desk and saying, "If we listened to environmentalist dingbats, we'd never get anything built." In the background, Teton Dam is bursting, houses and cows flying into the air.

A year later the Kelly Barnes Dam in Tocoa, Georgia, which supplied power to a nearby college, collapsed, killing 39 people, including 18 children. This disaster in his home state secured Jimmy Carter's opposition to hydropower. During his campaign, he had promised to get the Army Corps of Engineers—which, as the largest operator of hydroelectric power plants in the United States, controls about one-third of the nation's hydropower—out of the dam-building business. Carter insisted that our nation had built enough dams and he said that as president he would be extremely reluctant to build any more. He was not completely successful in this effort, but dam construction and hydropower became quite unpopular over time. In 1980, hydroelectric power accounted for 97 percent of our electricity from renewable sources. By 2008, hydroelectric's share was down to two-thirds. The fastest-growing renewable electricity source in recent years has been power from wind.

Environmentalists in the 1970s, however, were not especially enthusiastic about the production of electric power from wind either. Walter Liebold, chairman of the National Energy Committee of the Sierra Club, for example, said that concerns about air and water pollution sometimes conflict with "worries about aesthetics and preserving public lands." Wind was far from his favorite renewable energy source because wind turbines were neither small nor beautiful.

The windmills of this era were generally found in large "wind farms," mostly in California. One, consisting of a large number of windmills clustered in the California desert alongside Interstate 10 not far from Palm Springs, resembled a graveyard of giant propellers that had been stripped off airplane engines and lined up in rows on tall towers to amuse highway passers-by. The electricity the wind farms generated was sold to the local public utility, which under the 1978 Public Utility Regulatory Policy Act was required to purchase it at "avoided cost." California's Public Utilities Commission promoted wind and solar power through what it called "standard offer" contracts. In 1983, for example, it promulgated a rule that compelled California electric utilities to enter into 15- to 30-year contracts to purchase wind energy with a 10-year price guarantee based on high avoided-cost projections for oil and natural-gas prices.

Every once in a while an enterprising individual (almost always living in a rural area) would purchase a windmill to generate his own electricity. In late 1972, for example, Neil Welliver, an artist who described himself as an "ecology freak," spotted an advertisement in the *Whole Earth Catalog* for a two-kilowatt windmill made by an Australian manufacturer. For about $3,500, the windmill, the 50-foot tower it was mounted on, a 60-battery storage system to hold the electricity until it was needed, and an inverter to convert the direct current it produced into the alternating current necessary to run American appliances were delivered and installed at Mr. Welliver's farmhouse in rural Maine. The windmill saved the Welliver family money: not only what would have been their monthly electric bills, but also the $10,000 that the local electric company would charge to run a power line from the road to the house a mile and a half away. The Wellivers, however, insisted that it was not the money savings that motivated them. "We're really doing all this out of perversity," Polly Welliver said.

Technology was not the same kind of obstacle for wind power as it was with solar. After all, windmills have been generating energy since the Dark Ages. In both the 1970s and now, the principal technological challenges have been to design windmills to produce the maximum possible electricity from the wind and to distribute the power produced by wind through the electric grid to where it is needed. Although most of the wind farms of the 1970s were located in California, wind is plentiful both on and off shore along both the east and west coasts and also in the Midwest plains from Texas to the Dakotas and in the Great Lakes.

In the late 1970s and early 1980s, the main barrier to widespread wind power for the United States was economic. Wind power was reasonably competitive with solar or nuclear power but more costly than electricity from oil, natural gas, or coal. In California, for example, wind power in 1985 cost 9 to 12 cents a kilowatt hour compared to about 7 cents for electricity from oil and gas.

The federal government, along with some states, especially California, attempted to eliminate the price disadvantage of wind power by providing it large tax incentives. As mentioned earlier, federal tax credits for renewable energy were first enacted in 1978, and, as with solar, tax breaks for wind power were expanded in 1980 and 1981. But they were allowed to expire at the end of 1985. During the interval, California enjoyed a "wind rush" of unprecedented scope: the total capacity of its wind power equipment tripled in the period from 1981 to 1983 and quadrupled again in 1984. By the end of 1985, about 5,000 California wind turbines were generating more than 1,000 megawatts of electricity, roughly the equivalent of one of the nuclear reactors at Diablo Canyon. By the mid-1980s, California's wind power output was 60 times greater than its solar output.

As with solar, however, wind power became a victim of extravagantly optimistic predictions of electricity production. Thomas Gray, then executive director of the American Wind Energy Association, predicted that by 2005 wind power in the United States "could supply up to the equivalent of the three million barrels of oil a day, or around 7 percent of total . . . energy demand." That did not happen: in 2005, wind power accounted for less than 0.5 percent of total U.S. electricity consumption. (Since then, wind power has been growing rapidly, nearly tripling, for example, between 2005 and 2008, even though it still accounted for just more than one percent of U.S. electricity production.)

The utilities that the 1978 Public Utility Regulatory Policy Act forced to purchase wind power were not enthusiastic buyers. Both Southern California Edison and PG&E, which together then purchased about 95 percent of all U.S. wind power, complained about its reliability.

Influential members of Congress found another cause for complaint. Representative Fortney H. (Pete) Stark, a liberal Democrat from northern California who served on the House Ways and Means Committee, was one of many critics who regarded wind production as far more significant as a tax shelter than as a source of electric power. "They're not wind farms; they are tax farms," he frequently said. The tax savings were indeed enticing. From 1978 through 1985, California offered a 25 percent tax credit for wind power, which could be combined with both a federal investment tax credit of 10 percent and a special 15 percent energy investment credit. A high-income Californian, for example, who purchased a $100,000 wind turbine—typically for $10,000 down with a nonrecourse loan from the seller for the balance—could reduce his combined state and federal taxes by as much as $50,000. With only a $10,000 down payment, the income tax savings alone produced a return of five times his investment, whether the turbines actually produced any electricity or not.

As with other tax shelter investments of the period, scoundrels abounded. Overvaluations, inflated prices, and excessive estimates of earnings were common. The wind turbines claimed on taxes sometimes did not even exist. One notorious example was International Dynergy, a developer of a prominent wind farm in southern California, whose turbines never produced more than 12 percent of their projected output. At one point, its employees attached used helicopter blades to broken wind turbines in an effort to confound an IRS investigation.

When tax reform came in the mid-1980s, these tax breaks were eliminated. The federal tax credits expired in 1986, California's in 1987. (California also ended its regulatory wind purchase program in 1985 after oil and other energy prices began falling substantially.) Allowing the tax credits to expire let all the air out of the California wind rush. John Ekland, of Fayette Manufacturing, one of the large legitimate manufactures of wind turbines, said that without the tax credits there would be no wind industry. Others believed that by the mid-1980s the industry had reached a level and a maturity that it could survive the loss of the tax incentives. The naysayers were proved right, though.

However, in 1992 new federal tax incentives came into place, and the wind power industry has since revived, growing dramatically since 2005. Despite this recent growth, wind has yet to reach the potential as a source of electricity for the United States that its 1970s enthusiasts had hoped for and expected. Wind's future again looks promising, and billions of dollars of subsidies are being directed to it, but whether its promise will be realized depends on technological, economic, and policy choices discussed in subsequent chapters.

Supporters of wind power and especially solar energy in the 1970s were often linked to what the *New York Times* called the "alternative technologies movement." This movement, which was adamantly antinuclear, has been described as "groping toward new, smaller, cheaper institutions—food co-ops instead of supermarkets, community phone systems instead of AT&T." Its advocates boasted, "At this point in the 70s, you can see its manifestations everywhere." The alternative technologies movement was anticorporate, anti–big government, and fundamentally anti–economic growth. Its bible was E. F. Schumacher's *Small Is Beautiful: Economics As If People Mattered*. Schumacher, a British economist who had headed planning for Britain's National Coal Board from 1950 to 1970, insisted that continued economic growth was undesirable and that current methods of both consumption and production "ravish nature" and "mutilate man." He wanted people to return to an era when business enterprises operated on a much smaller scale. His book was a national best-seller and was translated into fifteen languages.

The families who moved from cities to rural areas—along with many others who were just bystanders—got much of their information about alternative energy sources from publications such as the *Whole Earth Catalogue* or the monthly journal *Rain*, published in Portland, Oregon. In 1977, the publishers of *Rain* gathered together much of their learning into the 250-page *Rainbook: Resources for an Appropriate Technology*, which contained, for example, 12 pages of references, phone numbers, magazine lists, and places to get plans for solar water heaters and solar energy. Anthony Maggiore, chairman of the board of the National Center for Appropriate Technology, headquartered in Butte, Montana, said that "appropriate technology can make neighborhoods more self-reliant and improve the quality of life." He pointed to a project financed by his center that had converted two tenement buildings on Eleventh Street in New York City to use both solar and wind energy. California's governor Jerry Brown created his own Office of Appropriate Technology.

The alternative technologies movement produced some creative products and inspired some optimistic community offshoots, but it was hardly upbeat. Pessimism seemed to run through its veins. At the three-day conference "People, Neighborhoods, and Appropriate Technology" held in Milwaukee in December 1977, one of the keynote speakers compared his fears to the dreadful visions of the end of the world depicted by the painter Hieronymus Bosch and worried aloud that fossil fuels would run out by the year 2000. Along with many others, he believed that a growing population and a concomitant shortage of essential products—especially food and energy—would soon wreak havoc on our nation if we failed to revise our lifestyles drastically. In the energy arena, this movement's antigrowth, small-is-beautiful advocacy hindered renewable energy's ability to grow over time to a level that might actually rival oil or coal use, a legacy we live with today.

One of the principal endeavors of the National Center for Appropriate Technology and other advocates for renewable energy sources was to promote fuels other than gasoline and replacements for the automobile. One cannot help but notice an irony that emerged out of the great push of the late 1970s by the small-is-beautiful movement to develop and use biofuels for transportation vehicles. As it turned out, that effort's greatest financial beneficiary was the agribusiness giant Archer Daniels Midland (ADM).

The same 1978 legislation that awarded tax credits for wind and solar power also instituted a substantial tax incentive to mix gasoline with methanol or ethanol or both. Methanol ultimately failed to work out, but the production and sale of so-called gasohol—a mixture of 90 percent gasoline and 10 percent ethyl alcohol or ethanol—soon became common. Ethanol is produced primarily from corn in the United States. (In Brazil, the second-largest ethanol-producing nation, ethanol is produced from sugar cane.) Some Americans recognize ethanol as 200-proof "corn whiskey," but many more now know it as a fuel additive. In 1979, ten ethanol facilities existed in the United States, producing 50 million gallons annually. In one of its rare spot-on predictions, the Energy Department that year estimated that ethanol's use would climb to 600 million gallons by 1985. According to the Renewable Fuels Association, a national lobbying and public-relations trade association (that now spends more than $3.5 million annually promoting the use of ethanol), in 2008 there were 168 ethanol distilleries in 26 states capable of producing just less than 10 billion barrels of ethanol a year. Measured solely in terms of substituting a domestic "renewable" fuel source for oil, the ethanol program has been a resounding success.

Barry Commoner, in his 1979 book *The Politics of Energy,* expressing his usual zeal for alternative energy sources, told the story of Minnesota farmer Archie Zeithamer, who in the summer of 1978 built an ethanol plant on his 500-acre dairy farm. Commoner explained how efficient and self-sufficient Zeithamer's ethanol-manufacturing venture was: it returned $12,000 of his $16,000 investment in the first year of operation. In singing ethanol's praises, Commoner insisted that alcohol fuel could be produced at little or no additional cost and at a commercially feasible price simply as a by-product of farming. As he put it, "The farmers who are investing in alcohol production can expect to turn a profit in 1979, which will increase year by year thereafter." Commoner also cited a Nebraska test suggesting that a mixture of 10 percent ethanol with 90 percent gasoline would get 5 percent better gas mileage than gasoline alone. Who could resist the appeal of ethanol? It supposedly would help small farmers without increasing their costs and at the same time produce a cleaner, more efficient automobile fuel as a by-product.

The idea of using ethanol to power automobiles was not new in the 1970s. In fact, the automobile's internal combustion engine in its earliest days could be made to run on either gasoline or ethanol. Henry Ford's Model T was a flexible fuel vehicle that could run on ethanol or gasoline or a mixture of the two. In the formative years of the automobile, the higher price of corn (ethanol's primary feedstock) compared to that of petroleum and federal taxes on alcohol (which from 1861 to 1908 were $2.08 a gallon) led to gasoline's becoming the favored automotive fuel. When corn prices fell dramatically after World War I and oil shortages seemed likely, Standard Oil briefly produced and marketed an ethanol–gasoline mixture, but a rebound in corn prices killed that venture. After that, automobile engine manufacturing focused on gasoline as its fuel, and gasoline stations sprang up nationwide to fuel this rapidly growing industry.

In the 1970s, in response to the oil crisis, along with rising concerns about the environmental effects of the lead products that had commonly been added to gasoline to stop engine knock, the idea of mixing alcohol and gasoline again came to the fore. In the early 1970s, the EPA issued regulations requiring reductions in the levels of lead in gasoline for older cars and requiring new cars to use unleaded gasoline. After some delays in implementation, lead additives were completely eliminated in 1986. As lead was being phased out of gasoline, ethanol became attractive both to boost gasoline's octane rating and to reduce our dependence on imported oil. Increasing corn prices to help farmers was also an important political plus.

In 1978, Congress enacted a 40-cents-a-gallon excise tax credit for ethanol used in gasoline. Unlike other credits for renewable energy sources that were allowed to expire, the ethanol credit has always been extended at a level between 40 and 60 cents a gallon. The Government Accountability Office estimated that Congress gave more than $19 billion of tax breaks to the ethanol industry between 1980 and 2000. Per unit of energy produced, ethanol subsidies during that period exceeded all other government energy subsidies.

In addition to these tax incentives, the federal government provided grants and loans for ethanol production and subsidies for growing corn. In early 1986, after oil prices dropped sharply, ethanol plants were losing money even with their tax credits and other subsidies. Dwayne O. Andreas, the chairman of ADM, the nation's largest ethanol producer, threatened to close some or all of the company's plants if the government did not provide immediate assistance. Notwithstanding a report by the Agriculture Department's Office of Energy concluding that gasohol "cannot be justified on economic grounds" and that "the industry had no long-term prospect for survival without massive new government assistance," the government responded with the largest one-time subsidy the Department of Agriculture had ever handed out: $29.23 million in certificates for free corn. ADM received more than half of the total amount of these corn certificates. What ADM did not use itself to acquire corn, it sold on the commodity markets for up to 135 percent of the certificates' face value. (That year ADM reported more than $230 million of profits on $5.3 billion of revenue). Most of the rest of the certificates went to four ethanol producers, each of which was half-owned by large oil companies. State governments also contributed with tax breaks and subsidies of their own.

In 1980, Congress enacted a controversial tariff on imported ethanol, and that tariff continues to this day, amounting to 54 cents a gallon in

2008. The tariff increases the price of ethanol, which benefits domestic producers and effectively keeps cheaper Brazilian ethanol produced from sugar cane out of the United States—a replay of our foolish oil import quota.

We accomplished the goal of substituting ethanol for a substantial amount of gasoline, but when one compares the costs of the program to its benefits, the applause must vanish. There have been numerous studies of ethanol policy over the years, and complaints have abounded. For starters, despite Barry Commoner's claims to the contrary, gasohol has a lower fuel economy than gasoline. According to a 1986 report by the U.S. Agriculture Department, "Each gallon of ethanol contains about two-thirds as much energy as does gasoline." And the Department of Energy concluded that gasohol-fueled vehicles average 4.7 percent fewer miles per gallon than gasoline-fueled vehicles.

Critics also claim that the amount of energy required to produce ethanol may be as great as the amount of energy the ethanol contains. For example, according to an oil industry study in 1980, one ADM plant was using 15 percent more energy to grow, transport, and process the grain used to distill the alcohol than the energy the plant would produce. ADM disputed this calculation, claiming that the ethanol it produced contained 39 percent more energy than used in its production. Some environmentalists have also claimed that the environmental damage from producing ethanol—its adverse effects on land use, biodiversity, soil erosion, acidification, and groundwater contamination—largely offset the environmental benefits of replacing gasoline with it. Moreover, because farm land is used to produce biofuels, forested land may be converted to agriculture to make up for the lost food production. Ethanol subsidies were not originally conceived in the United States as a response to the problem of greenhouse gas emissions, and, as subsequent chapters discuss, if reduction of these emissions is going to be a major goal of our biofuel policies going forward, substantial changes will be necessary.

No one doubts that the move to ethanol has resulted in higher corn prices. Indeed, helping America's corn farmers was one of the original purposes of ethanol subsidies and the tariff. This support has not made other farmers—such as poultry and cattle producers, who have had to pay more for their feed—happy, and food prices have generally risen as a result of the use of ethanol as a fuel.

Despite these shortcomings, ethanol subsidies, mandates, and tariffs enjoy great political support. This phenomenon is due to three factors. First, farm states are represented by a substantial, bipartisan, and aggressive cadre of influential senators. For example, both Tom Daschle, a South Dakota Democrat, and Robert Dole of Kansas, who ran for vice president on the Republican ticket with Gerald Ford, served as majority (and minority) leaders of the Senate and were vigorous supporters of ethanol subsidies. And probably the only thing that Iowa's liberal Democratic senator Tom Harkin and its conservative Republican senator Charles Grassley agree on is their support for ethanol. The special importance of Iowa as the earliest state to play a significant role in our presidential nominating process and the importance of corn to that state's economy have also led many a vocal ethanol opponent to reverse that position when running for president. Indeed, every president who has moved into the White House since the 1970s has made campaign commitments to support ethanol subsidies. Finally, Dwayne Andreas, ADM's longtime chairman, was a generous financier of political campaigns and of the pet projects of influential politicians—so much so that at one point he became the poster boy for campaign finance reform in the United States.

Confidently assessing the costs and benefits of our nation's ethanol policy is a difficult exercise. Some costs and benefits are very difficult to measure and value. More than one recent study suggests, however, that we have incurred—and will incur—far greater costs than benefits by continuing to subsidize ethanol. Some ethanol supporters are now pinning their hopes on wood and cellulosic ethanol (ethanol distilled from switch grass and other biological wastes.)

In 1979, when the subsidies were much more modest than they would soon become, the *Washington Post* reported: "The gasohol lobby says the $18.80 a barrel in available subsidies—compared with a $16.00 a barrel cost for foreign oil—is not enough." Ethanol subsidies have often exceeded the entire cost of the gasoline that the ethanol has replaced. And despite stories like Barry Commoner's of small farmers building their own ethanol plants, the ethanol industry was essentially invented by ADM. Even recently, after all the new plants built in response to government largesse have come online, that company produced about one-third of the nation's supply. In July 2010, CBO issued a study indicating that in 2009 corn ethanol incentives cost the taxpayers $1.78 for every gallon of gasoline saved; cellulosic ethanol cost $3.00 per gallon of gasoline. (CBO also estimated that ethanol cost about $750 in subsidies for every ton of carbon dioxide emissions saved. As chapter 13 describes, climate change legislation that passed the House of Representatives in 2009 would have priced a ton of carbon dioxide emissions at less than $30 in 2019.) It is simply not plausible that putting so many eggs into this particular biofuel basket has been wise policy.

All in all, it is difficult to count the 1970s efforts to move to alternative, renewable fuels a success. In 2008, just 7.4 percent of our nation's total energy supplies, including biomass, primarily ethanol in fuel, came from renewable sources—compared to about 5.5 percent when Jimmy Carter took office more than three decades earlier. Of that 7.4 percent, more than half is from ethanol and about 2.5 percent still comes from hydropower. Although growing industries, wind and solar power together accounted for about one percent of the total.

Indeed, the one unambiguous success story involving an alternative to imported oil in the 1970s was the effort to conserve energy. Chapter 3 has already related how the CAFE automobile fuel standards were enacted in 1975 and how, despite that law's shortcomings, gas mileage of automobiles increased. The 55-mile-per-hour speed limit enacted in late 1973 also no doubt served to conserve gasoline. Higher gasoline prices were an additional factor in reducing total consumption. But the transportation sector was only part of the story.

In 1973, approximately one-quarter of U.S. energy consumption occurred in the transportation sector. About 35 percent was consumed in residences, commercial stores, and offices. The remaining 40 percent was used in the industrial sector.

Any effort to conserve energy faces four challenges. First, inertia: absent large and obvious cost savings or specific legal requirements, it is difficult to stimulate people to make the kinds of changes that would substantially reduce their energy use. Second, the size and timing of costs: energy savings often require large up-front costs to achieve small amounts of cost reductions spread over a long period. Third, information and uncertainty: people do not know how much an investment in energy conservation will really save them. Fourth, the frequent mismatch between who will bear the costs of energy savings expenditures and who will ultimately reap the benefits of lower periodic costs: builders, for example, may not be able to recover the costs of energy-efficient features from their buyers.

The states and the federal government enacted a number of laws in the 1970s to encourage or mandate conservation, but the dramatic increases in oil prices probably had the largest effects on behavior. After the first oil crisis, in the period 1973 through 1975 energy consumption declined by about 5 percent. From 1976 through 1979, when oil prices were relatively stable, it increased by about 10 percent, but it declined again by about 6 percent in 1980 and 1981 following a second spike in oil prices in 1979.

In the industrial sector in the 1970s, about 85 percent of energy was used in manufacturing, with most of the balance in agriculture and mining. Absent direct mandates or subsidies, conservation decisions in these industries are straightforward. If the costs of saving energy are lower than the costs of using it, energy will be conserved. These business decisions are sometimes confounded by the fact that conservation usually requires capital improvements, but energy consumption is an operating expense. Some companies reaped great returns on conservation investments; American Can Company, for example, invested $73,000 at a New Jersey plant to reduce energy consumption, and it saved $700,000 annually. But the mid-1970s recession and the extraordinarily high interest rates of the late 1970s dampened all sorts of capital investments, and these economic difficulties no doubt kept capital investments for conservation below what they otherwise might have been. The economy's decline itself in turn, however, also reduced energy consumption.

The price controls on oil and natural gas of that decade affected cost trade-offs in a manner adverse to conservation. Nevertheless, after OPEC quadrupled the price of imported oil following its 1973 embargo and prices doubled again at the beginning of the Iranian Revolution in 1979, industrial energy consumption fell significantly.

Owners of commercial buildings, like industrial users, have financial incentives to conserve energy as oil and gas prices rise whenever they rather than their tenants pay the electric and heating bills. It's no surprise that these owners responded enthusiastically to requests by Richard Nixon and Jimmy Carter to lower thermostats in the winter and raise them in the summer because they could save money on heating and cooling while insisting to their tenants that they were just being patriotic.

Congress also passed laws during the 1970s providing subsidies and mandates for conservation. Tax legislation also gave incentives to homeowners to install energy-conservation devices, such as insulation and new tighter windows and doors. The National Energy Conservation Act of 1978 required electric and gas utilities to engage in "energy audits" and other activities (including the provision of financial incentives) to encourage energy conservation by their customers. Federal funding of R&D for energy-conservation technology more than quintupled from 1974 to 1980. But achieving conservation in the residential sector remained challenging. In rented units, for example, which have housed 40 percent of New England households, tenants who move frequently have little incentive to make energy-saving investments. Landlords are also reluctant to make capital improvements simply to lower their tenants' heating bills. Installing insulation, energy-efficient furnaces, or energy-saving windows and doors is rarely a landlord's priority. Likewise, builders are reluctant to incur the up-front costs of installing energy-efficient features for new homes because those costs are difficult to recover through higher prices. The 1975 Energy Conservation and Policy Act contained some regulatory requirements, such as elimination of space heaters in new construction, but energy efficiency in the residential sector made only slow progress.

Jimmy Carter's inconsistency did not help. For example, the Department of Energy estimated that a New England pilot program that gave thousands of people a small device to lower the water flow in showers, along with a booklet outlining 10 other inexpensive ways to reduce their household oil consumption, saved 2.6 million barrels of oil in 1978 at a cost of only $1 a barrel. (Oil was then selling for about $30 a barrel.) The department proposed expanding the program nationwide, but Carter eliminated it instead even though he was pushing conservation.

Importantly, despite the challenges, energy conservation and greater energy efficiency, especially in the industrial and commercial sectors, during the 1970s broke the "iron law" that had previously been thought to exist between increasing energy consumption and economic growth. Growing the economy by 3 or 4 percent was supposed to require a 3 or 4 percent increase in the demand for oil. Through the mid-1970s, that relationship had seemed to hold. A National Academy of Sciences panel, however, published a report in *Science* magazine in 1978 that showed that great variations in energy demand might produce similar levels of economic growth over time. An earlier study also had showed that the same U.S. living standards could have occurred using 40 percent less energy, an amount equal to almost as much as all the oil used that year. Energy analyst

Daniel Yergin estimates in his book *The Prize: The Epic Quest for Oil, Money, and Power* that in the period following the 1973 oil embargo, using 10 percent less energy had no adverse effect on U.S. economic growth, and he contends that 3 percent annual growth in GDP could then have been achieved with no increase in energy use. Yergin insists that public policy in the 1970s was not nearly as aggressive in stimulating conservation as it should have been, describing public policy in that period as "miserly" and "disheartening."

Nevertheless, even though we use large amounts of energy per capita compared to others around the world, our economy is now much more energy efficient than it was. And many further opportunities for conserving energy and improving our nation's energy efficiency remain.

There may be some skepticism about conservation among politicians—Ronald Reagan once joked that conservation means that we will be hot in the summer and cold in the winter—but there is no organized political opposition to energy conservation. The methods for achieving conservation, however, are highly varied and opportunities often quite diffuse. As with the automobile fuel standards and mandates for energy-efficient residential construction, when particular industries are targeted for energy conservation, they often balk. On the other side, with the paltry exceptions of insulation and window manufacturers, there is no real organized political support for conservation policies.

Jimmy Carter made conservation the centerpiece of his energy strategy and emphasized the immorality of energy wastefulness, but he also demonstrated the limitations of presidential exhortations to conserve. Writing for the *New York Times* editorial page, the influential columnist James Reston, in a June 1979 column entitled "O Jimmy! It's Easy," offered a long laundry list of "proposals" to conserve energy because, as he said, Jimmy Carter "obviously needs help." Reston's first idea was to have "all government employees report to work only every other day." "This odd–even system," he said, "would cut the government's consumption of gas in half, and double its efficiency, especially if you could fire the odds." Another Reston suggestion was to "recycle all the political gas on Capitol Hill into useful energy." You get the idea: Jimmy Carter's moral suasion was an easy target. More important, throughout the 1970s, whenever presidents sought greater conservation by directly increasing energy prices—either through decontrol or increased energy taxes—political barriers in Congress eviscerated or halted them. Political resistance to energy taxes still remains strong.

Overall, the sharp rise in the prices of oil and natural gas in the 1970s brought renewable energy sources, such as solar, wind, and biofuels, along with energy conservation, to the forefront of the nation's public-policy agenda. Renewable fuels became potentially viable, both technologically and economically, for the first time, but when oil and gas prices dropped precipitously in the 1980s, the country returned to its old habits. Energy usage picked up again. Along the way, overly optimistic predictions for alternative energy production actually hurt its cause, making the only incremental upticks disappointing and frustrating to its advocates.

The 1970s alternative technologies movement's vision of personal energy independence—with solar panels on every rooftop, windmills in every backyard—remained in an unresolved conflict with the need for massive subsidies and mandates from federal and state governments and for the cooperation of large businesses—especially electric utilities—to make such energy sources viable. The government came through with large increases in R&D dollars and a wide variety of subsidies, especially tax incentives, but—with the dramatic and unfortunate exception of ethanol production—these incentives were often misdirected and inadequate to the task, and they offered only off-and-on support.

The 1970s movement toward renewable energy sources and conservation ultimately required higher prices for traditional fuels, especially oil and natural gas, along with a committed effort by a government willing to incentivize and in some instances to mandate both conservation and the use of alternative fuels. Ronald Reagan was willing to oblige by decontrolling oil and natural-gas prices, but the market would soon thereafter produce a collapse in the costs of those very same commodities. And Reagan had no interest in mandating either alternative energy use or conservation. As a result, during the 1980s neither of these essential conditions could be counted on. Movement toward substantially increased reliance on solar and wind power then became largely moribund until a worldwide movement to limit greenhouse gas emissions gathered force, and oil prices once again spiked during the first decade of the new millennium.

As the 1970s were moving toward their close, however, Jimmy Carter would face a new and unanticipated challenge—and from a country whose friendship he had taken for granted. Ominous change was brewing in Iran.

8 A Crisis of Confidence

All of Jimmy Carter's efforts to encourage conservation and alternative energy did little to halt the crisis he faced in the summer of 1979. Inflation in the United States was then running at 14 percent. The nation was reeling from gasoline shortages and a wildcat insurrection by truckers frustrated by fuel shortages and high prices. Gas lines and odd–even day rationing had returned to America. Tempers were frayed. On May 31, a young pregnant woman's husband was shot to death in a New York gas line. America's truckers were especially aggrieved. In Levittown, Pennsylvania, a truckers' highway blockade set off riots that injured 100 people and led to more than 170 arrests.

These conditions were a direct offshoot of the 1978 revolution in Iran. Oil exports from Iran, then the second-largest exporter in the world after Saudi Arabia, had come to a halt that December. Although the lost Iranian production was largely compensated for by increases in output elsewhere, the Iranian Revolution created great fears of supply disruptions in the importing nations and an opportunity for suppliers to enhance their profits greatly by increasing prices. The upward pressure on prices was further exacerbated by oil companies' hoarding inventories in anticipation of further price increases. Perceived as price gouging, this behavior was one reason that the American people blamed the oil companies for the crisis. Motorists, fearing a repeat of the 1973 gasoline shortages, kept refilling their gas tanks to make sure they would not be caught short. One expert estimated that Americans burned up more than 600,000 gallons of oil daily waiting to fill up their gas tanks. The panic buying increased worldwide demand for oil by 3 million barrels a day and, when added to the reduced supply of about 2 million barrels a day, created a gap equal to about 10 percent of worldwide oil consumption. Oil prices shot up from $13 to $34 a barrel.

On July 3, 1979, Carter returned to the White House from an unsuccessful G-8 economic summit in Tokyo, where he had tried to convince the members gathered there to go along with a raft of suggestions to reduce Western dependence on OPEC oil. The Europeans had resisted almost all of the president's proposals. After the summit, Carter and his wife, Rosalynn, had planned to spend a few days vacationing in Hawaii, but the president's staff insisted that he return to Washington immediately to take charge of an increasingly disgruntled nation.

Within hours of Carter's arrival at the White House, his staff requested network television time for a presidential address to the nation two days later, on July 5. But he had previously addressed the nation four times on energy issues and was disinclined simply to do that once more. Carter's moody young pollster Pat Caddell had told the president that he had the approval of only about 25 percent of the public—a level as low as Richard Nixon's just before he resigned. Caddell convinced Carter that the American people believed that both the government and the oil companies were incompetent or dishonest or both. He said that another speech on energy would fall on deaf ears. He wanted Carter to give up his "transactional" presidency and embrace a "transformative" one. He convinced the president that the American people were losing their "faith in themselves and in their country." He urged Carter not just to talk about energy, but also to address the basic relationship between the people, their government, and other institutions.

On July 4, President Carter canceled the speech scheduled for the next day in order to give himself the time to tackle this much larger subject. Not knowing himself what he was going to do, Carter refused to explain the cancellation to the press. His vice president, Walter Mondale, who thought Caddell's approach was "sophomoric and silly," was apoplectic at this turn, which he believed would lead to political catastrophe (and sharply diminish his own prospects of ever becoming president). But Carter would not be dissuaded.

The president retreated to Camp David, where over the next 10 days he invited a stream of guests—including the young Arkansas governor Bill Clinton—to give him advice. His advisers included many Washington "wise men," along with some ordinary folks. Carter later would say that most of the public's doubts about him "had arisen from the struggle over energy, with my repeated exhortations and lack of final action by

Congress." "It was not pleasant for me to hear this," he added, "but I felt their analysis was sound." Carter wrote in his diary for July 9: "it's not easy for me to accept criticism and reassess my ways of doing things. And this was a week of intense reassessment."

Finally, on July 15, 1979, having returned from his sojourn at Camp David, Carter delivered to the nation one of the most remarkable peacetime speeches in American political history. One hundred million people turned on their televisions to hear Carter's somber remarks—triple the audience that had watched Carter's most recent speech on energy that April.

When the American people tuned in that Sunday evening, they did not find Jimmy Carter sitting in a rocking chair in front of a well-stoked fireplace dressed in a cardigan sweater. This was a suit-and-tie Oval Office occasion. President Carter looked straight at the camera and asked, "Why have we not been able to get together as a nation to resolve our serious energy problem?" He began answering this question by observing that the "problems of our Nation are much deeper—deeper than gasoline lines or energy shortages, deeper even than inflation or recession."

Carter then launched into the "crisis of confidence" part of his speech. Sounding more like a southern preacher behind a pulpit in church or teaching one of his Sunday school classes in Plains, Georgia, than a president sitting at his desk in the Oval Office, Jimmy Carter described an "invisible" crisis "that strikes at the very heart and soul and spirit of our national will," a crisis that "threatens to destroy the social and the political fabric of our nation," one reflected in "growing doubt about the meaning of our own lives and in the loss of a unity for our nation." Carter accused the American people of "losing faith," of accepting a "growing disrespect for government and for churches and for schools, the news media, and other institutions." Channeling both Christopher Lasch's surprise bestseller *The Culture of Narcissism* and Tom Wolfe's famous article on the 1970s, "The 'Me' Decade," Carter said: "In a nation that was proud of hard work, strong families, close-knit communities, and our faith in God, too many of us now tend to worship self-indulgence and consumption." "Human identity," he said, "is no longer defined by what one does, but by what one owns." Carter then scolded Americans about materialism, saying that "owning things and consuming things does not satisfy our longing for meaning . . . piling up material goods cannot fill the emptiness of lives which have no confidence or purpose."

When his sermon was done—the portion of his speech that would soon inspire the national media to label it Carter's "malaise" speech, even though he never used that word, the president returned to the problem of the day: energy. "We ourselves," he said, "are the same Americans who just ten years ago put a man on the Moon. We are the generation that dedicated our society to the pursuit of human rights and equality. And we are the generation that will win the war on the energy problem and in that process rebuild the unity and confidence of America." "Energy," he added, "will be the immediate test of our ability to unite this nation, and it can also be the standard around which we rally." Calling our energy problem the "moral equivalent of war," Carter predicted, "On the battlefield of energy we can win for our nation a new confidence, and we can seize control again of our common destiny."

The president then proceeded to finish his speech with 18 paragraphs (close to 1,000 words) describing his energy goals and current program point by point by point. This approach, no doubt, discouraged everyone who might have thought that Carter's ordeal during the previous two years to get the huge National Energy Act through Congress had resolved the issue. His laundry list of proposals included (1) imposing an oil import quota; (2) creating an energy security corporation and a solar bank to spend "a lot of money" developing alternatives to imported oil "from coal, from oil shale, from plant products for gasohol, from unconventional gas, from the sun"; and (3) mandating a major shift of electricity production from oil to coal. (Energy security was his motivation; concerns about climate change were not yet on the agenda.) Carter also urged Congress to enact a "windfall profits tax" on oil companies and to increase aid to "needy Americans" to help them deal with rising energy prices.

On conservation, Carter sounded positively schoolmarmish: "I'm asking you for your good and for your nation's security to take no unnecessary trips, to use carpools or public transportation whenever you can, to park your car one extra day per week, to obey the speed limit, and to set your thermostats to save fuel." "Every act of energy conservation like this is more than just common sense—I tell you it is an act of patriotism," he said.

For Jimmy Carter—despite the massive energy legislation he had signed in 1978—energy had morphed from a confusing, confrontational, and confounding policy issue into the path out of the nation's "crisis of confidence." "So," he said "the solution of our energy crisis can also help us to conquer the crisis of the spirit in our country." Moreover, "it can rekindle our sense of unity, our confidence in the future, and give our nation and all of us individually a new sense of purpose." The president concluded his speech with a warning, observing that there are "no short-term solutions to our long-range problems. There is simply no way to avoid sacrifice."

But the American people were in no mood to sacrifice. Carter (and Caddell) might have mistaken anger and frustration for a lack of confidence. Americans felt they were already sacrificing enough with high energy prices, ravaging inflation, a sluggish economy, stagnating wages, and now a president who seemed to be spinning out of control.

Thirty years later a consensus has emerged that it wasn't the July 15 speech that did Carter in, but rather the "massacre" of his cabinet soon thereafter. Immediately following his address, Carter's poll numbers actually jumped 11 points. But when two days later Carter asked his entire cabinet to resign and replaced a handful (including Energy Secretary James Schlesinger), he gave more credence to the idea of a malfunctioning executive than to the notion of a dysfunctional populace. His cabinet purge turned the finger Carter had pointed at the American people right back at himself. Many Americans no doubt thought that Jimmy Carter was proving right Tom Wolfe's later description of him as a "down-home matronly voiced Sunday school soft-shelled watery-eyed sponge-backed millennial lulu." By the end of July, his approval rating had sunk again to its previous low of 25 percent.

If Jimmy Carter's crisis-of-confidence speech and the cabinet purge that followed did not mark the beginning of the end of his presidency, a phone call he received early in the morning of November 4, 1979, from his national security adviser Zbigniew Brzezinski surely did. Brzezinski called to tell Carter that about 3,000 Iranian militants had stormed the U.S. embassy in Tehran and captured 50 or 60 Americans who were being held hostage. They composed the entire American staff at the embassy that day—a tiny fraction of the 1,100 who had manned the embassy when the shah was in charge.

It is difficult to know, even with the advantages of hindsight, how much of the anti-Americanism that accompanied the Iranian Revolution to lay at Jimmy Carter's feet. Like many U.S. presidents before him, Carter had cordial relations with the shah, whom he regarded as a "strong ally" and a "likeable man," "surprisingly modest in demeanor." The shah had visited Carter in the White House in November 1977, and Carter had then urged the shah (to no avail) to consult with the Iranian dissidents and to have his police force treat them more gently. Six weeks later, on New Year's Eve, Carter had stopped to visit the shah in Iran. They both had acknowledged the close relationship between Iran and the United States, and Carter had toasted Iran as "an island of stability in one of the more troubled areas of the world." Some island, some stability. Before the new year was over, neither the close relationship nor the stability would remain.

In September 1978, the shah had declared martial law in an effort to quell public demonstrations. Throughout that fall, the revolution had continued to gain momentum, and on November 10 Jimmy Carter had written in his diary: "The Shah is on very shaky ground," but "the Shah should know that we are with him." By December, a general strike was under way in Iran, and a few weeks later the shah had left the country, never to return. His departure marked the end of Iran's role as both a reliable ally and a stable source of Middle East oil for the United States. The country's new leader, the fundamentalist Ayatollah Ruhollah Khomeini, had declared Iran an "Islamic republic," condemned the United States almost as often as he railed against the shah, and blamed Iran's poor economic and social conditions on the United States. Anti-American demonstrations had become daily spectacles in Iran. During this time, the shah had been kept outside our borders, but on Monday, October 22, 1979, after much agonizing in the White House about whether to let him into the country, the shah was allowed to come to New York for medical treatment. Less than two weeks later, the U.S. embassy in Tehran was seized, and the Americans there were taken hostage. That same week both Massachusetts senator Edward Kennedy and California governor Jerry Brown announced that they would run against Carter for the Democratic nomination for president.

The Middle East had taken a turn that would prove very costly to us for a very long time. Islamic fundamentalists spurred violence in Saudi Arabia. A mob burned the U.S. embassy in Pakistan.

Two days after Christmas, the Soviet army invaded Afghanistan. In his 1980 State of the Union Address, Jimmy Carter announced that any attempt by the Soviets to gain control of the Persian Gulf "would be repelled by any means necessary including military force." Carter also responded to the Soviet invasion by canceling grain shipments, halting the Senate's discussion of ratification of the SALT II arms-reduction treaty he had negotiated with the Soviets, withdrawing the U.S. athletes from the Moscow Olympics that year, and indirectly supplying aid to Afghan fighters.

Despite his troubles, on January 21, 1980, Carter won the Iowa Democratic caucuses by a wide margin. All the while, the American public worried about the hostages. ABC television's *Nightline* sent pictures of Iranians burning flags and yelling "death to America" in a nightly spectacle it called "America Held Hostage Day ____." As the journalist Roger Wilkins said, "The whole world saw these people stomping on images of Carter, burning American flags, and the most rancid sort of disrespect and hatred of the United States, on television, around the world, all the time."

In late April, the president launched a military hostage rescue mission, sending eight helicopters to the Iranian desert, where they were to rendezvous with six large transport planes for refueling and to pick up assault troops. The helicopters were then to fly to Tehran, free the hostages, and transport them to a secure air force base nearby. The mission failed dramatically. Navigation equipment failed on one helicopter; another had a mechanical breakdown; and a third crashed into one of the transport planes in a blinding sandstorm. These mishaps left too few helicopters to carry out the mission, so President Carter aborted it. Eight servicemen died; three more were badly burned in the crash.

The mission's failure was a huge embarrassment to the United States, which looked remarkably weak. Carter's secretary of state Cyrus Vance, who had opposed the rescue attempt, resigned and was replaced by Maine senator Edmund Muskie. The failed rescue effort helped cement despondent Americans' view that their president was incompetent. The hostages were then scattered around Iran, where they were held until just after noon on January 21, 1981—the 444th day after their capture—when the new president of the United States, Ronald Reagan, announced their release. As Yale historian Gaddis Smith says about Iran, "President Carter inherited an impossible situation and made the worst of it."

Following the hostage rescue attempt, oil prices continued their rise. Writing in the summer 1980 issue of *Foreign Affairs*, Walter J. Levy, a well-known energy consultant, described oil-consuming companies and countries as willing to pay OPEC nations virtually any price they asked in order to get an adequate supply of oil. His article was entitled "Oil and the Decline of the West." "The year 1979," he said, "was one of grievous setbacks for the oil supply of the Western world, its economic and financial prospects, its strategic capabilities, and its political stability." Levy was right.

Levy's observations echoed fears expressed nearly a year earlier by James Schlesinger, shortly after he had been forced to resign as energy secretary. Schlesinger—an arrogant and supremely confident man who had never given himself a grade less than A+ on anything—said that he could give President Carter only a B on his handling of energy policy. Schlesinger was sure we would now face ongoing instability and unpredictability concerning the flow of oil from the Middle East. "Oil that has driven the vast economic expansion since World War II will no longer be available to fuel the further growth of the world's economy," he said in a speech in August 1979. He worried that our nation's near-term energy future now depended excessively on a "politically stable Saudi Arabia," a nation he described as "exposed to the ferment of social change." He was sure that, absent worldwide recession, limited supply and ever-increasing demand would "inevitably" and "inexorably" drive up the international price of oil. And, like Carter, Schlesinger complained that the issues he was speaking about "do not engage our people," even though "on our success depends the economic and political survival of the United States and her allies—and quite possibly of freedom itself." His conclusion: "The energy future is bleak and likely to grow bleaker in the decade ahead."

In September 1980, Saddam Hussein's Iraq attacked Iran, and the threat of another shock to oil supplies and another price spike loomed large. Supplies dropped by almost 15 million barrels a day, and spot prices soared to $42 a barrel. In December, after the initial shock of Iraq's invasion had worn off and the two Middle East nations had settled into what would be a very long war, OPEC set an official price of $41 a barrel. (By the end of January 1981, the average OPEC price would fall to $35 a barrel.)

Although Carter secured the Democratic nomination for president, both he and the Democratic Party were badly bruised by Ted Kennedy's challenge, which did not end until the convention that August (Jerry Brown had dropped out in April). In the fall campaign, Carter proved no match for the affable, optimistic Republican nominee Ronald Reagan. Reagan, who had served as governor of California and had nearly snatched the

1976 Republican nomination from Gerald Ford, personified the reconstruction of the Republican Party from the right that had been gaining force throughout the 1970s.

The year 1980 had been one of grievous setbacks for Jimmy Carter. With a weak economy, high inflation, exorbitant interest rates, and the ongoing Iranian hostage crisis, Carter would have faced an uphill battle against anyone. He came close to negotiating release of the hostages a few times before the November vote, but he failed. The Democratic Congress—fearful of higher gasoline prices in an election year—took away Carter's power to impose an oil import fee and, for the first time since the Truman administration, overrode a veto by a president of their own party. A media frenzy over dealings by Carter's alcoholic brother Billy with Libya also hurt the president. Election day marked the one-year anniversary of the hostage taking, and the news media talked about little else throughout the weekend before the election. In his home town of Plains, Georgia, folks would have described Carter as "snake bit."

The president employed a "Rose Garden" strategy, staying at the White House for most of the campaign and tried to portray Reagan—who he thought would be his easiest Republican opponent—as an over-the-hill, heartless, intellectual lightweight, extreme right-wing, militarist, B-movie actor. In his diary on July 31, 1980, Carter observed that the policy differences between Reagan and himself were "perhaps the sharpest divisions . . . of any two presidential candidates in my lifetime." He optimistically added: "we're on the right side and if we can present our case clearly to the American people, we will win overwhelmingly in November." Reagan supplied Carter some ammunition by making a few misstatements, such as "trees cause pollution," but Carter got little traction with the voters.

In their only televised debate, held in Cleveland just a week before the election, the 68-year-old Reagan came off as upbeat, humorous, and completely in control—an unmistakable counterpoint to the dour, moralistic Carter. Reagan's optimistic "vision of a shining city on a hill" provided a sharp contrast to Carter's era of limits, his crisis of confidence. Reagan's critical question "Are you better off than you were four years ago?" sounded Carter's political death knell. Although the election seemed reasonably close until the very end, Reagan won in a landslide. He outpolled Carter by nearly 10 percentage points and carried 44 states to Carter's six (plus the always Democratic District of Columbia). The Electoral College vote was 489 to 49, a crushing defeat for an incumbent president.

Jimmy Carter left the presidency a defeated man. Along with the Iranian hostage crisis, the dreadful combination of a stagnant economy and galloping inflation—"stagflation"—had overwhelmed him. During his presidency, this Georgia engineer had approached energy policy as a problem simply in search of the right technical solution. He radically underestimated both the political and cultural barriers to the drastic transformation he sought. Nor was he ever able to reconcile the contradictions between his environmental objectives and his energy policies. As Carter said, the man who replaced him in the White House could not have been more different—both in style and substance. Ronald Reagan entered the presidency exuding confidence both for himself and for his nation. Energy policy, which had bedeviled each of his three predecessors, was about to take a U-turn and move toward the bottom of the nation's agenda.

9 The End of an Era

When it came to energy policy, the differences between Ronald Reagan and Jimmy Carter were stark. They reflected not only their disparate policy preferences, but also their fundamental dispute over the proper role for the federal government. Energy had been the centerpiece of Carter's domestic presidential agenda. He had pushed to transform our nation's energy policies more than any president of the twentieth century. In addition to the energy legislation he secured in the first three years of his term, in June 1980 Carter signed six more pieces of legislation known collectively as the Energy Security Act. These laws were intended to stimulate the supply of alternative energy sources and to encourage energy efficiency through a complex combination of subsidies and tax breaks. At least one well-known energy economist, however, described this legislation as "probably the low point in contemporary U.S. energy policy." All in all, Jimmy Carter signed many thousands of pages of energy legislation during his single term as president.

During the 1980 campaign, Carter took credit for the fact that oil imports that year were down more than 20 percent from their level in 1970, but he failed to mention that they were down only 7 percent from their 1976 levels, the last year of the Ford administration. Nor did he mention that imports were lower largely due to a combination of higher prices for imports, lower U.S. demand because of the recession, and an almost tenfold increase in Alaskan oil production since 1976 to 1.6 million barrels a day.

Reagan, in contrast, who frequently said that "Washington invents cures for nonexistent diseases," told an enthusiastic black-tie audience at the Petroleum Landmen's convention in El Paso, Texas, that "the problem isn't a shortage of fuel but a surplus of government." He suggested that the United States had no energy problem until the government had gotten involved, and he urged the elimination of the Department of Energy. "It has a budget as big as the total after-tax profits of the entire U.S. oil industry," he said. "What the hell has it accomplished?"

In addition to his desire to abolish the Department of Energy, which was Jimmy Carter's pride and joy, Reagan had other major energy policy disagreements with the Democratic president. During the campaign, in the midst of a five-day swing through six key eastern and midwestern states, Reagan vigorously attacked Carter's policies. Reagan told 6,000 people at an outdoor rally in Cleveland, Ohio, that, according to reports of the U.S. Geological Survey and the Department of Energy, "America has a proven and potential 47-year supply of oil, including oil shale, a 27-year supply of natural gas [which has since grown dramatically], and at least a 321-year supply of coal." He accused President Carter of concealing these figures from the public and leading people "to believe that there is an acute shortage of energy resources in this country." Reagan blamed government regulations and "production barriers" for the energy shortages. He promised to increase U.S. energy supplies. "We will get America producing again," he said. "Coal, oil, natural gas, shale oil, solar, geothermal, and safe nuclear power. Every available resource we have must be used to free us from OPEC oil domination." Reagan used an increase in oil imports from Libya to take a shot at Carter's brother's troubles, and in an over-the-top effort to link energy policy to the nation's weak economy he accused Carter of "deliberately putting two million people out of work in a single year." "If that's Mr. Carter's idea of good energy policy, I don't want any part of it," he said.

In fact, Reagan wanted little of Carter's energy strategy, nor did he regard environmental concerns as serious objections to expanding domestic production. Moreover, he was quite congenial to using federal lands for domestic exploration and production. Reagan wanted to move the federal government out of the energy producers' way and laid the blame for the nation's energy problems on the government's "price-fixing, regulating, and controlling" of the industry. He proposed immediately decontrolling oil prices: eliminating regulatory and environmental barriers to coal and nuclear power; leasing more federal lands for energy exploration and development; limiting spending on synfuels, solar, and other alternative fuels, and supporting basic research; and significantly reducing mandates and regulatory requirements for energy efficiency in appliances, automobiles, and buildings. If conservation was the cornerstone of Jimmy Carter's effort to achieve energy security, increased domestic production served that role for Ronald Reagan. Government incentives and controls would be moved out; market decisions would take priority. Reagan's inauguration on January 20, 1981, thus marked the end of an era for energy policymaking in the United States.

Once in office, Reagan did not delay in taking action. Despite concerns by some of his aides that higher prices might cost him politically, on January 28, 1981—eight days after he was sworn in—Reagan used his unilateral authority to lift all federal controls on oil and gasoline prices. "For more than nine years," he said, "restrictive price controls have held U.S. oil production below its potential, artificially boosted energy consumption, aggravated our balance of payment problems, and stifled technological breakthroughs."

Lifting the oil price controls, whatever its political consequences, was clearly sound policy. Writing in the *Wall Street Journal* in January 1981, one former energy regulator described how when station owners and wholesalers had submitted 14,000 applications for greater allocations of gasoline during the period from March through August 1979, backlogs grew, along with arbitrary decisions, in an exercise that was "consuming thousands of hours of labor by station owners, oil-company lawyers, and government officials." "We just spread around the shortage—and the profits," the regulator said. "Despite reservations, I surrendered to my duty."

The Reagan administration estimated that gasoline prices would increase 3 to 5 cents a gallon immediately after decontrol and by 8 to 13 cents more down the road. Domestic crude oil, which was then selling for an average of $23 a barrel, with some price-controlled oil selling for as low as $7 a barrel, was expected to increase to the world price of $35 a barrel within a few days. However, a combination of increasing non-OPEC supplies from Alaska, Mexico, and the North Sea; increased Saudi production; dampened demand due to a weak economy; and better international coordination of the inventories that had accumulated the previous year averted the kinds of worldwide price hikes that had accompanied the Iranian Revolution. Some members of Congress threatened to block Reagan's action, but Democratic House Speaker Tip O'Neill said that there was "no way" they could succeed.

Decontrolling natural-gas prices was another thing altogether. The contentious compromise Congress had forged in the 1978 Natural Gas Act phased out price controls on most natural-gas prices by 1985. The new chairman of the Senate Energy Committee, Republican James McClure of Idaho, warned Reagan that "Congress isn't likely to plunge back into that thicket." McClure also expressed great skepticism about Reagan's proposal to eliminate the Department of Energy. In addition, he said that he wanted to continue to fund Jimmy Carter's Synthetic Fuels Corporation and reiterated his strong support for building a nuclear breeder reactor. Reagan was unlikely to fare much better with his Republican colleagues in the Congress than Jimmy Carter had with his Democrats.

In 1981, the new president formally asked Congress to shut down the Department of Energy. "[We] don't need an Energy Department to solve our basic energy problem," he said. "As long as we let the forces of the marketplace work without undue interference, the ingenuity of consumers, business producers, and inventors will do that for us." Congress did not bite. Many of the functions of the Energy Department would have had to be retained in other departments in any event, and Congress was in no mood for reorganizing deck chairs. The committees that would have had to advance such legislation would lose substantial chunks of their power. That was not going to happen.

Reagan had more success in diminishing the federal government's interest in solar energy. When the Energy Department in 1981 replaced the blazing sun that had sat atop the Solar Energy Research Institute's stationery with a blank, dull gray to save costs, Denis Hayes said, "It was an omen"—not a good one. On the morning of the first summer solstice after Reagan's inauguration, Hayes was asked to resign as the institute's director. He said, "Each day now, the sun falls lower; so does federal support for solar energy." Reagan's energy secretary, James B. Edwards, did not mince words. "[A] vote for Ronald Reagan" he said, "was a vote for a nuclear future."

In July of Reagan's first year in office, responding to a congressional mandate, his administration released its National Energy Policy Plan (NEPP), a report setting forth its philosophy and plans for energy policy. In his message transmitting the NEPP to Congress, President Reagan observed that it marked "a break" from the philosophy of the two previous plans submitted by Jimmy Carter. Reagan insisted that our nation's energy supplies were not nearly as grimly low as others claimed and said: "Our national energy plan should not be a rigid set of production and conservation goals dictated by Government. . . . When the free market is permitted to work the way it should, millions of individual choices and judgments will produce the proper balance of supply and demand our economy needs." In addition to eliminating price controls, Reagan also urged "reforms in leasing policies and removal of unnecessary environmental restrictions" as "part of this same effort to reduce bureaucratic burdens on all Americans." He then listed the three roles that he regarded as appropriate for the federal government: (1) exercising its responsibility for federal "lands which contain a major share of our resource wealth"; (2) fostering some long-term R&D related to energy production and distribution with the goal of developing "technological innovations to the point where private enterprise can reasonably assess their risks"; and (3) rapidly filling the Strategic Petroleum Reserve.

Public spending for energy-related purposes, the report said, "is secondary to ensuring that the private sector can respond to market realities." "Using public funds to subsidize either domestic energy production or conservation buys little additional security and only diverts capital, workers, and initiative from uses that contribute more to society and the economy." The contrast could hardly have been greater with the first principle of Jimmy Carter's National Energy Plan submitted to Congress on in April 1977, a few months after he took office: "The energy problem," Carter's plan said, "can be effectively addressed only by a Government that accepts responsibility for dealing with it comprehensively."

Reagan was determined to move energy away from the center of U.S. domestic policy and to replace it with a more general program aimed at restoring robust economic growth. He placed his faith in the "strength of the private sector." You can find no hint in his NEPP that Reagan might think that the American people were morally deficient because of what Carter had labeled their "profligate" energy consumption. He did not. The age of limits was over.

Congress was unsurprisingly not completely convinced to undo its legislative handiwork of the 1970s, and Ronald Reagan was unable to achieve nearly as complete a dismantling of 1970s energy policies as he desired. Congress, for example, increased the Energy Department's budget beyond what Reagan had requested. On October 1, 1987, that department celebrated its tenth anniversary. Despite its setbacks in Congress, however, the Reagan administration marked an effort to return to the "old-time religion" that had characterized energy policy before Richard Nixon's price controls and the 1973 oil embargo.

Most important of all, the oil market in the early 1980s broke sharply in favor of Reagan's laissez-faire approach. OPEC, holding a one-day meeting in Geneva on October 29, 1981, agreed to set its base price for oil at $34 a barrel through the end of 1982. This base price was an average increase in OPEC's price by $1 to $2 a barrel, and it was expected to produce an increase in the price of U.S. gasoline and heating oil of about three cents a gallon. But looking at the average conceals some important variations. For Saudi Arabia, this decision increased prices by $2 a barrel; however, for the 12 other OPEC Middle East nations it was a price reduction of about the same amount. Libya reduced its prices by $4 a barrel. This agreement was intended to alleviate the downward pressure on OPEC oil prices resulting from a decline in energy demand in the West and increasing non-OPEC supplies. The Saudis, in particular, were concerned about the damage that rapidly rising prices had been having on Western economies. The need for price stability had become particularly acute when the assassination of Egypt's president Anwar el-Sadat two weeks earlier threw world oil markets into turmoil. With a glut of oil available on worldwide markets, this October's OPEC agreement was really an attempt to head off a downward price spiral.

This would turn out to be the last time for a long while that oil prices might be said to rise. Less than two years later, on March 15, 1983, at the end of a contentious 12-day meeting in London, OPEC agreed to cut its base price by $5 a barrel to $29. It also agreed to reduce its output to 17.5 million barrels a day. In 1979, OPEC daily output had been 31 million barrels; in 1981, it was 20.5 million. Between 1979 and 1983, worldwide oil consumption had dropped by about 6 million barrels a day, and non-OPEC production had increased by 4 million barrels a day. With prices falling, oil companies were dumping onto the market inventories they had accumulated in 1979 as a hedge against price increases during the Iranian Revolution.

The worldwide oil market had changed dramatically. Even though the Iranian Revolution and the Iraq–Iran War had severely reduced the export capacity of those two oil-rich nations, the worldwide supply of oil now greatly exceeded demand. In 1982, non-OPEC production had become greater than OPEC production.

When Western leaders came together in May 1985 in Bonn, Germany, for their yearly economic summit, the U.S. economy was booming. The other nations there were also enjoying substantial economic growth. Iran and Iraq were still at war and blowing up each other's oil fields, refineries, and tankers, but oil and energy policy was not even on this summit's agenda.

By the end of that year, the price of oil had fallen off the table. West Texas crude was selling for $10 a barrel. A barrel of oil in the Persian Gulf could sometimes be bought for as little as $6. The turnaround was so stunning that Ronald Reagan sent Vice President George H. W. Bush to Saudi Arabia in an effort to keep prices high enough to protect a declining U.S. oil industry. The energy world had once again turned upside down. In December 1986 in Geneva, OPEC, with the cooperation and agreement of non-OPEC producers, managed to stabilize prices in a $15 to $18 range.

In the 1970s, dramatic oil price increases had contributed to both inflation and recession in the United States. The price drop of the 1980s produced opposite effects, helping to reduce inflation and aiding economic growth. Oil prices would not spike again until 1991, when Iraq invaded Kuwait, and the United States responded with its Desert Storm offensive. They would not reach their 1979–1980 peak again in real terms until 2008. Energy policy complacency was settling back in.

In 1986, the solar panels that Jimmy Carter had so proudly and publicly installed on the White House roof seven years earlier were taken off for "repairs." They were never reinstalled. Years later they could be found, not functional, on the roof of the cafeteria of Unity College in Maine. In 2007, two of them traveled to the Jimmy Carter library in Atlanta—supposedly driven there from Maine by two students in a vehicle powered by vegetable oil. Why Ronald Reagan had removed them was never completely clear. Reagan's press secretary told the Associated Press that putting them back would be "uneconomical." The panel installer told the *Washington Post* that Reagan "felt that the equipment was just a joke." Whatever Reagan's motive, the removal of the solar panels went unremarked by the press and unnoticed by the public. Energy was no longer at the forefront of America's policy radar screen.

All the efforts of three presidents to confront a crisis that had periodically disrupted and at times threatened to cripple the U.S. economy seemingly evaporated with the collapse in the price of the one fuel that had come to dominate our energy system. Price was king. And under Reagan, policy, always slower than the movements of a spot market, merely followed in its wake.

In the meanwhile, a new challenge was coming to the fore. In 1987 James Hansen, a NASA climatologist, who according to *The New Yorker* is often called "the father of global warming and sometimes the grandfather" and who had done influential work on climate change since the 1970s, testified before Congress on more than one occasion that increases in carbon dioxide emissions threatened temperature increases that could dramatically affect our way of life. The scientific community was reaching a consensus that the ongoing accumulation of greenhouse gas emissions from fossil fuels threatened uniquely dangerous changes in the world's climate—a risk that the price of oil and of other fossil fuels failed to reflect. But it would be a long time before this news would take hold in Congress.

10 Climate Change, a Game Changer

On December 10, 2007, former vice president Al Gore stopped briefly on the long red carpet leading up the steps to Oslo's city hall. He nodded, smiled, then waved to the cheering crowd gathered on the streets circling the building. Gore—who had won the popular vote for the U.S. presidency in 2000 but lost to George W. Bush in the electoral college after a controversial Supreme Court decision halted a recount of votes in Florida—was in Norway to accept the Nobel Peace Prize. He was the second American in a decade to do so, having been preceded five years earlier by Jimmy Carter, who had won this prize for his postpresidential work promoting democracy and human rights.

The Nobel Prize Committee had announced on October 12 that Gore would split the 2007 award with the Intergovernmental Panel on Climate Change (IPCC), a group of 2,500 scientists from more than 130 nations established in 1988 by the United Nations to study climate change and report what it found to the public. The Nobel Prize Committee praised the IPCC for creating "an ever-broader informed consensus about the connection between human activities and global warming" and Al Gore as "probably the single individual who has done most to create greater worldwide understanding of the measures that need to be adopted." Together, the IPCC and Gore, the committee said, had illuminated "the processes and decisions that appear to be necessary to protect the world's future climate, and thereby reduce the future threat to the security of mankind." Rajendra K. Pachauri, an Indian who as chairman of the IPCC was there to accept that organization's award, said that the prize represented a victory for science over skepticism.

Al Gore, who had been involved with environmental issues since he first came to Congress in the 1970s, had publicized the dangers of global climate change from emissions of greenhouse gases most successfully in his Academy Award–winning 2006 documentary *An Inconvenient Truth* and its best-selling companion book *An Inconvenient Truth: The Planetary Emergence of Global Warming and What We Can Do about It.* The movie consists mostly of Al Gore's "slide show" on the causes and dangers of global warming—which Gore calls "the greatest danger we've ever faced." It grossed about $50 million in theaters, an extraordinarily large amount for a documentary film, and was seen by millions of people worldwide. His companion book hit number one on the *New York Times* best-seller list in the summer of 2006. According to a 2007 Nielsen survey, two-thirds of the people who saw Gore's film said they had "changed their mind" about global warming. Carl Pope, president of the Sierra Club, has called Gore the "indispensable player in the drama of mankind's encounter with the possibility of destroying the climactic balance within which our civilization emerged and developed."

From the stage in Oslo's City Hall, Al Gore accepted the Nobel Peace Prize, saying that we need to act "boldly, decisively, and quickly" to respond to "a planetary emergency—a threat to the survival of our civilization that is gathering ominous and destructive potential." Comparing the climate change threat to the global threat posed by Adolf Hitler, Gore described a world now "spinning out of kilter," facing an unprecedented catastrophe. He said, "we must quickly mobilize our civilization with the urgency and resolve that has previously been seen only when nations mobilized for war." What is required, he said, is a "mighty surge of courage, hope and readiness to sacrifice for a protracted and mortal challenge." Gore called for a worldwide moratorium on the construction of any new coal-generating facility that does not have the capacity to capture and store carbon dioxide as well as for a tax on carbon and an alliance of energy-consuming nations to make "solving the climate crisis its first priority." Making clear that he viewed climate change as a moral issue and offering a prayer that we make the right choices, Gore said that we "have everything we need to get started, save perhaps political will, but political will is a renewable resource." After nearly a two-minute ovation, with many in the audience standing, Al Gore placed his open hand on his heart, nodded to his appreciative audience, and took his seat.

Despite overwhelming scientific evidence, the successes of political leaders such as Al Gore in making the problem known, and the best efforts of environmentalists and climate scientists to galvanize public support for addressing emissions of greenhouse gases, not everyone is convinced.

Speaking on MSNBC's *Hardball with Chris Matthews* in August 2009, representing a view firmly held by more than a few conservative Republicans and other doubters, the commentator Pat Buchanan, who had challenged the incumbent president George H. W. Bush for the 1992 Republican presidential nomination, characterized Gore's stirring up concerns for climate change as a "hoax." Buchanan did not deny that greenhouse gas emissions occur, but he said that the "idea that we're all gonna die of this is utter nonsense." "It's a power transfer to governments here and governments abroad," he insisted. As for Al Gore's arguments to the contrary, Buchanan did not challenge Gore's sincerity. "It's a religious belief," Buchanan said.

In the winter of 2009–2010, a number of relatively small but visible errors in the IPCC's work (including, for example, an overstatement of how rapidly the Himalayan glaciers might melt), the leaking of emails showing that certain prominent climate scientists sought to squelch publication of contrary views, and an extraordinarily snowy winter, especially in Washington, D.C., provided ammunition for climate change skeptics. Oklahoma's Republican senator Jim Inhofe built a six-foot igloo, which he labeled "Al Gore's New Home." His South Carolina colleague Jim DeMint predicted that it would keep snowing in D.C. until "Al Gore cries uncle." For his part, Al Gore said that he genuinely wished that "the climate crisis was an illusion," but that a few mistakes do not change the conclusions of the "thousands of pages of careful scientific work over the last 22 years by the Intergovernmental Panel on Climate Change." We are facing, he insisted, "an unimaginable calamity requiring large-scale, preventable measures to protect human civilization as we know it."

Despite overstatements and bad behavior on both sides, as well as, more important, the inevitable uncertainties about the magnitude and timing of temperature changes, the scientific community generally agrees that climate change is occurring as a result of human emissions of carbon dioxide and other greenhouse gases and that potentially such change will cause harmful consequences, such as rises in sea levels, ice melts, droughts, and more rapid spread of diseases. Those who reject this evidence believe that the climate scientists are not behaving in good faith and are motivated by their political desires rather than by their scientific observations. The deniers seize on every mistake, overstatement, or contrary anecdote to bolster their case.

It is inescapable that the risks of climate change from greenhouse gas emissions are real. It is also inescapable that we do not and cannot know with precision exactly what will happen or when it will occur. But we would be foolish not to take action to insure against the risks we face. After a detailed and careful review of the science and the climate change debate, *The Economist* concluded: "The fact that the uncertainties allow you to construct a relatively benign future does not allow you to ignore futures in which climate change is large, and in some of which is very dangerous indeed. The doubters are right that uncertainties are rife in climate science. They are wrong when they present that as a reason for inaction."

Studies by the U.S. Defense and State departments in 2009 described the potential geopolitical and military issues that climate change might cause. Armanda Dory, an Obama administration Defense Department official charged with incorporating climate change issues into that department's strategic thinking, said (without irony) that there had been a "sea change" in the Pentagon's attitude. The military must "wrestle with" these issues in thinking about its national security strategy. Food shortages, droughts, and catastrophic flooding might strain military resources and require humanitarian and perhaps military responses. The department's 2010 *Quadrennial Defense Review Report* stated that "climate change could have significant geopolitical impacts around the world, contributing to poverty, environmental degradation, and the further weakening of fragile governments." This review also suggested that climate change "will contribute to food and water scarcity, will increase the spread of disease, and may spur on or exacerbate mass migration."

Retired marine general Anthony C. Zinni said, "We will pay to reduce greenhouse gas emissions today, and we'll have to take an economic hit of some kind. Or we will pay the price later in military terms." In December 2009, President Barack Obama, accepting his own Nobel Peace Prize, said that the "world must come together to confront climate change," citing potential conflict and threats to security as a reason why. "It is not merely scientists and activists who call for swift and forceful action," he said, "it is military leaders in my country and others who understand that our common security hangs in the balance."

At the time of Gore's Nobel acceptance speech in December 2007, oil was selling for more than $90 a barrel on the New York Mercantile Exchange, nearly 150 percent of its price a year and a half earlier when his movie was released. In real terms, taking inflation into account, oil was once again reaching the price peaks of the end of the 1970s following the Iranian Revolution. And oil's price climb was not over: in July 2008, oil prices peaked at about $148 a barrel. Americans were paying more than $4 a gallon for gasoline just six months after it had broken through the $3 barrier that January. The era of low oil prices that had lasted for two decades since the mid-1980s had come to an end. During this interval, oil and gasoline prices had spiked up once in a while—for example, in January 1991 during the first U.S.–Iraq War, in September 2001 after the attack on the World Trade Center, and again in August 2005 when Hurricane Katrina hit the Gulf of Mexico on her way to New Orleans. As in the oil crises of the 1970s, these price spikes were responses to supply shortages, but, unlike in the 1970s, oil prices in these periods quickly returned to their preshortage levels. In 2008, however, the price increases were demand not supply driven, spurred by the burgeoning appetites of modernizing and growing economies in China and India. The price increases were also a response to the decline in the value of the dollar, which had lost 30 percent of its value relative to other currencies since 2002. Despite the volatility and short-term price swings on oil futures, most experts regarded these price changes as more likely to be permanent than temporary.

The rise in prices, coupled with growing concerns about the risks of climate change from fossil fuel emissions, once again, after a long hiatus, moved energy policy front and center on our nation's policy agenda. If the challenges of the 1970s were daunting, adding the risks from climate change to the mix massively multiplies their complexity and difficulty.

In 2001, the administration of the former Texas oilman George W. Bush, well populated by climate change skeptics and ideological opponents of environmental regulation, essentially abdicated any American leadership role in confronting the climate change challenge. That year President Bush rejected the Kyoto Protocol to the 1992 United Nations Framework Convention on Climate Change, which called for the United States to make legally binding commitments to reduce greenhouse gas emissions. (In an earlier 95-to-0 vote, the Senate had made clear that it would not ratify an international agreement that placed binding commitments on the United

States but not on developing countries, such as China and India.) This abdication would remain in effect for most of Bush's two terms in office. In 2007 and 2008, however, President Bush convened leaders of nearly 20 other countries in what he labeled the "major economic meeting on energy scarcity and climate change," a forum that provided a useful ongoing context for international discussion among major emitters of greenhouse gases. In April 2008, facing the dizzying rise of gasoline prices and an ever more overwhelming scientific consensus on climate change, Bush made a parting nod to the issue of global warming. Acknowledging "concerns" over climate change as well as "concerns" over the economic effect of policies designed to combat it, he outlined some modest new emission-reduction goals (stopping the growth of U.S. greenhouse gas emissions by 2025, which was very small compared to the Kyoto Protocol target of emissions 6 percent less than 1990 levels by 2012). And, as had many presidents before him, he promised new money for clean technologies—such as for "clean coal," wind and solar energy production, biofuels, and nuclear power—that, he hoped, would solve problems of both energy security and climate change without diminishing the country's prospects for economic growth.

Just four months later, speaking at Michigan State University, Senator Barack Obama, then the Democratic nominee for president, spelled out his energy policy vision. By that time, it was clear to everyone that the nation was in the midst of a sharp recession—one that would turn out to be the most difficult since World War II. Oil prices—having doubled between Labor Day and May—had peaked in July, and by August 4, 2008, the day of Obama's speech, they had dropped below $120 a barrel, their lowest level since May 6. But gasoline was still averaging $3.88 a gallon, and the nation's pain was palpable. Obama, who despite the political temptations had rejected his opponents' foolish calls for a summer gas tax holiday, began by focusing on what he called "our addiction to foreign oil." "We've heard promises about energy independence from every single President since Richard Nixon," he said. "We've heard talk about curbing the use of fossil fuels in State of the Union addresses since the oil embargo of 1973." "Back then," he added, "we imported about a third of our oil." But "now we import more than half. Back then, global warming was the theory of a few scientists. Now, it is a fact that is melting our glaciers and setting off dangerous weather patterns as we speak." Calling Earth a "planet in peril,"

Obama linked climate change and the domestic economic dislocations from rising gasoline prices to "instability and terror bred in the Middle East." He was marrying fears about national and energy security to those about climate change.

Obama listed a number of specific proposals. This list was of little import, however, compared to the fundamental goal he announced: "We must end the age of oil in our time." "Breaking our oil addiction is one of the greatest challenges our generation will ever face," he said. *"It will take nothing less than a complete transformation of our economy."* "If I am President," Obama continued, "I will immediately direct the full resources of the federal government and the full energy of the private sector to a single overarching goal—in ten years, we will eliminate the need for oil from the entire Middle East and Venezuela."

Although environmentalists, such as Al Gore, have focused on climate change, Obama was right also to elevate energy security on the policy agenda. No one disputes its importance or validity. If the 1970s teach us anything about energy policy failures, it should be the dangers of transferring enormous wealth to the oil-producing nations of the Middle East. But in the first decade of the new millennium, that is exactly what we continue to do. We have gone from being the world's largest creditor nation—which we were in the 1970s—to its largest debtor. And the United Arab Emirates, Saudi Arabia, Kuwait, and Libya are all now among the world's 15 largest creditors. Foreign investors now own one out of every three U.S. government bonds, one in four U.S. equities, and one in five private-debt securities—all triple 1990 levels. Capital exports by the petronations grew from well less than $100 billion annually in the 1990s to nearly $500 billion in 2006 alone. The need for less reliance on Middle East oil is greater than ever.

Al Gore had called for the end of coal; Barack Obama was now urging the end of oil. In essence, we are being asked to stop relying on the two carbon compounds that have fueled the world's economies for more than a century. Coal, which Gerald Ford called our "ace in the hole," now is a joker in the deck; oil, long the mother's milk of our economy, now is considered poison.

And the fact that natural gas is abundant domestically and considerably cleaner than coal or oil does not satisfy Al Gore. Having previously called for the end of coal-fired electricity in the absence of carbon sequestration,

Gore, speaking before the Netroots Nation in Austin, Texas, in the summer of 2008, urged an end for the use of both oil and natural gas, advocating that we get "100 percent" of our electricity from renewable sources and in the "longer term" expand our electricity use to power our automobiles.

Although dissenting a bit on natural gas, Obama concurred. "In just ten years," he said in Lansing, we "will produce enough renewable energy to replace all the oil we import from the Middle East." "[W]e will reduce our dangerous carbon emissions 80 percent by 2050 and slow the warming of our planet." "And," he added, "we will create five million new jobs in the process." He promised that his presidency would mark "a new chapter in America's leadership on climate change."

Lumping together the goals of energy security—freeing our nation from the need to import oil, especially from the Middle East—and a halt to climate change may help to stimulate the American people's willingness to support fundamental change, but it obscures crucial policy differences. In 1970, 15 percent of our nation's electricity was produced from oil; in 1980, 13 percent. But in 2008 only one percent of the petroleum consumed in the United States was used to generate electricity, producing only 1.6 percent of our electricity. So shifting electricity consumption to renewable sources, such as wind and solar power, may slow the emissions of greenhouse gases—because coal supplies more than half our electricity—but will immediately do little or nothing to lessen imports or use of oil.

Jimmy Carter's effort to develop synfuels from coal to power our nation's cars and trucks illustrates well the potential conflict between energy security and climate change goals. Had this program been successful and our vehicles were now powered by liquids produced from coal, we would, to be sure, be importing far less oil. Our energy security would have greatly improved. But because burning coal to provide a coal-based synthetic fuel to drive an automobile would release much more carbon dioxide than gasoline, we would be emitting more greenhouse gases, threatening more rapid climate change.

Today we use oil primarily for transportation: to fuel automobiles, trucks, buses, and airplanes. In 2008, more than 70 percent of all petroleum consumed in the United States was used to transport people and goods. Nearly an additional one-quarter was used in the industrial sector, and the remaining 5 percent was used to heat residences and commercial buildings (especially in the Northeast). Ninety-five percent of the fuel used in the transportation sector is a petroleum product, mostly gasoline and diesel fuel. In 1970, when our domestic oil production peaked at 11.3 million barrels a day, we were importing just more than 3 million barrels of oil a day. In 2008, we produced 6.8 million barrels domestically, and our imports were 11 million barrels a day. Nearly 6 million barrels of these imports were from OPEC countries. So if we want to shed our "addiction to oil," the transportation sector should take priority.

To get a sense of the scope of that challenge, consider the fact that in 2008 there were about 230 million passenger vehicles on our nation's roads. In a good year, automakers sell about 16 to17 million new cars, vans, SUVs, and light trucks; in a bad year they sell 10 to 11 million. Many Americans hold their cars a very long time or purchase used cars. Our automobile fleet turns over rather slowly.

Three approaches might be taken to reduce or eliminate the use of oil for transportation: a switch to the use of natural gas (or possibly hydrogen) as fuel; massive increases in fuel economy, for example, by widespread sales of hybrid vehicles, many of which are already in use; or the long-promised adoption of electric cars.

At present, the most prominent proponent of the first of these ideas is T. Boone Pickens, who made his fortune as an independent oil man and who Daniel Yergin describes as having "remade the corporate landscape of oil." In 2008, Pickens announced his so-called "Pickens Plan"—a proposal to shift electricity production to wind power to combat climate change and to expand the use of natural gas in transportation to reduce oil imports. The total number of vehicles fueled by natural gas today is fewer than 150,000. There is, however, no technological barrier to converting cars and trucks to using this fuel, and natural gas is now abundant domestically. The barriers, instead, are expense and infrastructure. Converting a passenger vehicle to natural gas can cost more than $10,000 and a diesel truck about $65,000. With 6.5 million trucks on the road, just converting them would cost about $400 billion; converting the autos would take another $2.3 trillion, not that anyone is suggesting the latter. Then there is infrastructure. There are about 500 publicly available natural gas refueling stations nationwide—compared to about 120,000 retail gasoline stations. As chapter 6 has described, a home natural-gas fueling system costs around $4,000, and the one commercially available natural-gas-powered car in the United States, the Honda Civic GX, has never been successfully marketed. Making a bet on new natural-gas automobiles to capture any significant portion of the private automobile market would be very risky.

In the past decade, Toyota and then Honda introduced to the U.S. and world markets hybrid automobiles that combine electric and gasoline power to enhance substantially the cars' fuel economy. American and other automakers soon followed with hybrids of their own. These cars frequently achieve 40 to 50 miles per gallon of gasoline, although some of the heavier models provide only half that much. A limited number of hybrid cars per manufacturer have been eligible for federal tax credits in excess of $3,000 per car, but (as chapter 12 details) those tax credits have declined and disappear for particular cars as they penetrate the market. Some hybrid models have been quite successful, with the Toyota Prius leading the pack. Sales of hybrids increased annually from 2000 through 2007, but the combination of declining gasoline prices, tight credit markets, and a recession resulted in a 10 percent decline in sales in 2008 to about 315,000 cars compared to 350,000 in 2007. Long waiting lists for the Toyota Prius that had existed when gasoline was selling for $4 a gallon disappeared when gas prices dropped to $2 a gallon. Despite their success, hybrids still compose a small fraction of annual U.S. auto sales—less than one percent—but the number of hybrids on our roads increases annually.

The third possibility is the long anticipated arrival of the electric car. Because there are electric outlets in every home in America, finding power should not be a problem, but perfecting the battery technology has proven extremely difficult. And as the 2006 documentary *Who Killed the Electric Car?* makes clear, virtually all the major players—oil companies, auto companies, states, and the federal government—had a hand in dooming the electric car's first commercial roll out in the 1990s. Nevertheless, it may well be on the verge of resurrection. Both the automobile companies and the federal government seem to be transforming from culprits to converts. In August 2009, GM, which due to the economic meltdown was then largely under government ownership and control, unveiled the Chevrolet Volt, which became available in 2010. Fritz Henderson, GM's spokesman, said that the car, which has a small gas engine to help augment the car's battery, will be able to travel up to 40 miles on a single electric charge and will get 230 miles per gallon in city driving. The 2011 Nissan Leaf, an all-electric car, is said to get 367 miles per gallon using a similar methodology. But these are overstated; however, the combined city and highway driving mileage for both cars may reach triple digits. The Volt sells for about $40,000, the Leaf for closer to $30,000. Both auto makers expect these cars to be eligible for a $7,500 federal tax credit. Mr. Henderson said that "having a car that will get triple-digit fuel economy will be a game changer for us." Tesla Motors, a Silicon Valley manufacturer of high-performance electric cars using lithium-ion battery cells and selling for prices in six figures, enjoyed great success in its initial public offering in June 2010, despite not yet making a profit, and soon thereafter Tesla announced a joint venture with Toyota to produce electric cars.

Of course, the final verdict on the electric car will come in the marketplace. The Nissan Leaf, for example, although roomy, comfortable, and equipped with a GPS system that will tell you where the nearest electric recharging system is located, is not an attractive car. Moreover, it will take 8 to 20 hours to recharge when the battery dies, depending on the strength of a homeowner's electric power. Whether consumers—who were less than enthusiastic for the earlier generation of electric cars—will have changed their attitudes in sufficient numbers is uncertain. In addition to price, there are other drawbacks. Even though some cities, most notably San Francisco, are modifying their building codes to require electric car chargers, and some companies, also predominately in California, are introducing workplace charging stations, city dwellers, in particular, without a garage with electricity to recharge their cars overnight, may be particularly reluctant to take the leap. "If you're going to park it on the street, I don't know what to do actually," said GM's Mr. Henderson. "I don't know how to address that situation." A Nissan vice president was even more blunt, "A lot of those people aren't going to be able to get an electric vehicle initially," he said. Some cities, however, are attempting to solve this problem. London, for example, which in 2009 had 250 publicly available electric charging stations, has announced an ambitious goal of 25,000 by 2015. New York City in January 2010 released a comprehensive study analyzing what might be done there to encourage more electric vehicles. More government subsidies for such vehicles seem both necessary for them to succeed and inevitable. And more improvements in battery technology remain essential before Americans can be expected to move to electric cars in any meaningful numbers.

An important factor to consider, however, is that although electric cars would allow us to import less oil, they will not help much with climate change until we shift electricity production away from coal or find a viable economic way to capture and sequester its carbon emissions. Shifting away from petroleum-powered vehicles and coal-powered electricity are Herculean tasks facing enormous technical, political, economic, and cultural challenges.

If one looks forward—over a very long time horizon—both climate change and energy security can be reconciled: by shifting our electricity production to cleaner fuels and our transportation vehicles to run on electricity (or some other non-carbon-based fuel). Then, neither will require imported oil, and neither will increase greenhouse gas emissions. We should not fool ourselves, though: such a shift will not be quick, easy, or painless.

After Barack Obama became president, he moved to demonstrate that he was serious about transforming our nation's energy production and consumption. He appointed Steven Chu—a Nobel laureate in physics who had been serving as director of the Lawrence Berkeley National Laboratory, the oldest of the Energy Department's national labs—as our nation's twelfth energy secretary. When he announced Chu's appointment, he said, "The future of our economy and national security is inextricably linked to one challenge: energy."

Steven Chu clearly shares Obama's concerns and aspirations, and he has made it clear that he regards reducing greenhouse gas emissions as our energy policy priority. In July 2009, in testimony to the Senate Committee on Environment and Public Works, Secretary Chu summarized the scientific knowledge concerning climate change:

> Overwhelming scientific evidence shows that carbon dioxide from human activity has increased the atmospheric level of CO_2 by roughly 40 percent, a level one-third higher than any time in the last 800,000 years. There is also a consensus that CO_2 and other greenhouse gas emissions have caused our planet to change. Already, we have seen the loss of about *half* of the summer arctic polar ice cap since the 1950s, a dramatically accelerating rise in sea level, and the loss of over two thousand cubic miles of glacial ice, not on geological time scales but over a mere hundred years.
>
> The Intergovernmental Panel on Climate Change (IPCC) projected in 2007 that, if we continued on this course, there was a 50 percent chance of global average air temperature increasing by more than 7 degrees Fahrenheit in this century. A 2009 MIT study found a fifty percent chance of a 9 degree rise in this century and a 17 percent chance of a nearly 11 degree increase. 11 degrees may not sound like much, but, during the last ice age, when Canada and much of the United States were

covered all year in a glacier, the world was only about 11 degrees colder. A world 11 degrees warmer will be very different as well. Is this the legacy we want to leave our children and grandchildren?

One might expect—given the concern about the potential apocalyptic consequences from climate change and the Obama administration's goal of completely transforming U.S. energy consumption—that the American people would be asked to sacrifice now in order to protect "our children and grandchildren." But not since Jimmy Carter left office has it been fashionable to ask the American public to sacrifice. Steven Chu did acknowledge in his testimony that there might be some costs in addressing climate change, but he insisted that they were trivial. He told the Senate committee that meeting targets for reducing greenhouse gases "can be achieved at an annual cost between 22 and 48 cents per day per household in 2020." "That's about the price of a postage stamp a day," he added. And if that weren't soothing enough: "History," he said, "suggests that the actual costs could be even lower." "The right clean energy incentives will start the great American research and innovation machine, and I am confident that American ingenuity will lead to better and cheaper climate solutions."

President Obama also exudes optimism. Although acknowledging in passing that the great energy transformation he envisions will be "costly," Obama, like many climate change warriors, emphasizes the many "green jobs" that "clean energy" will produce: "Millions of new jobs. Jobs that pay well. Jobs that can't be outsourced. Good union jobs."

The notion that responding to environmental challenges will be a job creator not a job killer dates back more than two decades. In June 1987, David Clark of Britain's Labour Party claimed that his party's environmental policies would create at least 100,000 and perhaps half a million "green jobs" by century's end. In the United States, the greatest proselytizer for the potential of green jobs has been the *New York Times* columnist Thomas Friedman. "This is an innovation problem," Friedman says, speaking, as he puts it, of "climate change, petro-dictatorship, biodiversity loss, energy poverty and energy resource supply and demand." "We don't need a Manhattan project," he says. "What we need is 100,000 Dean Kamens [inventor of the two-wheel, electric-powered Segway] in 100,000 garages trying 100,000 things, so maybe ten of them will come up with that holy grail of abundant, cheap, clean, reliable electrons."

On our fortieth Earth Day, April 22, 2009, Barack Obama visited Trinity Structural Towers in Newton, Iowa, a plant that manufactures wind energy equipment, to push for energy legislation and to address climate change. "The choice we face," the president said, "is not between saving our environment and saving our economy. The choice we face is between prosperity and decline. We can remain the world's leading importer of oil, or we can become the world's leading exporter of clean energy." The president claimed that wind power alone, which in 2008 accounted for about 1.25 percent of the nation's electricity, could supply 20 percent by 2030.

President Obama's visit to Iowa was just one piece of an ongoing coordinated effort by his administration to convince the public that addressing climate change would also be good for the nation's economy. An op-ed article, coauthored by Energy Secretary Chu and Labor Secretary Hilda Solis, ran in newspapers across the country the same day, claiming that with "the depletion of the world's oil reserves and the growing disruption of our climate, the development of clean renewable sources of energy is the growth industry of the 21st Century." Van Jones, then of the White House's Council on Environmental Quality, said that the administration is "committed that green jobs be good jobs, and there's a strong commitment to make sure that it actually happens." He said that this is "the kind of 'for everybody Earth Day agenda' that the Obama administration stands for," adding, "There's a wingspan on these jobs [that] goes from GED to Ph.D."

The plant that Obama visited had 91 employees. Another plant nearby, TPI Industries, which makes blades for wind turbines, employs another 300 workers, and its workforce may grow to 700 if demand increases as expected. Both of these plants are located in factory space that previously housed a Maytag factory, which had employed 1,800 people. If the idea is that "green energy" jobs will make up for the manufacturing jobs we have lost, it seems Pollyannaish.

A study issued by Senator Kit Bond (R–Mo.), ranking Republican on the Subcommittee on Green Jobs of the Senate Committee on Environment and Public Works, indicates that in addition to federal direct subsidies and tax breaks for wind power, TPI Industries had received $6.6 million in state and local subsidies, or $20,000 per worker, and Trinity Structural Towers had received $2,000,000, or just more than $14,000 per job. If that is all the government spent, these jobs were bargains. Senator Bond's report describes solar photovoltaic–manufacturing enterprises in Georgia, Michigan, and Oregon that have enjoyed subsidies ranging from $100,000 to $325,728 per job. His report also cites economic studies suggesting that some California and other state studies of green jobs have overstated their benefits and ignored their costs. He points to German and Spanish studies showing subsidy costs per job ranging from $281,000 to $700,000. The Spanish study also estimates that each Spanish wind industry job involved subsidies of about $1.4 million. In addition to the size and scope of taxpayer-funded subsidies, Bond expresses concern about low wages and uncounted job losses. His conclusion is that "these green jobs problems and pitfalls signal a yellow light urging caution with green jobs."

The Spanish study cited by Bond estimates that in addition to large government subsidies, each new green job costs the loss of 2.2 other jobs. Conservative columnist George Will reported the following exchange when President Obama's press secretary was asked about this contention:

Robert Gibbs It seems weird that we're importing wind turbine parts from Spain in order to build—to meet renewable energy demand here if that were even remotely the case.

Questioner Is that a suggestion that his study is simply flat wrong?

Mr. Gibbs I haven't read the study, but I think, yes.

"[O]ne can be agnostic about both reports," Will says, "while being dismayed by the frequency with which such findings are ignored simply because they question policies that are so invested with righteousness that methodological economic reasoning about their costs and benefits seems unimportant."

A hefty dose of skepticism regarding the job-creating force of shifting to cleaner sources of energy is further reinforced by an analysis by Sunil Sharan, who served as director of GE's Smart Grid Initiative, in which electric meters will be shifted to "smart meters" that constantly and automatically transmit data about energy consumption to its provider. Such "smart grid" technology allows much more efficient management of electricity production, distribution, and consumption, and it therefore serves as an important example of the kinds of technological advances that might play an important role in limiting climate change and helping to secure greater energy security. Writing in the *Washington Post*, Sharan questioned the job-creating potential of the more than $4 billion of 2009 stimulus funds for producing and installing nearly 20 million "smart meters" by 2015. He pointed out that manufacturing of the meters will occur predominately overseas and that domestic manufacturing jobs as well as supervisory domestic management, R&D, and information technology jobs are likely to number only in the hundreds or low thousands. Sharan estimated that it will take 1,600 new workers to install the 20 million smart meters over five years, but that eliminating the many meters that are now read manually each month will cost 28,000 meter readers their jobs over the same period. After expressing similar skepticism about the net new jobs likely to result from solar and wind energy and from a conversion from gasoline to electric cars, he concluded: "Those who take great pains to tout the 'job-creation potential' of the green space might end up inducing labor pains all around."

The reality, unfortunately, is that there is no guarantee that the United States rather than, say, Asia will be where the bulk of new green jobs occur. In 2009, for the first time, China manufactured and installed more wind turbines than the United States did. Despite the billions in direct subsidies and tax breaks now available to domestic wind and solar manufacturers and customers, much of the job growth seems to be occurring elsewhere. Suntech Power Holdings Company, which received a grant of $2.1 million to build a solar panel plant in Arizona, is hiring 70 workers there to assemble components made by the company's 11,000 Chinese employees. First Solar Inc., the world's largest producer of thin-film solar power modules, received $16.3 million to add 200 jobs in Ohio, but the company also employs 4,500 workers worldwide, mostly in Malaysia, where it expects nearly three-quarters of its expected factory growth to occur. Many multinational companies from Europe and the United States are building large factories in China to supply that nation's rapid growth in generating electricity from wind and solar sources. They are manufacturing equipment there both for China's large market and for export to the West. Although President Obama insists that these jobs should be in the United States, Kit Bond claims that the "good-paying, middle-class supporting manufacturing jobs" are occurring in Asia. He says that companies in his home state of Missouri are "creating low-wage jobs to install" the products manufactured abroad. No one wants even to whisper that we might ultimately end up substituting dependence on Asian clean-energy equipment imports for dependence on imports of Middle East oil, although doing so could represent a major improvement in terms of both energy-supply stability and national security.

Everyone, of course, wishes that reducing greenhouse gases will be a great spur to the economy, but it takes a leap of faith to believe that moving away from fossil fuels will create more jobs than it will destroy. Matt Taibbi, a writer (most often for *Rolling Stone* magazine), describes Tom Friedman's "massive investment in green energy" as "obvious" and "inoffensive," but he also attacks Friedman's credibility. "Where does a man who needs his own offshore drilling platform just to keep the east wing of his house heated get the balls to write a book chiding America for driving energy inefficient automobiles?" Taibbi asks.

Even if you are a Tom Friedman fan, Taibbi is on to something important here. Moving this nation away from fossil fuels may imply a massive change in our culture. Taibbi describes Friedman as living "in a positively obscene 11,400 square feet suburban Maryland mega-monstrous-mansion" and being married to the "heir of one of the largest shopping-mall chains in the world . . . whose family bulldozed 2.1 million square feet of pristine Hawaiian wilderness to put a Gap, an Old Navy, and even a [expletive deleted] Foot Locker in paradise." If so, Tom Friedman himself is a great exemplar of the cultural effects of our nation's longstanding reliance on inexpensive fossil fuels—especially oil and coal.

So is Al Gore. He too lives in a megamansion of 20 rooms and 8 bathrooms, and in 2006 his home (and pool house) consumed 221,000 kilowatt hours of electricity at a cost of $30,000—ten times the national average. In 2007, a spokesman for Gore said that he was in the process of installing solar panels, that he used energy-efficient compact florescent light bulbs, and that he had purchased carbon offsets to bring his carbon footprint down to zero. The inventor Saul Griffith, although acknowledging that Gore "has done a huge amount to help this cause," calls Gore "the No. 1 environmental hypocrite." Like Gore, *An Inconvenient Truth* producer Laurie David apparently flies regularly in a private jet. This practice prompted the *New Republic* writer Gregg Easterbrook to observe that one cross-country flight in her Gulfstream emits the same greenhouse gas as if she drove a Hummer for a year.

By describing Friedman and Gore's energy profligacy, I do not mean to insist or even suggest that our most visible, vocal, and effective advocates for transforming how our nation obtains and uses its energy must follow Mahatma Ghandi's famous admonition to "be the change you want to see in the world." Rather, it is to demonstrate starkly some of the barriers to such change.

A more ordinary American, Barnard P. Giroux, a real estate entrepreneur from Fall River, Massachusetts, elaborated his view of the cultural impediments to transforming our nation's energy in a letter to the editors of the *Wall Street Journal* in 2009. The letter, which the *Journal*'s headline writers entitled "When a Hybrid Can Pull a Boat, Then We'll Talk," was in response to a proposal for higher gasoline taxes. Mr. Giroux was blunt. "Americans don't want small vehicles," he explained. "We have great distances to travel, mountains and plains to cross in all seasons of the year. We tow our boats and other contrivances. We haul our children around and travel with them over the continent." In case readers missed his point, he added: "This is not Europe. This is the United States of America, a vast country with amazing distances and varieties of geography and climate."

Mr. Giroux's outrage may be excessive, but the sentiments he expresses are widespread. Many of us travel long distances to work everyday, often in bumper-to-bumper traffic. Most Americans don't have or don't like mass transit; we rely instead on our nation's highway system. We are unwilling to give up our individual privacy and flexibility to car pool. We don't even like commuting with the one additional person who might give us access to faster high-occupancy-vehicle lanes on the highway. The California Highway Patrol routinely pulls over drivers who have placed dummies in the passenger seat. Our children look forward to nothing more than the day they can get their driver's licenses: the automobile is our nation's freedom machine. Despite government subsidies, fast trains are as common as centaurs in the United States.

Moreover, as Mr. Giroux insists, many Americans like their automobiles large and powerful. U.S. automakers for many years owed most of their profits to SUVs and pickup trucks. The demographic shift that took hold in the 1970s with people moving away from the Northeast to the more wide open spaces of the South and West increased our nation's total automobile mileage, as did the growth of our suburbs and then exurbs. The destruction and longstanding opposition to mass transit by the auto industry and other business interests must also share some of the blame. In 1975, U.S. vehicles drove a little more than 1 billion total miles; three decades later that number exceeded 3 billion. And we live in larger homes than people in any other nation. If fuel switching and technology do not completely solve the problems, will the American public be willing to change longstanding preferences and habits in order to combat climate change and to cure our "addiction" to oil?

It is tempting to believe that the "small is beautiful" movement of the 1970s has passed on and that, notwithstanding new worries about climate change, it is not likely to be revived anytime soon. But not for Bill McKibben, whom in 2010 the *Boston Globe* called "probably the nation's leading environmentalist" and *Time* described as "the world's best green journalist." McKibben in 2007 founded 350.org, which according to *Foreign Policy* organized "the largest ever global coordinated rally of any kind" with 5,200 demonstrators in 181 countries calling for a limit of the worldwide concentrations of carbon dioxide in the atmosphere to 350 parts per million (ppm), about 10 percent less than current levels. In his 2010 book *Eaarth: Making a Life on a Tough New Planet,* following a pessimistic recitation of a large number of adverse consequences of climate change that he says have already occurred, McKibben endorses the findings of the Club of Rome's 1972 publication *Limits of Growth,* which he praises as "ahead of the curve." He says that it is "possible" that we have already "seen the peak of economic growth" and insists that "we can't grow." Instead, he urges that we "try to manage our descent," that we "aim for a relatively graceful decline," that we "adjust to the fact that we are not going to get bigger." McKibben, who practices what he preaches, lives in rural Vermont with solar panels on his roof. He is a true believer that small is beautiful. He would like all of us to live like he does. The kind of cultural transformation he urges could hardly be a more dramatic rejection of the very idea of a global economy, a call for returning to self-sufficient local economies. He quests for durability, sturdiness, and stability rather than expansion and economic growth. But the small-is-beautiful view failed in the twentieth century and will most likely fare no better in the twenty-first. Bill McKibben sounds much like Barry Commoner. And this time the environmental challenge entails emissions all over the world. Americans are not the only ones who will have to change.

It takes only a thimbleful of empathy to see why leaders of developing countries such as China and India have been rather reluctant to enter into international treaties committing their nations to limit greenhouse gases just as their people are gaining enough income to purchase electricity, automobiles, and central heating. Even though we use less energy per dollar of economic output than we did in the 1970s, the United States, with only 4 percent of the world's population, consumes one-quarter of the energy the world uses each year. And 1.6 billion people around the world still have no electricity.

In the summer of 2009, Hillary Clinton, on one of her first missions as U.S. secretary of state, went on a five-day visit to India accompanied by Todd Stern, the administration's special envoy for climate change. She traveled with Jairam Ramesh, India's minister for forests and the environment, to tour the headquarters of the hotel division of the Indian tobacco company ITC Ltd.—a "green building" in Gurgaon on the outskirts of New Delhi. After praising the building, which its owner claims has a 30 percent smaller carbon footprint than buildings similar in size, as a "monument to the future," she acknowledged that "developed countries like mine must lead on this issue." She then suggested that reducing greenhouse gas emissions might help, not hinder, India's economy. She said that since more than "80 percent of the growth in future emissions will be from developing countries," it is "essential for major developing countries like India to also lead."

Minister Ramesh did not take this admonition well. He complained about pressure to address climate change from the United States and said that India's view "is that we are simply not in the position to take legally binding emissions targets." The Indian government has suggested that a per capita emissions limit would be fair—something the United States obviously cannot live with. (In 2006, for example, India's per capita energy consumption was about 5 percent that of ours, China's was 16.7 percent. Americans thus prefer talking about "energy intensity," the ratio of energy use to GDP: India's energy intensity that year was 85 percent of ours; China's 156 percent.) Secretary Clinton responded by insisting that no one "wants to in any way stall or undermine the economic growth that is necessary to lift millions out of poverty." Indian officials have indicated that if the Western nations want their country to roll back emissions, they will have to help offset its economic costs.

China generally echoed India's position in November 2009 even as it announced that by 2020 it would cut its carbon intensity—the amount of carbon dioxide emissions relative to its economic output—by 40 to 45 percent from 2005 levels. China heralded this "voluntary action," which came shortly after a summit between President Obama and Chinese leaders in Beijing, but others pointed out that our Energy Department had previously estimated that China was already on a path to reach this target under its existing policies. President Obama told the Chinese leaders that the United States would reduce its total greenhouse gas emissions by about 17 percent from 2005 levels by 2020 and by more than 80 percent by 2050. The latter would return U.S. carbon dioxide emissions to roughly their level in 1905.

Tensions between the developing countries and richer nations remained manifest when more than 190 countries met and negotiated for two weeks in Copenhagen at the December 2009 United Nations Framework Convention on Climate Change. After two years of preparation, many days of round-the-clock negotiations, and more than a little posturing and grandstanding, that gathering of world leaders produced a three-page "accord" that was neither formally accepted by the convention nor made legally binding, although most of the nations attending said they would accept its terms.

The Copenhagen Accord, which had been forged in negotiations among the United States, Brazil, China, India, and South Africa, contained no immediate (or even midterm) emissions-reduction goals but set an objective that world temperatures would not rise by more than two degrees centigrade above preindustrial levels by 2050. It does provide a system for reporting and monitoring individual nations' progress toward their own emission-reduction goals and commits wealthier nations to transfer hundreds of billions of dollars to poorer countries vulnerable to climate change. The richer nations, including the United States and Europe, agreed to provide $100 billion annually, an amount that some observers and representatives of developing nations still considered inadequate. The next convention was scheduled for 2010 in Mexico, to be followed by another in 2011 in South Africa.

Evaluations of the Copenhagen Accord were unsurprisingly mixed. Ban Ki-moon, the United Nations secretary-general, described it as "a new step toward the era of clean energy and toward an era of green growth." The Sierra Club's Carl Pope described it as "a historic—if incomplete—agreement to begin tackling global warming." David Doniger of the Natural Resources Defense Council said that it "broke through years of negotiating gridlock to achieve three critical goals: first, it provides for real cuts in heat-trapping carbon pollution by all of the world's big emitters; second, it establishes a transparent framework for evaluating countries' performance against their commitments; and third, it starts an unprecedented flow of resources to help poor and vulnerable nations cope with climate impacts, protect their forests, and adopt clean energy technologies." But the climatologist James Hansen, head of the Goddard Institute for Space Studies, said, "The proposals discussed in Copenhagen were like the indulgences of the Middle Ages": "The sinners are the developed countries," and "they want to continue business as usual by buying off the developing countries." Wen Jiabao, China's prime minister, disagreed: "These are hard-won results made through joint efforts of all parties, which are widely recognized and should be cherished." President Obama demurred. After calling the accord "an unprecedented breakthrough," he said, "I think that people are justified in being disappointed in the outcome in Copenhagen." "We've come a long way," he added, "but we have much further to go." One key player in Copenhagen, Yvo de Boer, who had for more than three years served as the United Nation's top climate change official, was so discouraged that he resigned. By June 2010, however, more than 100 countries had endorsed the accord.

One question everyone was asking after Copenhagen, especially after de Boer's resignation, was whether the United Nations Framework process had run its course. Getting more than 190 nations to agree to anything may be an impossible task. And some of the world leaders used the 2009 forum simply as an occasion to rant publicly. The language of the accord was largely hammered out by the leaders of five nations, and many participants left Copenhagen certain that any further progress would be made through small groups meeting in different forums.

Despite the uncertainties about the effects of future emissions of greenhouse gases, there is now widespread agreement around the world that the risks are substantial and the consequences potentially dire. The challenge for policymakers is how to manage those risks. Nearly everyone who has studied the subject agrees that if the United States is to reduce its carbon emissions substantially, it will have to increase the price of using carbon as a fuel. Although the prospect of green jobs is certainly real and enormously appealing—seductive even—it is difficult to believe that shifting away from oil and coal will produce a *net* increase in jobs any time soon.

Thus, a conflict between economic and environmental goals seems as inevitable today as it was in the 1970s. But few are willing now to argue explicitly for the kinds of limits to economic growth that the Club of Rome and its supporters embraced then. Instead, politicians and pundits alike promote green jobs hoping that no one looks too carefully with green eyeshades. Politicians know what happened to Jimmy Carter when he called for the American public to sacrifice. No one wants to repeat his debacle.

In the 1970s we struggled with some but not great success to fashion energy policy that would be conducive to growing economically, protecting the environment, reducing our nation's vulnerability to the machinations of foreign governments, eliminating economically destructive supply interruptions, avoiding price spikes, and doing all of this in a manner fair to both energy producers and consumers in every region of the country and across the entire income spectrum. Today we face great threats from Islamic jihadists in the Middle East, remain vulnerable to supply interruptions, and need to confront the existential threat of climate change. We must find a way to limit greenhouse gas emissions in a manner that protects and even enhances the welfare of the American people and is simultaneously also fair to the billions of people who are just beginning to reap the benefits of the modernization of their nations and societies.

Absent some remarkable technological breakthrough, there will be many hard choices to be made, difficult truths to absorb. Substitutes for fossil fuels must become economically feasible and profitable in an ongoing basis without large, wasteful, never-ending government subsidies. We will either have to find an economically viable way to capture and sequester carbon dioxide emissions from coal or have to transform at least half of our electricity production to cleaner fuels. We will also need to restructure our transportation sector. If we are to reduce dramatically our consumption of energy without an equally dramatic transformation in the quality of our lives, we must overcome significant cultural barriers. At the same time, effective international agreements must be forged and then enforced—an enormously difficult task. As we have seen, relying on our own domestic political process to solve long-term energy problems has long been a dicey enterprise at best. For starters, raising the price of fossil fuels is something that American people have long resisted.

Harvard sociologist Daniel Bell, who wrote the influential 1970s book *The Cultural Contradictions of Capitalism,* said about Jimmy Carter's crisis-of-confidence speech: "I do not think one can yoke a theme that is primarily moral and cultural to a 'cause' or 'crusade' that is so complex as energy." But that is exactly what we *must* do now. And not just for reasons of morality, but in order to forestall the risks of a genuine disaster.

11 Shock to Trance: The Power of Price

On November 17, 2008, the second Sunday after his election, Barack Obama sat down for an interview with Steve Kroft of CBS's *60 Minutes*. Here is what was said about energy in that conversation:

Kroft When the price of oil was $147 a barrel, there were a lot of spirited and profitable discussions that were held on energy independence; now you've got the price of oil under $60.

Obama Right.

Kroft Doing something about energy, is it less important now that . . . ?

Obama It's more important. It may be a little harder politically, but it's more important.

Kroft Why?

Obama Well, because this has been our pattern. We go from shock to trance. You know, oil prices go up, gas prices at the pump go up, everybody goes into a flurry of activity. And then the prices go back down and suddenly we act like it's not important, and we start, you know, filling up our SUVs again. And as a consequence we never make any progress. It's part of the addiction, all right. That has to be broken. Now is the time to break it.

No one can deny the truth of President Obama's description. When oil prices spike, politicians rush to act; the public scurries to respond to the new situation. When oil prices decline again—as they always have, so far—sighs of relief reverberate through the nation. Old consumption patterns reemerge. During the 1980s and 1990s, the American people entered a long energy policy trance. Relatively low energy prices removed energy from the front policy burner for both presidents George H. W. Bush and Bill Clinton.

The only federal energy legislation of much consequence in the 1990s was the Energy Policy Act of 1992—a response to an oil price shock after Iraq's invasion of Kuwait and the subsequent U.S. attack on Iraq.

There is a difficult dilemma here: higher energy prices may be what we need to achieve energy security and limit climate change, but neither the public nor their political representatives really want these higher prices. On the one hand, higher prices encourage us to consume less energy and emit less carbon dioxide and other greenhouse gases and pollutants, thereby lessening our need to import oil and helping to avoid climate change. On the other hand, they have been associated with recessions, and in the 1970s they were accompanied by a dreadful "stagflation," the debilitating combination of recession and inflation. In addition, high energy prices are especially burdensome for those on fixed incomes or with low incomes. What we want, what we may really need, is clean, reasonably priced, secure domestic energy, but if we are ever going to get that, we will need to learn to live with higher prices for carbon-emitting and imported energy sources, most notably coal and oil. Fortunately, as we have moved from a manufacturing-intensive economy to one based more on services, we have been able to reduce our energy intensity, and we have become better able to cope with higher energy prices.

Prices of oil have long refused to behave as the experts have predicted. Virtually every time oil prices have spiked, energy experts have predicted the increase to be permanent, but then prices have tumbled. Energy projects are typically both capital intensive and long term. Changing our pattern of energy consumption takes time. Porous buildings still leak when heating costs are high. Gas guzzlers get no better mileage as gasoline prices soar. We can cut back the electricity we use by only a little—at least in the short run. When oil prices are high, increasing exploration and investment in alternatives seems wise, but such investments may rapidly turn sour when oil prices fall. And oil price volatility and uncertainty is greater now than ever, due not only to ongoing conflict in the Middle East, but also to the speculative market in oil futures.

The quantity of trades in oil futures on the New York Mercantile Exchange every day is ten times the world's daily oil consumption. Oil is no longer just a fuel; it may also serve as financial protection against declining value in the dollar. Around the world, the total daily trades in oil on paper are often as much as thirty times greater than the consumption of oil in barrels. Fluctuations in oil prices are quick and frequent. They often have little or nothing to do with the fundamental relationships of oil supply and demand. Dramatic price swings and uncertainty about the future of both supply and demand, coupled with the necessity of large long-term capital investments, exacerbate the problems of creating sound policy. Given, then, the enormous power of price to drive patterns of consumption and the private-sector activity that arises to cater to that consumption, how can we harness the price dynamic to create a better, more stable energy future?

One of the few policies virtually all economists agree on is the need for substantially higher taxes on gasoline and other products from fossil fuels. The purpose of such taxes would be to decrease both consumption and carbon emissions, not necessarily to increase government revenues. *Congress could return the revenues to Americans either on a uniform basis to each person or based on income or wages, such as by reducing payroll taxes.*

Economists have long maintained that when the price of a good fails to take into account important "externalities," by which they mean the effects of transactions on people other than just the buyer and the seller, imposing a tax may be appropriate or even necessary. The classic case for such a tax is on a product or process that pollutes. The market price for fossil fuels fails to reflect the costs of its attendant pollution, including not only but especially emissions of carbon dioxide. A tax on carbon would require the producer of the emissions to take into account the costs he is imposing on society.

Market prices also fail to reflect the costs of our dependence on oil imports. In addition to the burdens on the overall economy from sharp price increases at the hands of the OPEC cartel, there are national security costs, such as deploying our military to protect and secure the flow of oil from the Middle East. We also pay a price for transferring enormous sums of dollars to nations that may not always be friendly to us. One researcher has estimated that oil dependence cost our economy between $700 and $800 billion in 2008 alone.

In addition, underpriced commodities tend to be overused. If you want people to consume less of something, taxing it can be a sure-fire tool. With emissions of carbon dioxide becoming the main focus of our policy concerns, a carbon tax should be an obvious policy instrument. Increasing the price of emitting carbon dioxide by taxing carbon would incentivize the public to consume less fossil fuel, entrepreneurs to invest in alternatives, manufacturers to produce more energy-efficient appliances and automobiles, and builders to construct more energy-efficient buildings. One reason that Europeans drive smaller cars and fewer miles than Americans is that they face much higher gasoline taxes. Gradually increasing the price of using fossil fuels through a tax on carbon would also make cleaner sources of energy more economically appealing, thus spurring innovation and development of these sources. As Al Gore states in *An Inconvenient Truth*, "The easiest, most obvious, and most efficient way to employ the power of the market in solving the climate crisis is to put a price on carbon."

A carbon tax would essentially impose a fee for each ton of carbon dioxide emitted by taxing each ton of carbon contained in fossil fuels. By being imposed directly on fossil fuels, a carbon tax can build administratively on existing excise taxes on coal and petroleum and can be limited to about 3,000 companies, while capturing about 80 percent of U.S. emissions. For modest additional costs, coverage can be extended to almost 90 percent of emissions. Emitters would generally invest in carbon dioxide control measures until they cost more than the tax rate, so setting the rate becomes important. Congress might set the tax rates to rise over time to specified levels and delegate to the EPA the authority to revise the rates if emission-reduction targets fail to be met or as new information becomes available, subject to congressional oversight and review. In Canada, both Quebec and British Columbia have adopted such a tax. In British Columbia, the rates rise from an original rate of $10 a ton of carbon content to $30 by 2012 (equivalent to about 30 cents per gallon of gasoline) with all the tax's revenues returned to residents, in part per capita and in part based on income. Although several members of the U.S. Congress have introduced carbon tax bills, these bills have not gained any real traction. A substantial tax on fossil fuels is one of the few energy policy strategies that we as a nation have consistently failed to employ.

In the 1970s, the Nixon administration announced it was considering a substantial gasoline tax increase, but it quickly dropped the idea. Gerald Ford rejected any increase in gas taxes (although he did propose an oil import fee), despite support from Alan Greenspan, the chairman of the Council of Economic Advisers. And President Ford fired his energy secretary John Sawhill when Sawhill publicly suggested that gasoline taxes be hiked up to 30 cents a gallon. In Congress in 1975, the House Ways and Means Committee chairman, Oregon Democrat Al Ullman, proposed a substantial gasoline tax increase, but his plan was soundly defeated on the House floor and never even considered in the Senate. In 1977, Jimmy Carter proposed, as part of his comprehensive energy plan engineered by James Schlesinger, increasing the gasoline tax by a nickel a gallon each year for ten years, up to a 50-cent ceiling, for every percentage point that the nation's gasoline consumption exceeded specified national goals. He would have refunded most of the revenues from this "standby gasoline tax" to low-income automobile owners. Despite support from the Democratic leadership in both houses of Congress, this proposal went nowhere. In March 1980, President Carter exercised the authority he had been given by Congress to impose a fee on oil imports, designed to function similarly to a tax on gasoline, but Congress voted overwhelmingly to stop this import fee from ever taking effect. John Anderson, an independent candidate for president in the 1980 election, urged an increase in gasoline taxes of 50 cents a gallon, but he got only about 7 percent of the popular vote. Ironically, in 1983 the famous tax cutter Ronald Reagan signed a gas tax increase of a nickel a gallon to provide additional funds for highway construction and mass transit.

In both 1990 and 1993, Congress came close to imposing a substantial tax on energy consumption, but the motivation then was deficit reduction, not energy policy. In 1990, many observers blamed Congress's failure to enact an energy tax on the fact that oil prices nearly doubled (from $14 a barrel to $24 a barrel between July and September) after Iraq invaded Kuwait. This price spike made it difficult for politicians to pile additional costs onto their constituents. When all was said and done, Congress in 1990 simply increased the federal gasoline tax by another nickel a gallon.

In 1993, when oil prices were again low (having fallen back to about $14 a barrel), President Clinton urged Congress to enact an energy tax—a so-called BTU tax. After much presidential arm twisting, the House of Representatives passed this provision—without garnering a single Republican vote. The BTU tax never got out of the Senate. The legislation finally enacted in 1993 limited the tax to gasoline and diesel fuels and increased the amount by 4.3 cents a gallon, bringing the total federal tax to 18.3 cents for a gallon of gasoline, 24.3 cents for a gallon of diesel. The next year Republicans won a majority in the House of Representatives for the first time in a generation, and many House Democrats who voted for the BTU tax were not reelected.

In the Senate, as usual, regional politics inhibited sound policy. Higher energy taxes were opposed by a variety of regional interests, ranging from northeastern liberals worried about low-income constituents who burn home heating oil to western conservatives worried about voters who drive long distances. A number of senators, particularly from the Midwest, were concerned about the potential impact of an energy tax on the international competitiveness of energy-intensive manufactured products, such as steel and chemicals. The BTU tax also foundered on the opposition of key senators from the oil-producing states of Louisiana and Oklahoma. It became clear from the Senate's discussions that even if such a tax were enacted, it would be riddled with exceptions and special rates.

Following the attack on the World Trade Center on September 11, 2001, George W. Bush might have rallied public opinion and Congress to support a substantial increase in gasoline taxes, an oil import fee, or perhaps even a broad-based energy tax to fund the military operations he launched in Afghanistan and Iraq. He apparently never even considered such an option, though, preferring to fund those ventures through additional borrowing.

Nor has President Obama any intention of proposing a carbon tax, a gasoline tax, or any other tax to advance his energy policy goals—whatever the merits. On April 16, 2008, debating Hillary Clinton in Philadelphia at a crucial moment in their campaign for the Democratic presidential nomination, Obama pledged not to raise taxes on Americans earning less than $250,000 a year. Hillary Clinton made a similar pledge. One of their interlocutors, ABC's George Stephanopoulos, described this promise as "an absolute, read-my-lips-pledge." By using that phrase, Stephanopoulos was alluding to George H. W. Bush's promise never to raise taxes. When Bush accepted the 1988 Republican nomination for president, he suggested, making his biggest political mistake, that the Democratic Congress would push him to sign a tax increase and his response was: "Read my lips, no new taxes." Two years after that, facing a pressing need to take control of spiraling budget deficits and a Democratic Congress insisting that both tax increases and spending restraint be part of any budget deal, Bush negotiated and signed the 1990 Budget Reconciliation Act, which curtailed government spending and raised taxes. An additional two years later, with a weak economy, the voters no longer trusted him and denied him reelection. Barack Obama repeated this pledge many times after the Philadelphia debate during the 2008 campaign and even after he took office: no tax increase for any family making less than $250,000 a year. This, of course, would rule out a gasoline tax increase, any broader tax on petroleum fuels and products, or a new carbon tax.

That two liberal Democrats running for president in 2008 felt it necessary to rule out any tax increases except on the "rich" demonstrates the continuing potency of the antitax politics that took hold in our country in the late 1970s. Very few days have turned American politics upside down, but June 6, 1978, was surely one of them. That day—by a nearly two-to-one margin—the people of California added a measure known as Proposition 13 to their state constitution. Proposition 13 limited state property taxes and made it very difficult for the state legislature to raise other taxes. In rejecting calls to be "responsible" by virtually all of their elected representatives, California voters transformed tax increases from merely a dangerous political suggestion into an act courting political suicide. The antitax movement—reflecting widespread antigovernment sentiments—soon struck fear into Democratic politicians and became the linchpin of Republican Party politics. Opposition to tax increases has since been the glue that has held the Republican coalition together, and it has frightened many Democrats into acting much like Republicans on this issue. Grover Norquist, an outspoken leader of the Republican antitax movement, describes taxes as "the central vote-drawing issue." "You win this issue," he says, "you win—over time—all issues." Barack Obama's no-tax-increase pledge deprived his Republican opponent of any advantage on the tax issue; in making it, however, Obama may also have deprived himself of a critical tool for governing—or he made a promise he will have to break.

If you want people to conserve energy, produce energy-efficient products, and move from fossil fuels to less-damaging sources of energy, and if you want, out of political fear, to take energy taxes off the table at the same time, only two other policy options remain: (1) subsidize or mandate the behavior you want to encourage or (2) limit through regulations or mandates the behavior you want to curtail. Let's start with subsidies.

In theory, if the goal were only to substitute more benign fuels for carbon-emitting fuels and not to curtail energy use, a subsidy for the favored fuel substitutes could work as well as a tax on disfavored fuels. Congress, for example, might either increase the gasoline tax by a dollar a gallon or subsidize alternatives by a dollar for every gallon of gasoline they save. Likewise, Congress might impose a tax of $25 a ton on carbon-emitting fuels or grant a subsidy based on an equivalent amount of carbon dioxide emissions avoided. Either the tax or subsidy approach would decrease the costs of alternatives relative to the prices of oil, coal, and natural gas. In practice, however, taxes and subsidies operate quite differently.

The burdens and benefits created by taxes and subsidies will be different. A tax imposed on the carbon content of fossil fuels, for example, would burden the producers and consumers of carbon-intensive products. It would raise the price of coal-fired electricity, for example, compared to solar, wind, hydro, or nuclear power, which are carbon free. A tax would reduce demand for carbon-emitting products so that people, for example, would drive fewer miles or lower their thermostats, and producers might tend to earn smaller profits. Consumers would face higher prices for much of the electricity, gasoline, or home heating fuels they use (although the revenues from the tax can be returned to the public so that low- and middle-income consumers would not have less money to spend or save). In contrast, the costs of subsidizing alternative sources of energy would be financed by the public at large, and the subsidies would increase the profits of those who produce the favored products and perhaps also lower the costs for those who use them.

Importantly, imposing a tax on disfavored fuels does not create any favorites among cleaner alternatives or among particular technologies. When it comes to subsidies, however, Congress is very much in the business of picking winners and playing favorites. In response to a tax on energy, people may change a wide range of behaviors—such as turning off lights, lowering thermostats, driving less or more slowly, properly inflating tires, and better maintaining their automobiles. As a practical matter, Congress would have a hard time subsidizing each of these activities in anything like an efficient manner. *It is also virtually impossible to design a subsidy so that it does not provide an unnecessary benefit for behavior that people would have done anyway without the subsidy.* Some folks, for example, are going to buy insulation or a more energy-efficient air conditioner or furnace (at least when the old one wears out) without a government subsidy. If half the amount of the favored activity would have occurred without any subsidy, that doubles the costs of a subsidy without any additional benefits. It is extremely difficult and usually impractical to limit a subsidy to genuinely incremental activity.

Since the 1970s, U.S. policy has been to subsidize the production and consumption of fuels we would like to encourage rather than to tax the use of fuels that we want to discourage. Those subsidies, however, have not been concentrated on R&D of new technologies. For the subsequent three decades, federal spending on energy R&D never again reached its late 1970s peak. In the 1980s, federal R&D spending was dominated by defense. And since the 1990s, health R&D has been the government's priority, although in recent years spending on energy and climate change has rebounded from its 1980s lows. In 1979, energy research funding accounted for 23 percent of the federal R&D budget; in 2008, it was less than 4 percent. Federal R&D spending on renewable energy, for example, increased from less than $1 million a year in 1970 to more than $1.4 billion in 1979, then declined to about 10 percent of that amount ($148 million) in 1990 before starting to grow again. In real dollars, federal energy R&D in 2006 was half its level in 1980. And public and private energy R&D spending combined fell from 10 to 2 percent of federal R&D spending.

Since the turn of this century, Congress has greatly expanded total federal subsidies for energy. According to a comprehensive analysis by the Energy Information Administration, total federal energy subsidies more than doubled (in inflation-adjusted dollars) from $8.2 billion in 1999 to $16.6 billion in 2007, growing by more than 9 percent annually. Subsidies for renewable sources of energy increased from 17 percent to 29 percent of the total, whereas subsidies for natural gas and petroleum declined. Coal subsidies changed very little, accounting for about 6 percent of the total in 2007. The amount of energy subsidies provided in the form of tax breaks more than tripled during the same interval, with $37 billion added in 2008 and 2010, and they now account for about 60 percent of the total. During this same period, total energy consumption in the United States grew by less than 5 percent, and energy production remained relatively flat. The "stimulus bill" of 2009 (the American Recovery and Reinvestment Act) added an additional $60 billion of energy spending, which was intended to be used in a four-month period to bolster economic recovery, making available a total of about $100 billion in federal tax and other incentives for renewable energy and energy-efficiency projects.

Federally financed R&D has a crucial role to play in helping to identify, develop, and induce the private sector to adopt the kinds of technological improvements that may ultimately enable us to shift from oil and coal to more climate-friendly fuels. But the government's spending priorities are not set by scientists and engineers. The essential point is this: any way you compare our subsidies for energy, you will find wide variations in their amounts relative to the fossil fuel savings they yield. Government subsidies are certainly not neutral across products or technologies.

Members of Congress frequently insist on their own personal priorities—directing funds to "individual projects, locations, or institutions" by "earmarking" projects. Between 2003 and 2006, congressional earmarks in Department of Energy programs for energy efficiency, renewable fuels, and electricity production tripled from $46 million to $159 million, with earmarks accounting for about 20 percent of the total 2006 budget. By 2008, congressional earmarks totaled $180 million, with an additional $46 million directed to specifically identified energy projects, including particular biofuel plants and specific green buildings. Earmarks that year took up one-half of the total R&D budget for biomass, one-third for wind, and more than one-quarter for hydrogen projects. The American Academy for the Advancement of Science has lamented that "earmarks eat up whatever increases there are for most energy programs and cut deeply into core R&D programs." Many members of Congress are obviously concerned more with rewarding politically well-connected constituents and contributors than with advancing science or promising technologies.

The tax system, as described earlier, has become more difficult to use for curtailing undesirable behavior. At the same time, however, it has become our politicians' favorite mechanism for transferring benefits and providing subsidies. The political advantages of enacting tax breaks to encourage this or that activity or expenditure are obvious: Republicans in Congress virtually never see a tax cut they will not embrace, and Democrats often view tax benefits as the only way to achieve their policy goals without being tarred as big spenders. As a result, presidents and the Congress have come to use tax breaks the way my mother used chicken soup—as a cure-all for any ill American society faces.

A close look at the tax incentives for changing our nation's energy production and consumption patterns is, however, quite discouraging. In April 2009, the staff of Congress's Joint Committee on Taxation (JCT) published a 120-page pamphlet describing the tax breaks that subsidize energy production and conservation. A 19-page table summarizes roughly 40 different provisions, including tax credits for electricity produced from certain renewable sources; tax benefits for alternative transportation fuels, including ethanol and other biodiesel fuels; tax credits for various means of energy production; a variety of tax breaks for homeowners and businesses to install or produce energy-saving devices; tax credits ranging from $400 to $40,000 for vehicles powered by alternative fuels, including fuel-cell, natural-gas, "clean diesel," electric, and hybrid vehicles; a special credit up to $200,000 for refitting property for alternative fuels; as well as a host of benefits for fossil fuels. The JCT estimated that these provisions would cost a total of about $40 billion for the five-year period from 2008 to 2012.

In an effort to assess how well these subsidies have performed in stimulating alternatives to fossil fuels and the extent to which they involve picked favorites, JCT calculated the value of this mélange of subsidies per million BTUs of fossil fuels that each tax benefit saved. It concluded that the benefits ranged from $1.01 to $8.45 saved per million BTUs. Wind power, for example, saved $2.12; ethanol $5.92. The JCT also compared the tax benefits received by two solar energy plants, one getting eight hours a day of sunlight and the other getting five hours. It concluded that the former costs taxpayers $9.64 per million BTUs of electricity, the latter $15.42.

JCT also estimated the dollar value of tax credits for specific hybrid motor vehicles per million BTUs of fossil fuel saved. It found that in 2006 a GMC pickup truck got $5.60 in credits per million BTUs, a Honda Accord $6.59, and a Toyota Camry $11.49. Economist Martin Sullivan also looked at the tax breaks for hybrid cars and compared tax benefits per gallon of gas saved. He found that the tax benefits ranged, for example, from zero for the Toyota Prius (which gets nothing because of its relative success with consumers), to $5.59 per gallon for the Chevrolet Tahoe, which gets less than half the gas mileage of the Prius.

Tax credits that may appear to be neutral are often not. Current law, for example, provides a tax credit for the production of electricity from renewable carbon-free sources equal to 2.1 cents per kilowatt hour for the first 10 years of production. Tufts University economist Gilbert Metcalf has calculated that this subsidy equals $7.74 per ton of carbon dioxide saved for geothermal energy but $12.28 per ton for wind. Given their different capacities and production locations, geothermal-generated electricity is more likely to substitute for coal-generated electricity, and wind is more likely to replace natural gas. Because the same amount of electricity emits less than half the greenhouse gas when produced from natural gas than from coal, an equal subsidy based on the amount of electricity produced has quite different effects in terms of greenhouse gas emissions saved.

Using a different methodology, Professor Metcalf also estimated the costs of new break-even investments in 2007 across various energy sources and calculated their effective tax rates as a way of comparing the tax benefits for various energy sources. He found that the effective tax rates ranged from a high of 39 percent for certain coal operations to a low of *minus* 245 percent for production of solar electricity. Generating electricity from natural gas was taxed at an effective rate of 34.4 percent; wind at *minus* 164 percent. In looking at direct federal subsidies that year for various sources of electricity production, the Energy Information Administration also found wide variations—ranging from 25 cents per megawatt hour for natural gas and petroleum liquids to 44 cents for coal, $23.37 for wind, and $24.34 for solar. Coal produced about half of the electricity generated in the United States, whereas solar and wind combined produced less than one percent, so the total federal dollars spent look quite different: $854 million for coal, $742 million for wind, $14 million for solar. Per BTU of energy produced, the subsidy for coal was nearly double that for petroleum liquids and natural gas and about one-sixth that for renewable fuels. This is indefensible. Regardless of who does the calculations or what methodology they use, irrationality abounds—expensive irrationality at that.

The specific numbers reported here are far less important than the fundamental trouble they reflect. It is clear that Congress has chosen to award subsidies—whether in the form of direct spending or tax breaks—in such a way that their costs are often unrelated to the benefits they are intended to produce. At best, decisions about what to subsidize and by how much seem arbitrary and capricious. At worst, they are wasteful and sometimes even nefarious.

Waste occurs whenever technologies that are subsidized or mandated do not match the objectives established for them. Twenty-eight states, for example, have mandated that a specified amount of their electricity must be produced from renewable sources. In 2009, electricity from wind was generally the least expensive way for electric utilities to comply with such mandates. Wind turbines, however, tend to generate more electricity at night when demand is low than on hot humid afternoons when electricity use is greatest. Electricity does not store readily or at low cost; there are limits to utilities' ability to handle intermittent sources of electric power, especially from small scattered sources, and to distribute power to large cities on the electric grid. Moreover, shutting down coal and nuclear power plants when demand is low is impractical. So the utility simply drops the wholesale price it pays producers to zero when it cannot sell what it is required by law to purchase. During the twelve months ending May 2009, for example, wind electricity prices in West Texas dropped to zero 11 percent of the time. This same thing happened, although less frequently, in California, Illinois, New York, and Ohio.

Even with a price of zero, however, wind producers still get 2.1 cents per kilowatt hour from the federal government. According to *Forbes* magazine, the grid operator in Texas, the Electric Reliability Council of Texas, told wind power developers a few years ago that it could not handle more than 4.5 gigawatts of wind power, even at peak demand. The developers built 8 gigawatts anyway. (A $5 billion upgrade in this transmission system is due to come online in 2013, and it should then be able to deliver wind power to Dallas.) So the wind producers often simply settle for their government subsidies even though no additional electricity is going to homes or businesses.

The "black liquor" scandal is far worse, and it became the most notorious recent instance of the pitfalls of congressional efforts to pick winners and subsidize them. Black liquor is a fuel by-product from the chemical production of wood pulp used in manufacturing paper. It has been used as fuel to power paper mills since the 1930s. In 2007, Congress expanded the definition of alternative fuels eligible for a 50 cents per gallon tax credit to include a wide range of petroleum fuels containing biomass products. Paper companies soon thereafter discovered that by adding some diesel fuel to their black liquor, they could become eligible for billions in tax credits. Instead of reducing the amount of petroleum fuel by substituting a biomass product, they added diesel fuel to the biomass simply to obtain tax credits. The subsidy in this case increased rather than reduced the consumption of products refined from oil. International Paper Company alone received as much as $1 billion in tax credits for the black liquor it used in 2009. Benefits to the U.S. paper industry as a whole may total $8 billon—all due to the industry's inadvertent eligibility for a tax benefit that Congress expected to cost only a total of $100 million when enacted.

One would have thought that once Congress learned of this gambit, it would immediately have stopped it. But when the chairman and ranking member of the Senate Finance Committee threatened to repeal this loophole, a representative of the American Forestry and Paper Association said that revoking the paper companies' eligibility for the credit "could have serious consequences for our companies and our one million employees at a time of unprecedented economic challenges." The International Steelworkers Union agreed. Finally in 2010, Congress ended this game—just a particularly costly and egregious example of Congress's inability to target subsidies to genuinely incremental productive and beneficial activities as well as its inability to respond promptly even when it learns that a subsidy has gone badly astray.

Despite the shortcomings of current subsidies—whether in the form of direct spending or tax breaks—government-sponsored or favored *research* will be necessary to assist in the development and commercialization of clean sources of energy. Opportunities for progress abound, including, for example, the potential for algae or other plants to remove carbon from the air, carbon capture and storage at coal-fired electricity plants, new fuels for transportation vehicles, fuel cell technologies, a "smart grid" for electricity, increased efficiency for wind and solar power, lighting from semiconductors (light-emitting diode, or LED, lighting), and a new generation of nuclear power plants, to name just a few. The federal government will clearly have to provide substantial resources for these efforts, although experts do not agree on the total amounts that will be needed.

I do not describe here these technologies in detail or evaluate their relative potential. Some are long shots; for others, the technology already exists, but the underlying economics are quite adverse absent major technological improvements. One important lesson from our 1970s experience is that pursuing a diverse portfolio of technological options and seeking more neutrality among them are essential.

Consider, for example, carbon dioxide capture and storage. We currently have the technological ability to do this, and a handful of small carbon-sequestration projects already exist. As of the summer of 2010, however, no commercial-scale power plant in the world captured and sequestered its carbon emissions. Doing so is very expensive and requires additional energy to accomplish. The carbon must first be "captured," often through a gasification process, then transported, usually by pipeline, to an underground facility where it will be stored, one hopes, without the potential to escape. In 2003, the U.S. government announced the bold $1 billion FutureGen program—a joint public-private endeavor between the Department of Energy and the FutureGen Industrial Alliance, a nonprofit organization of large coal producers and electric utilities, to build and operate the world's first coal-fired, zero-emission electric power plant. By mid-2007, the Department of Energy, concerned about the rising costs for the plant, retreated and announced a major restructuring of the program to demonstrate carbon-capture and storage facilities at a number of existing and new coal-fired power plants around the country. In June 2009, the new energy secretary, Steven Chu, said that the first commercial-scale, carbon-capture and sequestration plant would be built, after all, in Mattoon, Illinois, and would be funded with $1.1 billion of federal money and the remainder from private sources.

Coal now produces half of the electricity in the United States, and a move away from coal would be potentially devastating to the industry—an industry that has great sway in Congress and that seems to enjoy political protections that outstrip its real value to the country. Putting carbon-capture and storage technology in place might keep up to half of all emissions from electric power plants from reaching the atmosphere, so commercialization and international development of such a process might have a large potential worldwide beneficial impact for coal-fired power plants. But getting to such a result will not happen quickly, if it happens at all. The stakes of such decisions—especially for the coal industry—are quite high. There is a real danger that carbon capture and storage may become the synfuels misadventure of this era, so we should not go "all in" on this bet. We need to pursue many other avenues simultaneously, including developing alternative sources of energy supply and increasing the energy efficiency of the buildings we inhabit and the products we use. Surely we are better off when the allocation of funds to such projects is determined by an independent panel of experts rather than by our political representatives.

During the 1970s, presidents viewed committing the nation to a particular major technological project as proof of their vision and determination. But the most comprehensive analysis of government energy R&D

efforts in the 1970s, a book aptly titled *The Technology Pork Barrel*, concludes that the biggest R&D efforts of that period—the breeder-reactor and synfuels projects—were "unambiguous failures" and that our overall energy R&D effort was "hardly a success." Only the efforts to develop better and more economical photovoltaics for solar power garnered even passing marks from the authors. The biggest problems have been the tendency for Congress to put geographic considerations above technological and economic prospects—along with a pattern of boom-and-bust financing, characterized by a debilitating combination of excessive optimism about technological developments, impatience for results, and a process of haste and waste. The synfuels program, for example, favored eastern over western coal for political, not technological, reasons. The more recent FutureGen project suffered stops and restarts mainly because of undue optimism at the outset about both its costs and performance. Political power in the Congress has long been a critical determinant of where the public's money is spent.

Analysts of energy R&D efforts, whether past or present, agree that major institutional changes are required if we are to be more successful this time around. Elimination of earmarks would be a useful first step, and multiyear budgeting for greater funding stability a second. Congress's diffuse and overlapping committee structure is a fundamental problem, some say perhaps even "dooming the enterprise to failure," but that structure will be very difficult to change. Even within the executive branch, new arrangements are necessary, including payments linked to progress, peer review by independent scientists and engineers, and perhaps new ways of scientifically evaluating research proposals and for financing and assessing the progress of the implementation of new technologies. Encouraging more private investment and production will also be important to bring new technologies to scale. We must be able to pursue a number of paths simultaneously.

At all events, much greater neutrality in the incentives for technological innovations and commercial development is necessary. Trying to pick winners and avoid losers has proved to be a fool's errand. Because the problem of climate change is global, new international arrangements for the development of promising technological advances and cooperation among nations, including especially China and India, will surely be needed. We can certainly do much better than the hodgepodge of subsidies and tax benefits Congress has enacted so far to encourage energy technologies and alternatives.

It is clear, however, that funding incentives to increase the supply of alternatives to fossil fuels will not suffice alone either to achieve greater independence from imports or to decrease greenhouse gas emissions. Public policy must also curb our demand for fossil fuels. Such a curb will inevitably require higher energy prices for them. A 2009 report by the Energy Technology Innovation Group at Harvard said that "the most important single step toward commercialization of low-carbon technologies is to put a price on carbon emissions." A substantial tax on carbon and on gasoline and other petroleum products—with the tax's revenues automatically returned to the American people to ease the potential burdens on low- and moderate-income families and to minimize any adverse economic effects from imposing the tax—would be neutral toward alternative technologies and surely would help us to achieve our goals. Our experience in the 1970s teaches well the importance of prices to economic and policy developments. And we have long been aware of taxes' ability to amplify price signals and encourage producers to take into account the full range of their actions' costs. Yet we have eschewed taxes and instead employed virtually every other policy tool imaginable. Handing out tens of billions of dollars in subsidies annually is far more seductive to politicians. But our political representatives know that subsidies alone are not likely to accomplish major reductions in either oil imports or greenhouse gas emissions, so they will also attempt to reduce our demand for fossil fuels and control our behavior through those other great levers of government: mandates and regulations.

12 The Invisible Hand? Regulation and the Rise of Cap and Trade

Before the 1970s, the federal government played only a bit part in regulating energy policymaking. The most important agencies were state entities, such as the Oklahoma Commerce Commission and the Texas Railroad Commission, which had the power to regulate oil production in their states. Their job was largely to manage abundance. In effect, they limited output so as not to exceed domestic consumption, although there were times when efforts by the states fell short, and the Interior Department had to intervene to prevent supplies from outstripping demand. Nevertheless, federal policy consisted mostly of the FPC's regulation of interstate natural gas, the AEC's insistent promotion of nuclear power, the licensing and building of dams for hydroelectric power, and the leasing of federal lands for exploration of oil and natural gas.

Oil was inexpensive and plentiful. Virtually no one gave a thought to potential scarcity. Recall Richard Nixon's refusal to accept the shah of Iran's 1969 offer to sell the United States a million barrels of oil a day for a decade for $1 a barrel. Our most conspicuous policy toward oil was the import quota adopted by Dwight Eisenhower to keep foreign oil out of the country. Limiting imports kept oil prices high enough to satisfy the oil companies but, because of oil's abundance and low price, not so high to make consumers fret. Along with a series of tax breaks for oil and gas exploration and production (of which the oil-depletion allowance was the most famous but not the most important), the quota kept our domestic oil producers happy—some might say fat and happy. The main effect of the quota policy was that we used up our own oil when it was cheap rather than devouring Middle East and Venezuelan oil when it was much cheaper. We have been paying the price for that mistake ever since.

Following the 1973 OPEC embargo, all this changed. As previous chapters have described, facing an ongoing national trauma, the federal government, under three presidents, threw itself into the field of energy policy and regulation. By 1980—after promulgating many thousands of pages of new laws and regulations—the national government had entered into every nook and cranny of our nation's energy policy, and federal regulations affected virtually all aspects of energy production and consumption.

Nevertheless, states continued to play an important role in further regulating our various sources and uses of power, sometimes to standards higher than those of the federal government. California, for example, in the 1970s began requiring energy-efficiency minimums for appliances and certain other products. After Florida and some other states joined in or threatened to do so, manufacturers became fearful that they might have to meet 50 different state standards and, seeing that they could not thwart this movement, began to press for uniform federal standards that they could handle. In 1974, Congress required the Department of Energy to set specific energy-efficiency standards for 13 household appliances and heating and cooling equipment, but the Energy Department dithered, and the Reagan administration refused to implement any rules. Congress responded in 1987 by passing the National Appliance Energy Conservation Act, which not only set national standards for appliances, but also imposed deadlines for the Energy Department to promulgate specific rules. In 1992, Congress extended energy-efficiency mandates to some lighting products and certain industrial and commercial technologies. Efficiency standards were further extended and tightened by more recent legislation. And the states continue to be active, with California remaining the most aggressive—now regulating flat-screen televisions, to give just one example.

The CAFE standards are merely the best known of the many energy mandates that firms and families face. Just walk around your house. The energy efficiency of virtually every large appliance and many small ones is now dictated by the federal government: refrigerators and freezers, clothes washers and dryers, dishwashers, hot-water heaters and gas furnaces, to name just a few. And commercial and industrial businesses also now face numerous specific federal energy requirements.

Proponents of such standards often claim very large benefits for them by assuming that no progress in energy efficiency would have occurred if the government had not required improvements. Detractors, however, claim that most of the energy-efficiency increases would have occurred anyway after energy prices spiked in the 1970s and that the mandates result in unnecessary costs, especially to low- and middle-income consumers. Standards have clearly had some effect. Refrigerators, for example, became 75 percent more efficient between 1972 and 2005. And because of tough requirements, California's per capita electricity use has remained relatively level over the past three decades, whereas the rest of the nation's has risen by nearly 50 percent. But California's higher electricity prices, smaller houses, and favorable climate have no doubt also contributed to that result.

Efficiency standards have been criticized for dealing with specific appliances (such as electric hot-water heaters) rather than with equipment that performs similar functions (all hot-water heaters). And updating standards over time is a slow, painstaking process that often involves much negotiation between manufacturers and regulators. Nevertheless, mandatory standards have become commonplace throughout the world and have frequently been buttressed by "strong voluntary programs," such as the EPA's Energy Star labeling program, which requires, for example, power-management "sleep" programs for computers when they have been inactive for a while.

As previous chapters have shown, energy production and consumption in the United States was dramatically affected when in the 1970s the environmental movement found its national voice, and Congress responded with a burst of environmental legislation, followed by many regulations and much litigation. The 1970 enactment of NEPA and the Clean Air Act as well as the creation of the EPA that year—all adopted at a time when the economy was strong and both inflation and unemployment were low—forged an unbreakable marriage of energy policy and environmental policy. This marriage has not always been a happy one. Stresses and strains between the two have been common, but energy policy and environmental policy are—and forever will be—closely bound together. Since climate change moved to the forefront of our nation's environmental and energy policy agendas, the overlap has become greater than ever.

In the beginning, much environmental authority was left to state governments: indeed, the 1970 Clean Air Act explicitly said that air pollution control was the "primary responsibility" of states and local governments. But although the states, especially California, have continued to exercise power over such issues, more and more control has over time gravitated to the federal government. As we have seen, Richard Nixon, who might not be mentioned in the first rank of environmentalists, elevated environmental protection to the forefront of the national agenda. And even after the economy had deteriorated, amendments to the Clean Air Act in 1977 reaffirmed and strengthened federal regulatory power over our nation's quest for cleaner air.

Virtually all of the federal and state regulations of the 1970s were of the "command-and-control" sort. Congress, the Energy Department, the EPA, and state authorities told producers and manufacturers exactly what level of emissions would be permissible and frequently also what kind of technology must be employed to attain the regulatory goals. Under the 1970 Clean Air Act, for example, federal regulators set air-quality standards for particular regions of the country, with state and local authorities then generally required to tell individual polluters what they must do so that the region's goals would be met. (In some circumstances, the federal regulators told polluters directly what limitations applied to their emissions.)

Throughout that decade, such "command-and-control" regulations came to be increasingly criticized as wasteful, expensive, and ineffective. Delays in updating and ossification complaints became commonplace. And litigation flourished, though with decidedly mixed results. Congress and the EPA also frequently loosened and postponed the standards they had originally set. For example, under the 1970 Clean Air Act, automobile emissions of carbon monoxide and nitrous oxide were required to be reduced for new cars by 90 percent of their 1970 levels within five years. Automobile manufacturers soon insisted that this goal was impossible to meet, and neither the EPA nor environmentalists could demonstrate that the manufacturers were wrong. By 1977, Congress had lowered the standards to achieve about 50 percent reductions and had twice extended the deadline so that even this reduction was delayed until 1981. EPA enforcement actions frequently resulted in promises by industries to comply "sometime" or "pretty soon."

In the meanwhile, academic economists and other commentators were urging alternatives to command-and-control regulation. As the previous chapter has described, one of these alternatives was to impose taxes on the offending pollutant, but politicians rarely endorsed such taxes. A politically more palatable alternative to taxes—and to command-and-control regulation—then emerged: a market-based regulatory regime known as "cap and trade."

Here's how cap and trade works. Congress (or the EPA) determines what volume of emissions of a particular pollutant will be permitted. In the case of climate change, the emissions concerned would be emissions of specific greenhouse gases in the United States, most important carbon dioxide. The government then issues transferable allowances to emit a specified quantity of the restricted substance or substances. For example, it might issue permits to emit one ton of carbon dioxide in any particular year, with the total number of permits adding up to that year's total permissible emissions. These emissions permits or allowances may either be sold—auctioned—by the government or given away, and they can be used by their owners or sold to others. The fundamental idea is that sales (or "trades") of the permits will operate to concentrate their ownership in those companies that find it most expensive to curb emissions. Companies that can reduce or eliminate their emissions more cheaply than the price of the permits will do so and will sell their excess allowances to others who would otherwise have to spend more than the permits' price in order to curb their own emissions. In this way, market transactions allow emissions to be reduced in the least costly way and avoid the wasteful additional costs that would occur under command-and-control regulations that require each company to limit its own emissions to a government-specified level. Allowances will typically apply only for a specific time period, so this process of buying and selling permits will be repeated over time. By providing financial incentives for the largest reductions in emissions to be made by those firms that are able to reduce their emissions most cheaply, a cap-and-trade system should allow whatever level of total reduction in emissions is specified by the government to be achieved at the lowest total costs.

A carbon tax would have similar cost-saving attributes because each firm would reduce emissions to the point where it becomes cheaper to pay the tax. As chapter 11 describes, the challenge is to set tax rates at the right amount to get the desired overall level of emissions. Taxing greenhouse gas emissions, therefore, might require the taxing authorities to change the tax rates over time if and when total emissions fail to fall to the desired level. With a tax, the price of emissions is fixed by the government, but when the quantity of emissions is set through a cap-and-trade regime, price changes will occur through the emissions permit market. With a tax, if new information about the science of climate change or its potential consequences emerges, the tax rate may need to be changed; in a cap-and-trade system, the government would instead have to vary the quantity of emissions permits—the cap. If the government changes its targets or the number of permits because of new scientific evidence, new technologies, or changes in the economy, the price of emission permits will also change. Under a carbon tax, emitters would limit emissions whenever the costs of doing so are lower than the tax rate; under cap and trade, these reductions would occur until their costs exceed the market price of purchasing allowances.

Many economists prefer controlling prices through taxes rather than controlling quantities through cap and trade because of taxes' greater predictability, lower volatility, and lesser administrative costs. (The cost of a tax is predictable for the upcoming years; the prices of emission allowances in the open market are not.) Many businesses, despite their general aversion to taxes, also interestingly prefer a carbon tax over cap and trade because of the greater price certainty and smaller compliance costs the tax provides. But environmentalists, regulators, and politicians approach environmental problems through their concerns with a given quantity of emissions, so they find regulatory caps more congenial, especially because they are unsure how firms will react to specific levels of taxation. Close analysis, however, reveals that a carbon tax and a cap-and-trade system can be designed to reach very similar results.

In principle, the older model of direct government regulation of emissions can achieve results similar to a carbon tax or cap-and-trade system by requiring specific but different firm-by-firm emissions reductions. The regulator, however, would need detailed information about the emission-reduction costs for every firm and the lowest-cost technological changes to employ. In practice, this kind of information is not available to the government, and, without it, command-and-control regulation may be unnecessarily expensive (as the Clean Water Act has amply demonstrated).

For example, it became clear not long after the Clean Air Act's passage that meeting the legislation's goals might prove very costly to the nation's economy. In those regions of the country where the improvements of air quality mandated by the act were not occurring—"nonattainment" regions in the language of the law—the EPA sometimes prohibited entry by any new businesses that would emit the specified pollutants until the standards had been satisfied. Needless to say, the prospect of turning away new businesses did not sit well with the nonattainment areas' residents or their political representatives. Nor would prohibiting new businesses from entering serve innovation, competition, or economic growth. So the EPA began to look for alternatives.

Confronting this problem of the inefficiency of command regulation, academic economists, beginning in the late 1960s, began advocating "market-based" alternatives—such as pollution taxes and cap and trade. By the mid-1970s, the cost-reducing advantages of such alternatives had become clearly demonstrated in the economics literature and rather widely known in policymaking circles. But the EPA moved quite cautiously. Environmentalists were reluctant to embrace the idea of tradable permits because of moral qualms about creating a market for licenses to pollute. Some observers viewed creating a market for pollution rights as equivalent to the Catholic Church's discredited practice of selling indulgences—paying to pollute as forgiveness for bad behavior. Nevertheless, in 1974 the EPA instituted an emissions-trading program that would become a precursor to cap and trade and by doing so initiated a revolution in regulatory practices.

Beginning in 1974, EPA rewarded firms that had voluntarily reduced emissions below the level required by the Clean Air Act with "emissions-reduction credits." The firms could use these credits to allow greater emissions at another regulated source, or it could "bank" them for use in future years. Three years later the EPA began allowing firms to trade these credits to other firms within their region. New firms were then allowed to enter nonattainment areas if they could acquire enough credits from other firms so that they could show that the total emissions in the region would be lower than would otherwise have been the case. In other words, new firms had to buy more credits than their emissions would require. One economist has estimated that by the mid-1980s these emissions credit programs had saved more than $400 million in aggregate compliance costs.

The EPA soon fully embraced a market-based pollution-control methodology in connection with a program to phase out the use of lead as a gasoline additive. The agency allowed those refineries that had outperformed their legal requirements to sell "credits" to refiners for whom removing lead was more costly. Subsequent analyses showed that refiners' ability to trade credits in this manner resulted in less costly and faster removal of lead from gasoline than would have occurred without the trading option. Although extremely successful on a nationwide basis, the trading of credits produced significant regional differences. Some areas of the country, where removing lead was less costly, achieved greater lead removal—along with its accompanying health benefits—sooner than had been anticipated; others, where less-efficient refiners purchased permits, got the lead out more slowly.

A version of cap and trade was again employed in the case of ozone-depleting chemicals. Pursuant to an international agreement to reduce and eliminate many of those pollutants, the United States in the late 1980s instituted both a tradable permit system and an environmental tax. In a move that prefigured the climate change debate, the EPA decided to allocate permits to use such chemicals without any fee to the small number of domestic producers of these chemicals. It was estimated that this decision produced billions of dollars of windfall profits for these companies relative to what they would have earned had the use of the offending chemicals simply been phased out. In reaction, Congress enacted a complex excise tax on all offending chemicals as a way both to encourage the switch to less-harmful products and to recover some of the profits resulting from using the permits during the phaseout of the pollutants. In retrospect, some analysts have concluded that the tax was ultimately more effective than the tradable permit system in achieving the desired reductions.

Even as cap and trade, with its cost-saving advantages over command-and-control regulation, has emerged over time as the favored regulatory approach for addressing environmental problems, there has been considerable reluctance to transform preexisting regulatory structures. Take, for example, the CAFE fuel-efficiency standards enacted in 1975 and phased in during the following decade. Even though the automotive industry took massive advantage of the "light truck" loophole (read "SUV"), resulting in the number of light trucks growing by two and a half times between 1979 and 1999, from 22 percent of the nation's motor vehicle fleet to 37 percent, Congress wasted 30 years before finally revising the CAFE rules in the Energy Independence and Security Act of 2007. The new rules prescribe fuel standards covering both light trucks and automobiles, and, beginning in 2011, they require average fuel economy to increase to 35 miles per gallon by 2020. In 2009, President Obama accelerated the required fuel mileage improvements, announcing that a new standard of 35.5 miles per gallon must be reached by 2016.

This recent attention to CAFE reflects both the renewed interest in reducing oil imports and the increased political salience of the climate change risks from greenhouse gas emissions. But serious shortcomings remain. Unlike a gasoline tax, the CAFE standards impose no burden on increases in miles driven (total passenger vehicle miles have tripled in the United States since 1975). Nor has Congress ever taken seriously the idea of making gas mileage allowances tradable among automobile manufacturers as a way to achieve our overall fuel-economy goals at lower costs. A cap-and-trade CAFE regime would permit automobile manufacturers who are most efficient at increasing gas mileage to sell excess credits to firms that find increasing the mileage of their vehicles more costly, thus bringing down the total costs to manufacturers in complying with the standards. In the 1970s, when CAFE was first instituted, for example, such a system would have greatly reduced the incentives for Japanese car companies to develop larger cars, such as the Lexus and Infiniti. And now, at a time when U.S. automobile companies face serious economic difficulties and challenges, lowering the costs of complying with CAFE would be no small improvement. But not even the U.S. automobile companies have ever urged such a change, nor has any enterprising member of Congress. This kind of inertia no doubt reflects the tendency of institutional structures, once enacted, to persevere. Today's requirements depend largely on how legislation was fashioned more than three decades ago. The status quo is difficult to overcome, and the tendency of policies to stick with their initial institutional arrangements greatly raises the stakes of initial enactments.

A different piece of environmental legislation that did adopt a cap-and-trade system has proved quite successful. In 1990, Congress amended the Clean Air Act to institute a pollution permit–trading program beginning in 1996 to tackle the problem of acid rain caused by emissions from coal-fired electric utilities. This law was designed to reduce sulfur dioxide levels over time by 10 million tons and nitrogen oxide by 2 million tons. Congress allocated the tradable permits—again without any fee—to facilities at 110 power plants, owned and operated by 61 electric companies, mostly east of the Mississippi River. After 1995, these utilities could emit only the amount of sulfur dioxide that they had permits for. Fines of more than $2,500 per ton of emissions in excess of those allowable are imposed on plants not in compliance. In accepting the market-based cap-and-trade technique, Congress broke a legislative logjam that had prevented it from dealing with the acid rain problem for more than a decade. The Government Accountability Office has estimated that cap and trade has saved business more than half the costs (up to $3 billion a year) of command regulations. Questions have arisen recently, however, concerning whether too many permits have been issued, a common occurrence in cap-and-trade programs.

This was the first cap-and-trade program in the United States to employ public sales of emissions allowances (run by the Chicago Board of Trade) in addition to allowing private sales. Anyone can purchase these permits. Tom Tietenberg, an economics professor who has long been concerned with environmental issues, reports that many of his students have purchased allowances as Christmas or birthday gifts. Such purchases, by removing permits from utilities, may further reduce total emissions in a given year. The acid rain program clearly has been the most important, sophisticated, and successful cap-and-trade program in the United States. But it is small potatoes compared to what would be required under a cap-and-trade limit for greenhouse gases.

Nonetheless, regions and governments have begun to experiment with the market-based regulatory model to take on this much larger task. In 2008, for example, ten northeastern and mid-Atlantic states joined together in the Regional Greenhouse Gas Initiative to create a cap-and-trade mechanism to restrict carbon emissions by electric utilities. These ten states have capped carbon dioxide emissions from power plants at levels designed to reduce emissions by 10 percent over the following decade. They also allow buying and selling of permits among electric utilities in any of the member states; an electric utility can use carbon dioxide allowances issued by any of the member states to comply with its state's requirements. Each state auctions nearly all of the Regional Greenhouse Gas Initiative's allowances quarterly. By the end of 2009, more than 170 million allowances (each permitting one ton of carbon dioxide emissions) had been auctioned at prices ranging from $1.86 to $3.51 per allowance, yielding nearly $500 million in revenues for the member states. Similar alliances are being launched in the West (the Western Climate Initiative, involving half a dozen states and four Canadian provinces) and the Midwest (the Midwestern Greenhouse Gas Accord, involving six states).

To date, however, the most extensive effort to contain greenhouse gasses with a market system has occurred in the European Union (EU). The EU's emissions-trading scheme was created to address Europe's obligations to reduce greenhouse gas emissions pursuant to the Kyoto Protocol to the United Nations Framework Convention on Climate Change. As I have described, the United States is a signatory to the Framework Convention and signed but did not ratify the Kyoto Protocol, which was agreed to by 190 countries, including all EU member states. The Kyoto Protocol set binding emissions targets for 37 developed countries for carbon dioxide and five other greenhouse gases, including methane, nitrous oxide, and hydroflourocarbons. Under the protocol, specified targets must be reached between 2008 and 2012.

In its first phase, which ended in 2007, the EU cap-and-trade system covered carbon dioxide emissions from about 12,000 firms in the electric power sector and certain energy-intensive manufacturing industries in 25 countries. These firms together account for about half of total EU emissions. In 2005, this program produced a $12.4 billion market in EU allowances sold through several markets, including three exchanges in the United States, and by 2007 the World Bank estimated the total value of EU allowances to be about $50 billion. The allowances—each of which permits the emission of one metric ton of carbon dioxide—were initially allocated to industries and firms by each member state, but they can be traded to any company in the EU.

This phase of the EU system encountered substantial difficulties. The member states allocated total allowances in excess of the actual emissions then occurring. This excess of course resulted in uncertain and minimal effects on total emissions and very little positive impact on technology innovation and development. In fact, the Environmental Audit Committee of the British Parliament concluded that the level of Europe-wide emissions actually increased by 38 million tons between 2005 and 2007. In addition, there was significant volatility in the prices of permits, with allowances selling for €10.40 per ton in 2005, €37.48 in April of 2006, and zero in 2007.

Phase II of the program began in 2008, and the European market was estimated in 2010 to total about €130 billion (which would expand to an estimated worldwide $3 trillion market in 2020 if Japan and the United States also were to participate). Price volatility remained an issue, with allowance prices ranging from nearly €30 a ton to less than €10 in 2008 and 2009. In January 2010, using a computer fraud technique known as "phishing," thieves obtained access codes from companies and traders to obtain permits, which they then sold, netting at least $4 million. The biggest concern, however, according to Larry Lohmann of Corner House, a British environmental advocacy group, "is measuring whether the so-called reductions in emissions that are claimed by participants in these markets actually happen at all." If not, that would "turn out to be the far bigger fraud."

In 2010, the British Parliament's Environmental Audit Committee raised serious concerns about the effectiveness of the EU's cap-and-trade system for the future. Tim Yeo, the committee's chairman, said, "Emissions trading should be helping us to combat climate change, but at the moment the price of carbon simply isn't high enough to make it work." He added, "If the government wants to kick-start serious green investment, it must step in to stop the price of carbon [from] flatlining." He also suggested that Parliament "seriously explore the possibility of a carbon tax." Sarah-Jayne Clifton of Friends of the Earth said, "Not only is trading failing to drive down emissions, banks are growing fat developing ever more complex trading systems and this risks another financial crash."

Adopting a greenhouse gas cap-and-trade system in the United States requires Congress (or the EPA) to resolve a number of inevitably controversial issues. As Europe's experience has shown, the details will be critically important in determining the extent to which environmental goals are met, the effects on the economy, and the distribution of benefits and burdens. Hundreds of learned articles and many books have been written analyzing the pros and cons of the various decisions that must be made, and I do not intend to plumb all their depths here. It is nevertheless instructive to examine briefly some of the more important issues.

First, the government must decide which pollutants will be capped at what levels over what time period and for which emitters. It must also decide how allowances will be initially distributed and whether one year's allowances will be permitted to be "banked" and used in a subsequent year or "borrowed" to be used in an earlier year. It must also establish rules and regulatory structures governing the buying and selling of allowances, including whether to restrict potential increases or decreases in allowance prices. Furthermore, it must determine to what extent and under what circumstances to allow "offsets"—the ability to pay someone else for lowering emissions, such as by revising agricultural practices or foregoing deforestation. Finally, because the problem of global warming is global, we must determine how to coordinate our efforts with those of other nations around the world. None of these issues is easy.

Let's begin with the difficulty of setting the cap. The main reason environmentalists prefer cap and trade to a carbon tax as the principal method to address the risks of climate change is that the latter focuses on increasing the price of carbon emissions, whereas the former directly limits the quantity of emissions. But setting the emissions caps and how they will be phased in is hardly a simple or straightforward exercise. Uncertainty abounds.

The IPCC has estimated that by the end of this century, without major policy changes, ongoing emissions of greenhouse gases will likely double the concentration of greenhouse gases in the atmosphere relative to their preindustrial level of 280 ppm. Concentrations currently are about 385 ppm and are growing by about 2 ppm a year. Many scientists believe that levels of 450 ppm are dangerous; some, including James Hansen, believe that the IPCC estimates are too optimistic and that we need to lower levels to 350 ppm. Very few scientists regard levels higher than 550 ppm as safe. Summarizing 22 scientific studies, the IPCC in 2007 estimated that doubling preindustrial levels would produce an increase in global mean temperatures "most likely" ranging from 1°C (1.8°F) to 4.5°C with an expected value of 2.5 to 3.0°C. The IPCC suggests that this range reflects a 60 to 90 percent confidence level, but that the probability of a temperature increase higher than 4.5°C is between 5 and 17 percent. These IPCC estimates thus imply that there is a 5 percent probability that conducting business as usual would lead to an increase in global mean temperature of 7°C or more and a one percent probability that it would produce a 10°C temperature increase.

The less likely but more dangerous estimates of temperature increases might be catastrophic worldwide. In addressing a United Nations meeting on climate change in 2009, Barack Obama called on nations to act now to reduce the prospect of a worldwide "irreversible catastrophe." It is, of course, the prospect of genuine disaster that grabs the public's attention and has galvanized the world's and our nation's political leaders into action. As Nicholas Stern has put it, "The issue for policy is how to manage risk, taking account of strong scientific evidence that the risks are potentially very large."

But not only are estimates of the magnitude and scope of the risks of climate change from greenhouse gas emissions controversial among scientists; these estimates will change over time as new facts and new scientific studies emerge. Temperature changes are not the only uncertainty; also uncertain are the potential effects of temperature increases on economies and ecosystems around the world. The likely effects will vary greatly by regions. If the IPCC's most probable estimates prove true, Canada and our northern states will mainly enjoy warmer winters, but low-lying islands may disappear entirely. In October 2009, Douglas Elmendorf, director of the Congressional Budget Office (CBO), told the Senate that with warming of about 4°C (7°F), a "relatively pessimistic estimate," is that the U.S. GDP would decline by about 3 percent. He added: "There is such great uncertainty about how a given quantity of emissions would ultimately affect global temperature that there is very little additional certainty to be gained from choosing a fixed emissions goal . . . rather than a price path." Elmdendorf's essential point was that there is little, if any, reason to prefer cap and trade's focus on quantities over a carbon tax's focus on prices. Experience so far in Europe's and our states' regional cap-and-trade efforts supports his view.

Having overlearned the lessons of Jimmy Carter's political problems from calling on Americans to sacrifice, our politicians now insist that any sacrifices by the public resulting from a dramatic transformation of our nation's energy production and consumption must be minimal. The politicians' unwillingness to confront the potential costs straightforwardly necessarily limits the potential range of legislative answers to the central questions that our democratic process must ultimately answer: How much change will be required and over what time period? In Congress, this question becomes transformed from a scientific to a political inquiry. In 2009, President Obama and Democratic congressional leaders essentially agreed that our nation's emissions of greenhouse gases should be reduced by 3 percent from their 2005 levels by 2012 (the nearest presidential election year), by about 17 to 20 percent by 2020, and by more than 80 percent by 2050.

These goals generally reflect the consensus among environmental advocacy groups that we should endeavor to limit worldwide greenhouse gas concentrations in the atmosphere to less than 450 ppm and worldwide temperature increases to 2°C (3.6°F). But the goals were also set to keep low the short-term costs to the public. This slow phase in of emissions reductions has provoked criticisms from some environmentalists, who believe that by going so slowly the United States is foregoing an opportunity for global leadership. In contrast, many economists and virtually all Republican politicians argue that abatement of carbon dioxide emissions should begin slowly or even be delayed while over the next decade or two we learn more about the sensitivity of the climate to greenhouse gases and about the potential human, ecological, and economic impacts of climate change and so we can see how technology advances can lower the costs of reducing the level of our greenhouse gas emissions. This split, of course, will affect the political compromises of any energy legislation.

Another key decision policymakers must make is which polluters shall be subject to regulation. Because greenhouse gases escape from virtually every automobile, every building, every manufacturing facility, every farm, and most electric utilities, the potential difficulties of monitoring emissions and enforcing limitations are daunting. Such monitoring and enforcement would be a far more difficult problem than enforcing the cap-and-trade program for constantly computer-monitored sulfur dioxide emissions from U.S. power plants.

Carbon dioxide is the greenhouse gas emitted in the greatest quantities, and about 80 percent of all greenhouse gas emissions occur from the combustion of fossil fuels (the bulk of the rest comes from some industrial gases, agriculture, and deforestation). Cap-and-trade legislation in both the United States and Europe has focused initially on utilities that generate electricity and on importers and refiners of petroleum or other fossil fuels. Industrial sources that annually emit more than a specific threshold of carbon dioxide, such as 25,000 metric tons, should also be covered, along with natural-gas local distribution companies. All in all, in such a regime about 7,400 facilities would be subject to limitations that would have to be monitored by the EPA, and at least 60 percent of greenhouse gas emissions should be covered. (A separate cap-and-trade regime would apply to hydroflourocarbons.)

Once the class of businesses to be regulated is decided, there are questions of how to design the market these firms will be required to participate in. One key aspect of any cap-and-trade legislation is whether firms subject to emissions caps would be permitted to purchase "offset credits" to reduce their costs further. Most cap-and-trade legislation allows such offsets despite ongoing questions about whether they will be genuinely additional, permanent, and real. For example, utilities might pay landowners for such offset credits if what they plant absorbs carbon dioxide from the atmosphere or if they forego deforestation. Carbon sequestration and various alternative energy projects may also qualify for offsets. Cap-and-trade proposals generally would allow from one-third to two-thirds of emission-reduction requirements—up to 2 billion tons of greenhouse gases annually—to be satisfied by purchasing offsets. As much as one-half or two-thirds of those offsets might be purchased abroad. As the Congressional Research Service has said, there are "almost an infinite number of possible scenarios for offsets."

Because reducing emissions through offsets may be much cheaper than the transformations required by utilities and industries directly subject to the caps, the potential cost savings from allowing offsets are large. There is widespread agreement that offsets may reduce allowance prices by at least half. The CBO has estimated that the use of offsets might reduce allowance prices by about 70 percent and that even more liberal use of offsets would generate even greater cost savings. It is also worth bearing in mind that as emissions caps tighten and the price of allowances increases over time, the advantages of purchasing offset credits will also grow.

The theory behind offsets is that they will encourage firms that are not subject to the emissions caps to participate in reducing emissions—and to profit from doing so—whenever it is cheaper for them to reduce emissions. Because climate change is a global phenomenon and reducing emissions abroad may often be considerably cheaper than in the United States, it also makes sense—in principle at least—to allow firms to purchase such offset credits anywhere in the world. By buying them, however, firms subject to emissions limitations will be allowed to exceed their caps. And whenever offsets are purchased internationally in lieu of reducing emissions domestically, the United States will exceed its stated goals for emissions reductions (which has happened in Europe under the Kyoto Protocol). By making the United States seem to be behaving much worse than it actually is, offsets may diminish U.S. credibility in international negotiations, and we might forfeit any claim to global leadership that we wish to assert. More important, ensuring that such offsets are real, permanent, verifiable, enforceable, and additional (i.e., they would not have occurred in any event) will be no easy task.

The Kyoto Protocol established such an offset credit program, known as the Clean Development Mechanism, which allows companies in the 37 countries subject to binding emissions caps to purchase offset credits for emissions reduction projects in developing countries. This experiment has been subjected to fierce criticism on the grounds that approval of projects is slow, costly, and cumbersome. More important is the concern that actual reductions of emissions beyond what would have occurred anyway are questionable and perhaps temporary.

There has also been considerable concern, albeit not from the financial sector, about the likely volatility of prices for emissions allowances (and for offset credits if they become tradable). Volatility in permit prices complicates utilities' ability to evaluate long-term investment prospects. Cap-and-trade legislation may employ a variety of techniques intended to reduce potential fluctuations in prices. One way to decrease such volatility is to permit owners of emissions allowances for one year to "bank" them to be used in future years. Because the effects on climate change from greenhouse gas emissions depend on their aggregate levels, not on the annual flow, this approach seems appropriate. "Borrowing" allowances from future years to be used earlier is also possible, but it creates concerns that it may cause pressure on the political willingness to maintain stringent limits on emissions in the future. The government may also hold back some allowances to be sold on the open market if and when the prices of emissions allowances seem high enough to raise concerns for the health of the U.S. economy. One variation sometimes suggested is to permit the president to impose price ceilings or a price collar (a floor and a ceiling on prices) whenever he or she deems the price increases or decreases or the volatility of allowances to be excessive. A cap-and-trade system with a price floor and a ceiling to limit the range of prices would provide increased price certainty to investors and, if allowances were auctioned, would move closer to a carbon tax in its overall economic effects.

Such a market in the United States will likely involve hundreds of billions of dollars of transactions annually and will inspire all sorts of derivative contracts similar to those now present in other commodity markets. The market for emissions permits will presumably be subject to regulatory oversight similar to that of other commodity or derivative markets. In recent years, however, such oversight has left much to be desired. New financial regulations have been enacted, and their effectiveness will impact the integrity of any cap-and-trade program.

If properly designed and regulated, cap-and-trade legislation, like a carbon tax, should raise the prices of carbon-based fuels. For this reason, it becomes important to protect from excessive burdens low- and moderate-income families who spend a disproportionate share of their income on energy consumption. There are a number of ways to accomplish this protection—for example, by providing increased cash benefits to current recipients of food stamps, unemployment or disability insurance, or Social Security or by expanding earned-income tax credits. Any or all of these mechanisms are preferable to endeavoring to minimize price increases for electricity or petroleum products because the latter would undermine the price effects that a cap-and-trade system is intended to effectuate. It will be necessary, however, to monitor price increases and such benefits on an ongoing basis to make sure that low-income families are protected over time.

In order to understand the full implications of enacting a cap-and-trade system for greenhouse gases, one must also take into account our federalist system. Any national legislation might preempt state or regional cap-and-trade programs but would permit other forms of state, regional, or local regulations—for example, California auto emissions regulations and energy-efficiency requirements—to coexist alongside federal laws. States might also enact their own carbon taxes, technology or renewable energy standards, or emissions limits that are more stringent than the federal cap for any particular year. If they did so, they might induce price changes in the national market for emission permits. If that happens to any widespread extent, businesses may petition the federal government to halt the state actions.

For all the enormous changes that cap and trade would bring to the regulation of carbon emissions, Congress and the states will not entirely abandon the older form of command-and-control regulations. Indeed, any cap-and-trade or other energy legislation will also contain a long list of energy-efficiency mandates affecting commercial and industrial buildings, a host of appliances, and other electric products in an effort to induce greater efficiency in U.S electricity consumption. Expanding on an idea from legislation of the 1970s, there may also be requirements that electric utilities obtain a specific portion—perhaps as much as 20 to 30 percent—of their total power output from specified renewable sources. (It would also be possible to allow "trading" among utilities of renewable energy requirements, much as the EPA did with Clean Air Act requirements in the mid-1970s.)

Because climate change is a global problem attributable to greenhouse gas emissions regardless of source, any effective response must reduce emissions worldwide over time. Improvement in everyone's climate prospects turns on efforts made throughout the world. International coordination and cooperation is essential, if difficult. Some have suggested that one of the advantages of cap and trade over carbon taxes is that the former makes reaching international agreements easier. As an MIT assessment of U.S. cap-and-trade proposals put it, "A major strategic consideration in setting U.S. policy targets should be their value in leading other major countries to take on similar efforts."

Because of U.S. cap-and-trade legislation's slow phase-ins of emissions reductions, we will not know of its success or failure for quite a long time. As Lincoln Moses, the first administrator of the Energy Information Administration, once said, "There are no facts about the future." Given, as we shall see in the next chapter, the enormous political and corporate forces acting to shape legislation in Congress, however, it is difficult to believe that cap and trade will actually achieve the emissions-reductions goals established by President Obama and Congress. Here's how a former Democratic congressional leader privately explained the political popularity of cap and trade: "No one really understands it." But in the wake of the recent financial crisis, opaque markets may have lost some of their appeal.

NASA's James Hansen views cap and trade as "essentially a sham," calling it the "Temple of Doom." He has urged instead a worldwide moratorium on new coal-fired electricity plants and a phaseout of all existing coal plants over the next two decades. We better hope that Hansen—who was apparently right about climate change in the 1970s, 1980s, and 1990s—is wrong this time; his suggestions are gaining little traction. Even with our discovery of ways to obtain large domestic sources of natural gas, coal-fired electricity is not about to disappear anytime soon.

Taxing carbon and returning the proceeds to the public would be more straightforward, more predictable, administratively easier, and perhaps even more effective than cap-and-trade legislation. Some EU member states are tellingly considering enacting carbon taxes as a backstop to its cap-and-trade program. Indeed, Ireland in 2010 enacted a carbon tax on petroleum products, coal, and natural gas. As in the United States, however, political barriers to a Europe-wide carbon tax loom large.

It is to those political barriers here at home that we now turn. A gradual but quickly tightening cap-and-trade system as the preferred method for the government regulation of energy may make sense to environmentalists, some academic economists, and certain believers in markets. It may even make sense in practice if designed and implemented in a world of rational policymaking. Alas, that is not the world we live in. We live in a world where any major reform affecting significant economic interests—whether they be based on profits in health care, finance, or energy—is subject to a legislative process so deeply influenced by the very industries that such reforms aim to control that the results can end up seeming more like a mockery of the original idea than a solution to the problem. The greatest barrier to making a cap-and-trade system work may not be the concepts, but the sclerotic system though which it must be enacted.

13 Government for the People? Congress and the Road to Reform

In 2009 and 2010, the most significant energy legislation in a generation was being debated in Congress. Hundreds if not thousands of interest groups from business, the environmental movement, organized labor, and citizen action committees, together with all their lobbyists, jammed the halls of the Capitol to press their case for myriad different policies and provisions or simply for their share of the pork. Think tanks and academics and the talking heads who spin their work into sound bites geared up for what advocates anticipated might be a landmark bill reshaping energy policy for decades to come. In 2009, the House passed its bill, which included a cap-and-trade system for carbon dioxide along with other regulatory mandates and subsidies. By the end of summer 2010, however, the Senate had all but given up. Cap and trade still seemed to hold center stage, but several alternatives were vying to garner the necessary support. Opponents of climate change legislation now had the upper hand. The battle echoed the 1970s' struggle over energy legislation, especially the pricing of natural gas, but the stakes this time were much higher: imposing a price on carbon emissions would affect all of our fossil fuels and the products they produce. In November 2010, voters swept the House Democratic majority out of office and in the process took comprehensive climate change legislation off the table—at least for a while. But, despite cratering in 2010, energy does not seem likely to disappear from Congress's agenda anytime soon.

The big question is, how effective will any energy legislation be? Or, in other words, will cap-and-trade bills of the sort that have traveled under the short hand of the Waxman–Markey bill in the House and its variations in the Senate—or any major new energy legislation—forge a path to economically sound and environmentally and politically sustainable energy production, consumption, and development?

To answer that question, we need to examine the forces that shaped the 2009 House bill and influenced the Senate in 2010 as well as the process that shaped this legislation. And to understand these things, we need to understand how power in Congress functions differently now than it did in the 1970s.

Although it has always been true that our federalist government was designed to diffuse political power and inhibit dramatic change, the difficulty of passing major legislation has reached new heights in Washington. Within both the House of Representatives and especially the Senate, where the rules permit unlimited debate absent a 60-vote majority to cut off a filibuster, powerful individuals who serve in leadership positions, including party leaders, chairs of relevant committees or subcommittees, and lawmakers in a position to supply or withhold critical swing votes, serve as "veto players" who are able to stymie or shape legislation. As the *Wall Street Journal*'s David Wessel has observed, "The ranks of congressional leaders with the skill and desire to fashion compromises instead of talking points are depleted."

Moreover, enhanced ideological divides between Republicans and Democrats, coupled with aggressive partisanship for short-term political advantage, have turned fashioning legislation into an evermore rancorous and difficult exercise. After observing that "members of Congress vote on and pass bills without any knowledge—or sometimes even an opportunity to know—the language in the bills under consideration," Georgetown law professor Richard Lazarus, who has studied the history of environmental law in detail, says that since 1970 "Congress's ability to serve a constructive role in the ongoing process of environmental lawmaking has virtually disappeared." Congress, he says, "displays no ability to deliberate openly and systematically in response to changing circumstances and new information." In their 2006 book *The Broken Branch,* longtime Washington observers Thomas Mann and Norman Ornstein describe Congress now as a "contentious" and "partisan" body that is "failing America." They add that the "decline in deliberation has resulted in shoddy and questionable policy—domestic and international."

This decline did not happen overnight. The energy crises of the 1970s occurred during a transformative period in our nation's modern political history. At the beginning of that decade, faith in our government's ability to solve our nation's economic problems—even to "fine-tune" our economy—loomed large. Since World War II, median real income had doubled. Economic growth had averaged 5 percent between 1965 and 1969 and, following a small downturn in 1970, was about 6 percent in 1972 and 1973. Americans took robust economic growth—fueled by cheap and abundant energy—for granted. We had recently fulfilled John Kennedy's 1961 promise to put a man on the moon within the decade, and faith in our nation's technological prowess had reached a zenith. The public also still believed that Lyndon Johnson's massive Great Society legislation of the 1960s promised success in the "war on poverty." Even a Republican president such as Richard Nixon was prepared to deploy enormous government power over the nation's economy, as his disastrous 1971 wage and price controls vividly demonstrated. Few Americans then thought it at all odd that Congress had granted him such broad power over the nation's economy.

Not long thereafter, however, the energy crisis, a continuing Vietnam War, and Watergate convinced the nation that its abiding faith in government was misplaced. By the mid-1970s, these events had also inspired drastic revisions in the relationship between Congress and the president and within the legislative branch itself.

Responding to Richard Nixon's abuses of power, Congress flexed its muscles vis-à-vis the president in both foreign and domestic policy: foreign policy in the War Powers Act of 1973 and domestic policy through the Budget and Impoundment Control Act of 1974. The latter was Congress's response to Nixon's refusal to spend by "impounding" money that the legislature had appropriated. This law limited such actions, created new House and Senate budget committees, and revised the congressional process for enacting budget legislation. It also created the Congressional Budget Office, an important counterweight to the president's Office of Management and Budget.

When NEPA and the massive expansion of the Clean Air Act were enacted in 1970, the number of important congressional players was quite small. Two Democratic senators—Washington's Henry "Scoop" Jackson, chairman of the Senate Interior and Insular Affairs Committee and that chamber's leader on energy policy, and Maine's Edmund Muskie, chairman of the Air and Water Subcommittee of the Public Works Committee (later renamed the Environmental Pollution Subcommittee) and the Senate's environmental policy leader—were largely responsible for shaping these two major statutes. (The Clean Air Act passed the Senate in September 1970 by a vote of 73 to 0.) In 1970, that legislation was referred to only one House committee, and five representatives served on the conference committee with the Senate that hammered out its final details. Senators and representatives with significant responsibility for legislation took a hands-on role, delegating far less control to their staffs than they do now. The 1990 Clean Air Act amendments, by contrast, were referred to several different committees and numerous subcommittees. In the conference committee alone, the House was represented by 130 members from seven different committees.

The virtually complete domination of the legislative process by a relatively small handful of powerful committee chairs—predominantly conservative Democratic southerners—had broken down by the mid-1970s. The days when "junior members were expected to be seen and not heard" were over. The details of how that occurred have been recounted elsewhere and need not be repeated here. But the transformation was stark. Through the early 1970s, committee chairs, selected based on their seniority alone, completely controlled committee agendas, staff personnel, and resources. In the House, they fashioned legislation behind closed doors, controlled the floor debate, which permitted few, if any, floor amendments, and decided who would represent their chamber in conference committees. Beginning in 1970, however, some liberal Democrats in particular began to chafe, and by January 1975—when 75 new freshmen Democrats took office following Nixon's Watergate debacle—a procedural revolution was consummated.

New power centers in the form of autonomous subcommittees were created. Power also shifted away from committee chairs to the Speaker of the House and to party leaders in both the House and Senate. In a stunning and unprecedented challenge to the seniority system, in January 1975 three Democratic House committee chairmen—Louisiana's Edward Hebert of the Armed Services Committee and Texans Wright Patman, head of the Banking Committee, and William Poage, chairman of the Agriculture Committee—were ousted and replaced by considerably more junior (and more liberal) representatives. In the House Energy and Commerce Committee, Harley Staggers, a Democrat of West Virginia, barely managed to hold on to his chairmanship; he responded to a minirevolt among committee members by creating six new largely independent subcommittee "fiefdoms."

Washington's superlobbyist Tommy Boggs (son of the powerful Louisiana Democratic congressman Hale Boggs) described the transformation: "Instead of 10 committee chairmen, you now have 70 people running the House and a hundred people running the Senate. In the past, a lobbyist needed to know only about 10 people on the House side. . . . [A]ll that has changed as power has dispersed." By the mid-1970s, responsibility for energy and environmental legislation was scattered across a "tangle of competing committees and subcommittees" in both Houses.

This diffusion of power, with many new Democrats finding themselves occupying positions of authority for the first time, ushered in a new era of partisanship. Heretofore, House Democratic chairs had worked closely with their committee's ranking Republicans, but such cooperation across the aisle began to break down after the 1974 elections produced a two-to-one Democratic majority. New subcommittee chairs found it much more rewarding to cater to their younger liberal Democratic colleagues than to the dwindling band of more senior Republicans. After 1978, partisan splits became even more tendentious when Newt Gingrich (R–Ga.) came to Congress determined to achieve a Republican majority in the House—a result he successfully fulfilled many years later in the 1994 elections. Gingrich and his allies deployed post-Watergate ethics rules to bring down Democrats, including House Speaker Jim Wright—a tactic that was subsequently turned against Republicans and no doubt over time has exacerbated tensions and enhanced partisan distaste and distrust. Partisan gerrymandering of House districts by state legislatures also contributed to this atmosphere. In the early 1970s, Democrats and Republicans voted against each other about one-third of the time; by the mid-1990s, that number had doubled. Enacting legislation had become more cumbersome and contentious.

By the time Jimmy Carter entered the White House, Congress had become determined to exercise its legislative power in a manner less deferential to the president and less beholden to executive branch expertise. But it had simultaneously made legislating anything coherent more difficult by diffusing its own power. Adding new powerful players to the legislative process created additional veto points, making it even more difficult to enact new laws and increasing the number of interests that had to be satisfied. As a result, legislative compromises became more complex and convoluted.

How have these changes in congressional practices and procedures manifested themselves in relation to energy legislation? In the House, the Energy and Commerce Committee has primary jurisdiction over environmental and energy legislation. Nothing better illustrates the 1970s transformation of congressional prerogatives and practices than the 2008 battle between Democrats John Dingell and Henry Waxman for that committee's chairmanship.

Although both Dingell and Waxman had long been powerful Democratic liberal lions of the House, they represent very different constituencies—Waxman: Beverly Hills, West Hollywood, and Malibu; Dingell: the southern outskirts of Detroit. Although Dingell describes himself as an environmentalist, he has vigorously and effectively defended the interests of the Michigan automobile industry and the United Auto Workers in Congress. The Californian Waxman has unsurprisingly been far more concerned with reducing air pollution and increasing fuel efficiency. Dingell largely got his way in the 1975 battles over the CAFE fuel-efficiency requirements (and the relatively small penalties for failures to meet them) when his Michigan neighbor Gerald Ford occupied the White House. But Waxman triumphed over Dingell in tightening automobile emissions requirements under the subsequent amendments to the Clean Air Act. Dingell has long viewed Waxman as believing that blue-collar Michigan constituents should pay for Beverly Hills shoppers to breathe cleaner air on Rodeo Drive.

The day after Barack Obama was elected president in November 2008, Waxman, then 69, announced that he would challenge John Dingell, age 81, for the Energy and Commerce Committee's chairmanship (since 1981, Dingell had been either the committee's chairman or its ranking Democrat). The House Democratic caucus, following the procedures introduced in 1975, met a couple of weeks later to decide this contest by secret ballot. When the votes were counted, Waxman had won 137 to 122.

Waxman's bid was significantly aided by the fact that his legislative priorities on health care, climate change, and stronger consumer protection coincided with President Obama's and House Speaker Nancy Pelosi's (D–Calif.) agendas. Although remaining ostensibly neutral, both of these leaders implicitly signaled their support for Waxman's replacing Dingell. When Waxman finished his speech at the caucus meeting, Speaker Pelosi initiated a standing ovation for him.

The Dingell–Waxman battle not only illustrates the changes in internal congressional power and procedures but also reflects the central importance that regional influence plays in the creation of national energy policy.

Critics of Congress appropriately lament today's bitter partisanship. But when energy and environmental issues are at stake, regional differences can dominate partisan ones. To be sure, both ideology (which is often reflected in partisan differences) and concern for constituents motivate the positions that members of Congress take on legislation. Much empirical work in political science has endeavored to separate and evaluate the relative roles played by ideology, political party, regionalism, and constituent interests. The empirical work unsurprisingly finds a mix of these factors at play, but it has not been particularly successful at disentangling them. Most of the studies view ideology as most important, but also find that regional interests affect ideology and play an important role in energy and environmental legislation. As issues become more important to constituent groups, members of Congress become more likely to listen to them. Representatives and senators are also apt to support legislation that imposes costs on districts or states other than their own.

Large differences among the states characterize energy production and consumption. The states along the Atlantic and Pacific oceans, with the exception of California, have large populations of energy consumers but relatively little production of fossil fuels. They import substantial quantities of energy. Texas, Louisiana, Oklahoma, Alaska, and several western states, including California, have historically produced most of our domestic oil and natural gas. Coal is found in the eastern states West Virginia, Pennsylvania, Kentucky, Tennessee, Indiana, and Illinois and in the western states Wyoming, Montana, New Mexico, Utah, and Arizona. (These coal states account for twenty-two senators.) Most hydroelectric power is produced in the West.

Previous chapters have illustrated the role of regional interests in fashioning energy legislation. The difficulties of removing price controls on oil and natural gas, for example, often turned on contests between legislators from oil-producing states and legislators from oil-consuming states. Enterprising and ambitious senators, such as Scoop Jackson and Edward Kennedy, were quite happy to script televised public hearings to stoke public anger at the large oil companies, which became villains when oil prices spiked. A majority of Americans were convinced that the companies were lining their pockets at the public's expense and had engineered shortages simply to get their way in increasing prices. For much of the 1970s, this attitude sufficed to thwart powerful oil-state committee chairmen and their allies from decontrolling prices.

During that decade, coal interests fared considerably better than the previously all-powerful oil companies. Eastern coal was especially well treated in the government's funding of technological innovation and deployment. As noted earlier, most of the money for the synfuels program went to technologies that could use eastern coal, and, as chapter 5 notes, a coalition of environmental groups and eastern coal interests also played a decisive role in shaping the 1977 amendments to the Clean Air Act.

Because regionalism has played a key role in environmental and energy policy, one might think that such regionalism reflects a healthy response on the part of senators and representatives to their constituents' concerns. But one would be mistaken to view the energy bills before Congress as a response to public pressures regarding climate change.

Gauging what the public thinks is, of course, a notoriously treacherous business. Public-opinion polls are often little more than a snapshot that changes depending not only on the day's headlines, but also on how questions are worded. An ABC/*Washington Post* poll, for example, asking the question whether temperatures have been increasing slowly over the past century, found that 85 percent believed so in 2006, but only 72 percent did in 2009. Of those who believe that global warming is happening, about 35 to 45 percent regard it as a serious problem. When asked whether they support "cap and trade" to address climate change, small majorities of those polled usually say they do, but an October 2009 *ABC News/Wall Street Journal* poll, taken when a Senate committee was on the verge of considering cap-and-trade legislation, found 55 percent saying that they had heard nothing about cap and trade and another 30 percent that they had heard "only a little." In one 2009 poll of 1,000 likely voters, 29 percent thought cap and trade has something to do with Wall Street (which it actually does!), 17 percent said it is related to health care policy, and 30 percent admitted they have no clue.

Unsurprisingly, the proportion of people who say they support doing something to address global warming typically declines when people are asked if they are willing to pay higher taxes or energy prices. For example, one poll found that support for a greenhouse gas cap-and-trade program dropped from 56 to 44 percent, and another poll found that it dropped from 58 to 39 percent when people were told it would raise their electric bills by $25 rather than $10 a month.

The most comprehensive 2009 study of the public's knowledge and attitudes regarding climate change summarizes its findings this way: "climate change remains a relatively low priority among the American public, many of whom perceive it as a distant problem in both time and space." The authors of this report (from the Yale University Project on Climate Change and the George Mason University Center for Climate Change Communications) were clearly distraught at what they found. "Throughout human history," they said, "never before has so much rested on the need to change so many so fast."

In a September 2009 Gallup poll, only one percent said that the environment was America's most important problem. So even though concern with climate change had made it onto the public's radar screen of potential problems, it reaches the top of the list for only a very small minority. American voters are obviously not clamoring for cap and trade or any other legislation to address climate change, but the polls do give Congress enough running room to enact major changes if it wants to.

With public support so tepid for congressional action on climate change, one must look elsewhere to find the impetus for such far-reaching legislation. Environmental organizations acting alone are not sufficiently powerful to push through Congress such a dramatic change in our nation's pattern of energy production and consumption. And although there are business beneficiaries of any pollution-control legislation—witness the creation of an entire pollution-control industry beginning in the early 1970s—those businesses that would *naturally* gain from this transformation, such as wind and solar electricity developers and producers, might be overwhelmingly outmanned by business forces that potentially have much to lose from climate change legislation, including coal and oil companies, coal-fired electric utilities, railroads that transport coal, major industrial energy consumers, and the automobile industry. In fact, the National Association of Manufacturers, the U.S. Chamber of Commerce, and the American Petroleum Institute have opposed cap-and-trade legislation. With many business groups opposed to such legislation and the general public ranking it low on the priority list, what accounts for the fact that Congress was actually spurred to move forward?

One answer is that the environmentalists received an important boost from the courts, as they did in the 1970s. Federal courts have again played a crucial action-forcing role on their behalf. In 2007, the Supreme Court decided that the Clean Air Act requires the EPA to regulate emissions of greenhouse gases from new motor vehicles if that agency makes a finding of "endangerment" to human health from such emissions. This lawsuit was brought on behalf of 12 states, 4 local governments, and 13 renewable energy, environmental, and other organizations claiming that the EPA had "abdicated its responsibility under the Clean Air Act to regulate the emission of four greenhouse gases, including carbon dioxide." In his opinion for a five-to-four Supreme Court majority, Justice John Paul Stevens stated: "Under the clear terms of the Clean Air Act, EPA can avoid taking further action only if it determines that greenhouse gases do not contribute to climate change or if it provides some reasonable explanation as to why it cannot or will not exercise its discretion to determine whether they do."

During George W. Bush's remaining years in office, his administration essentially ignored this decision, but shortly after Barack Obama entered the White House, the EPA took action. In April 2009, EPA administrator Lisa P. Jackson signed a proposed finding that the current and projected mix of six greenhouse gases, including carbon dioxide, "threaten the public health and welfare of current and future generations." This is the "endangerment" finding. The EPA administrator also proposed to find that motor vehicle emissions of greenhouse gases contribute to this threat. Then, on September 30, 2009, the same day that Senators Barbara Boxer (D–Calif.) and John Kerry (D–Mass.) introduced their version of cap-and-trade legislation in the Senate, Jackson announced that her agency would also begin regulating greenhouse gas emissions from power plants and large industrial facilities by requiring them to demonstrate that they have installed the "best technology available" to reduce greenhouse gas emissions. In what she insisted was a "coincidence," the EPA finalized its "endangerment" finding just as world leaders were gathering for a two-week meeting on climate change in Copenhagen in December 2009. These announcements sent a clear signal to Congress and to businesses: if Congress does not act to curb greenhouse gas emissions, the EPA plans to do so by regulation. Lisa Jackson made it clear that her agency's announcements were intended to spur Congress to act: "The EPA's ready to work with Congress," she said, "but we're not going to continue with business as usual while we wait for Congress to act." (This comment prompted some in Congress, including Senators Jay Rockefeller [D–W.Va.] and Lisa Murkowski [R–Alaska] to promote legislation that would halt the EPA's ability to go forward with regulations.)

The move to regulate U.S. industry and power plants was further bolstered by the prospect of additional protracted, expensive, greenhouse gas litigation against businesses. In September 2009, the Court of Appeals for the Second Circuit (which includes New York, Vermont, and Connecticut) may have opened a new route for environmental organizations and local governments to attack emitters of greenhouse gases. The court held that municipal and private plaintiffs could sue six electric companies that operate fossil fuel–fired power plants in order to seek abatement of a "public nuisance"—the emission of greenhouse gases. Although it is far from clear that such lawsuits will be successful, prospects of this kind of litigation, along with the EPA's endangerment finding, have made key businesses more willing to support legislation limiting greenhouse gas emissions. As one analyst put it, "Environmentalists and industry seem to agree that the courts are not the place to develop a solution to global warming. [T]his [Court of Appeals] ruling is certainly a prod to both the EPA and Congress to get on with regulating greenhouse gas emissions."

If these legal and regulatory developments were not enough to send businesses to the halls of Congress looking for a more favorable outcome, action at the municipal, state, and regional levels supplied additional impetus. After the failure of the Kyoto Protocol, 23 states joined together to institute regional carbon-trading schemes. As chapter 12 indicates, ten northeastern states created the Regional Greenhouse Gas Initiative, a mandatory cap-and-trade system for electric power plants, and seven western states and four Canadian provinces concurrently announced their own plans for a regional cap-and-trade system known as the Western Climate Initiative. California had previously adopted a carbon emissions cap in 2006, and in 2009 Governor Arnold Schwarzenegger issued an executive order requiring one-third of that state's electricity to come from renewable sources by 2020. In May 2009, 30 governors released a "statement of principles" requesting Congress to cap greenhouse gas emissions "at levels guided by science to avoid dangerous global warming." And in October 2009, 1,000 mayors had signed a pledge to meet the Kyoto Protocol targets for reducing greenhouse gas emissions in their cities.

These domestic pressures and international efforts to make progress on limiting greenhouse gas emissions spurred Congress toward enacting climate change legislation. A convenient, if somewhat surprising, alliance of environmentalists and business interests who prefer congressional action to federal regulation, ongoing litigation, and a patchwork of state and local requirements has supported that endeavor. Environmentalists have viewed cap-and-trade legislation as their best opportunity to enact a federal law addressing climate change, and businesses have hoped to secure national uniformity to provide greater long-term certainty in making their investment choices—and to grab some windfalls in the process.

Books will no doubt be written about the final legislative outcomes, and it is not my purpose here to delve too deeply into this sausage-making endeavor, but a glimpse into the forces moving this lawmaking process, at least on some of the crucial decisions, is quite a revelation. The dollars at stake are potentially enormous: hundreds of billions of dollars in both costs and potential price increases that might result from command-and-control regulations, a carbon tax, or a cap-and-trade program; a new market in the trillions of dollars if carbon emissions allowances (or perhaps offsets or renewable energy requirements or both) are tradable; and untold billions in government aid in the form of direct subsidies, special tax breaks, and loan guarantees. Businesses are eager to avoid new costs being imposed on their products, to capture a share of the hundreds of billions of dollars of tradable permits allowing them to emit greenhouse gases, and to obtain subsidies for "green jobs," alternative energy production and use, and R&D. As Daniel Gross, writing in *Slate,* has said, "Above all, the new business progressives are pragmatists. They know Obama is going to be around four, possibly eight, years. In wartime, there are no atheists in the foxhole. And during this deep recession, there are few global warming–denying libertarians in the boardroom." Or, as the well-worn cliché goes, American businesses would rather be at the table than on the menu.

Given the enormous scope and magnitude of the potential sugars and spices on this particular table, the size and diversity of the crowd that has gathered in the hopes of feasting—and avoiding being chewed up—from the decisions Congress makes are hardly surprising. Nor is the fact that seemingly unusual coalitions pushing one position or another have come together, pulled apart, and reconfigured as legislation has moved through Congress. To understand why, consider a simple game: the divide-a-dollar game, in which three people agree that they will divide a dollar by majority vote. The critical insight is that however they decide to split the dollar, a majority will have an interest in changing the distribution. Difficulties, therefore, always persist in putting and keeping together any stable coalition to support any particular legislation. Any combination of interests, any coalition, is inherently and inevitably unstable. And when, as here, the amounts at stake are so great and the subject of the legislation and its details are so arcane and complex, the ability of the public, the media, or public-interest watchdog groups to monitor Congress and curb its give-aways to constituents or contributors is quite limited.

Business interests clearly learned from their significant legislative failures in the 1970s not only on environmental legislation, but also on health and safety issues. The setbacks to business interests then were largely due to three important counterweights: labor unions, newly emergent consumer advocacy organizations, and environmentalists. In the first half of that decade, when these groups secured important victories, accountability to the public had moved front and center on the nation's political agenda. And the public's belief that ongoing economic growth was inevitable meant that members of Congress felt little need to kowtow to business interests. Large segments of the prosperous, well-educated middle class were disdainful of business interests, and the titans of the news media tended to side with the goals of public-interest entities, such as Energy Action (a group opposing decontrol of natural gas and urging the dismantling of the large oil companies) and Ralph Nader's numerous consumer organizations. The result was an outburst of congressional regulation of business operations and activities. Despite some ups and downs, environmental organizations have demonstrated staying power since then, but by the 1980s both organized labor and the consumer movement were losing sway.

The worm began to turn once stagflation set in the mid-1970s. At first, business interests were unable to gain traction, especially in the wake of post-Watergate revelations of illegal corporate contributions to Richard

Nixon's 1972 reelection campaign. Ironically, however, congressional efforts to clamp down on business-related campaign contributions and to limit campaign spending had the opposite effect. The Supreme Court declared that limits on individual contributions and candidate spending were unconstitutional, and although the Court upheld limits on corporate contributions, it also upheld a decision by the Federal Election Commission permitting corporations to form multiple political action committees (PACs). After that, using multiple PACs to funnel campaign contributions to political candidates, in particular powerful incumbents, became commonplace. (A controversial 2010 Supreme Court decision struck down as unconstitutional more recent legislative barriers to direct participation in political campaigns by unions and corporations.)

At the end of 1974, only 89 companies had created PACs, but by July 1980 that number grew to 1,204 . In the 1972 election, corporate and business trade association PACs contributed $2.7 million to candidates for Congress; in 1976, they gave $10 million—outspending labor unions for the first time. In 1980, they contributed $19.2 million. Throughout the 1970s, the business community, including large and small businesses, became far more active, better coordinated, and much more effective. What Ralph Nader labeled the "energy establishment"—the oil, natural-gas, and coal companies and their suppliers, the automobile industry and the railroads, the atomic energy industry, utilities, and major industrial fuel consumers, such as the steel and chemical industries—was becoming a "virtually irresistible political force."

The business community was simultaneously becoming much more adept at mobilizing "grasstops" support for its legislative agenda. The National Association of Manufacturers, the Business Roundtable, the U.S. Chamber of Commerce, and the National Federation of Independent Business learned to garner support for their legislative efforts by calling for help from business owners or executives who have a personal relationship with their senators or representatives. Visits by these people were—and still are—particularly effective. Formed in 1974, the Business Roundtable, an organization of chief executive officers from large companies, has been particularly good at such sit-downs: nothing a congressman likes better than a conversation with a CEO. Mark Green, director of Ralph Nader's Congress Watch, said Washington in 1978 was "a completely different city" than just a few years earlier, adding: the "Business Roundtable is the most powerful lobby in Washington. You lose bills in the districts, not in Washington," Green said, "and business is very good at what they do."

By then, the U.S. Chamber of Commerce, using computerized databases to match constituents to their Congressman, could generate 12,000 phone calls to Congress in 24 hours. These calls came mostly from owners of small businesses, but larger companies also became adept at galvanizing their shareholders and employees into political action. The Atlantic Richfield oil company, for example, in 1975 organized 53,000 shareholders and 8,000 current and retired employees into regional committees to advance the company's political agenda. When the 1978 energy legislation was being considered, utilities in California, Michigan, and Ohio also formed shareholder groups to pressure members of Congress. Neither Jimmy Carter nor the environmental organizations were nearly as effective as businesses in mobilizing local constituencies.

By the mid-1970s, business groups had also started forming ad hoc coalitions to support or oppose particular legislation, coalitions that dissolved once the fate of the legislation was resolved. Business interests also changed Washington's political landscape in the 1970s by funding new, vibrant, conservative think tanks, most notably the American Enterprise Institute and the Heritage Foundation. These think tanks and their allies routinely produce and disseminate analyses and position papers that arm members of Congress and their staffs with intellectual ammunition on behalf of probusiness positions, and they have over time come to play an increasingly important role in the legislative process.

Just as business interests were becoming more adept at influencing legislation, unions were commencing a long decline from which they have yet to recover. From 1971 through 1975, unions were influential in maintaining price controls on oil, and they strongly supported new environmental laws. Until the mid-1970s, union PAC contributions generally equaled or exceeded those from business, but not since then. Once the economy soured, unions became more concerned with rising unemployment and shifted their allegiances. The United Auto Workers, for example, in 1977 joined forces with the automobile industry against environmentalists, successfully supporting legislation to relax and postpone auto emissions requirements. Union membership has subsequently diminished substantially, especially in the private sector, and except for legislation directly affecting their ability to organize, unions have generally become less active and effective in the legislative process.

Fast-forwarding to how these now well-entrenched business interests have responded to the pressures for climate change legislation and endeavored to influence its content, it's worth examining, by way of example, the role played by Jim Rogers, the CEO of the Duke Energy Corporation, an electric utility serving 11 million consumers in North Carolina, South Carolina, Ohio, Indiana, and Kentucky. Duke Energy also happens to be the third-largest emitter of greenhouse gases among corporations in the United States. Nevertheless, Rogers is a leading advocate for federal legislation limiting greenhouse gas emissions. As early as 2001, he urged fellow energy industry CEOs to support federal cap-and-trade legislation regulating greenhouse gas emissions, and he has testified in favor of it before House and Senate committees.

Eileen Claussen, head of the Pew Center on Global Climate Change, has praised Rogers: "It's fair to say that we wouldn't be where we are in Congress if it weren't for him." "He helped put carbon legislation on the map." Not all environmentalists, however, have welcomed Rogers's enthusiasm for cap and trade; some have questioned his motives in pursuing carbon emissions limits. They view Rogers as attempting to "greenwash" his company's polluting practices by appearing environmentally concerned without altering his firm's behavior. Bruce E. Nilles of the Sierra Club commented, "Among the utility guys he's the most dangerous because he talks a good game, but his actions are among the worst." Frank O'Donnell, president of Clean Air Watch, described Rogers as "the master of double-talk." "Under his leadership," O'Donnell added, "Duke has one of the worst environmental records in the nation." "What he's really trying to do is boost the value of his stock." That, of course, is exactly what Mr. Rogers is paid to do.

Jim Rogers does not deny that self-interest is the basis for his climate change positions. He insists that cap-and-trade allowances should be allocated free of charge to corporations (a subject to which I return later) in proportion to their prior levels of greenhouse gas emissions. He claims this approach is only fair, pointing out that most of Duke Energy's coal-fired power plants were constructed in the 1970s, when U.S policy was to wean the nation from foreign oil by increasing our reliance on coal. Rogers also argues that workers and families in the Midwest and South should not be penalized for helping fulfill this new national policy. Sounding eerily like John Dingell, Rogers says forcing low-income people in the "heartland" to subsidize folks in California would be untenable. (He ignores the higher prices Californians generally pay for their electricity and their better record on electricity consumption.)

Rogers has also argued that the United States must seize its competitive advantage as the world's "leader in researching and developing nuclear technologies." He claims that nuclear plants "can be located close to growing demand centers and next to existing transmission lines." Needless to say, construction of new nuclear power plants—much less locating them near population centers—is a position not welcomed by many environmental organizations and is fraught with controversy.

By the end of 2009, Duke Energy had spent $50 million installing solar panels on customers' rooftops and pledged to purchase all of the electricity generated by the nation's largest solar energy farm, which generates 16 megawatts of electricity—a tiny fraction of the three 2,200 megawatt nuclear plants Rogers plans to construct. Duke Energy also purchased a wind power company for $320 million. Although Rogers supports installing massive transmission lines to transport wind energy from the windy Midwest to population centers, he has expressed doubts about the role of renewables, observing that they "produce power intermittently and must often be sited far from cities and the grid."

Duke Energy also wants to construct a new coal-fired power plant in southwestern North Carolina to replace two less-efficient coal plants. Despite considerable skepticism among scientists and engineers, Rogers is convinced that Duke Energy—supported by boatloads of federal dollars, of course—can ultimately meet any strictures of a cap-and-trade regime through carbon capture and sequestration. With federal financial support, Duke Energy already has invested in an experimental carbon-capture and sequestration effort at its coal plant in Edwardsport, Indiana. That plant's goal is to capture "about 18 percent of its carbon dioxide emissions within four or five years, and an additional 40 percent a few years after that." Rogers seems to expect that Duke Energy will meet any emissions caps through advances in carbon capture and sequestration and through construction of new nuclear power plants.

But Jim Rogers is just one particularly adept player in a much larger story of corporate influence on congressional production of a new energy bill. Electric power companies that now produce energy from cleaner sources are also enthusiastic about increasing the price of fossil fuels through cap and trade, a carbon tax, or other regulation. For example, Exelon Energy, one of the largest retail energy suppliers in the United States, a marketer of electricity in Illinois and Pennsylvania and of natural gas in Illinois, Michigan, and Ohio, told investment analysts in November 2009 that enactment of the House version of cap-and-trade legislation would increase its revenues by more than $1 billion annually. This number was based on a quite low estimate of $15 per ton for the value of carbon dioxide emissions permits. Exelon generates 132 million megawatt-hours of nuclear-based electricity annually, and if electricity prices were to rise by only half the costs of those carbon allowances, the company would be able to charge its customers an additional $1 billion a year. Florida Power and Light, another company with large nuclear plants, likewise stands to gain. Its chairman and CEO Lewis Hay has said, "The sooner we can establish a price on carbon dioxide, the sooner we can tackle climate change and begin the transition to the clean-energy economy of the 21st century"—and, he failed to mention, the sooner his company can raise prices for the electricity it supplies.

Exelon, Florida Power and Light, and Duke Energy are all members of the United States Climate Action Partnership (USCAP), the most influential entity advancing climate change legislation. This coalition of 25 large U.S. businesses and five leading environmental groups originated from discussions among GE's CEO Jeffrey Immelt; a small group of "like-minded" CEOs, including Jim Rogers; and leaders of the Environmental Defense Fund, the World Resources Institute, and the Pew Center on Global Climate Change. (GE is a major manufacturer of natural-gas-fired electric power turbines and wind power towers and has a major financial subsidiary; it, too, expects to gain significantly from curbs on greenhouse gas emissions.) Early in 2009, USCAP released a specific proposal detailing its desired legislative outcome, a document it called *A Blueprint for Legislative Action*. Of the signatories to USCAP's original call to action, only the National Wildlife Federation refused to sign the *Blueprint*, saying that the federation favored stronger goals and more stringent curbs.

USCAP's *Blueprint* proposes specific targets for emissions reductions: down to 58 percent of 2005 carbon dioxide levels by 2030 and to 20 percent by 2050. The *Blueprint* also describes the scope of coverage for a cap-and-trade program that USCAP prefers, specifies the nature of regulation it desires, articulates policies arguing for offsets and the banking of allowances, and offers a series of measures to expand funding for technology transformation and energy efficiency. Importantly, USCAP strenuously argues that rather than the government auctioning the emission allowances as President Obama has proposed, they should be given away for free. Newly minted Energy and Commerce Committee chairman Henry Waxman praised USCAP, saying that the legislative blueprint put forth by this coalition "became the model that we used as the basis for our starting point." Indeed, much of USCAP's *Blueprint* was embraced by the Waxman–Markey cap-and-trade legislation (and by some Senate bills).

The CBO has estimated that a government auction of allowances might produce between $30 billion and $300 billion annually between 2010 and 2019 depending on the details of cap-and-trade legislation. It estimated, for example, that an early version of cap-and-trade legislation approved by the House Committee on Energy and Commerce would create allowances worth a total of $32 billion in 2011 (or $15 per metric ton of carbon dioxide emissions), rising annually to more than $207 billion in 2019 (or $26 per metric ton of carbon dioxide emissions). In total for the decade following enactment, the CBO estimated that the federal government would raise nearly $850 billion in revenues if the allowances were auctioned. It could then return this cash to taxpayers per capita or disproportionately to low- and moderate-income families or by reducing taxes. *Because the allowances may immediately be sold for cash in the market, giving them away—or, in the CBO's phrase, "freely allocating" them—is pretty much the same as selling the allowances and transferring cash to the recipients.*

Nevertheless, if Congress enacts cap-and-trade legislation, it seems determined to give most of the allowances away largely but not exclusively to producers of coal-fired electricity. Several analysts have estimated that only 15 to 20 percent of the allowances are needed to compensate industries for their costs of complying with the caps. And the value of the permits in the early years is far greater than the costs of limiting emissions. According to the CBO, nearly $700 billion worth of allowances—or about 82 percent of the total allowances—would have been given away between 2011 and 2019 under the House bill.

Needless to say, competition in both the House and Senate to obtain the allowances was fierce. In the first round, electric utilities fared especially well; oil and gas companies did much less well. (Electric utilities got about 35 percent of the House allowances, oil and gas companies about 2 percent.) Although coal interests captured many allowances, they still have complained that too few allowances were passed out based on a business's historic emissions. Their reasoning: the worse you have been in the past, the more free allowances you should be given now. Under the House bill, supposedly by about 2030—if the law doesn't change before then—a majority of allowances will then be auctioned by the government.

President Obama made it clear in his first budget that he wanted the emissions allowances to be auctioned (with most of the money returned to low- and middle-income workers in the form of tax credits). His budget estimated that auctioning the allowances would generate $646 billion in the coming decade, and he proposed that $120 billion of that should be used to stimulate clean-energy technology with the remaining $526 billion transferred to low- and moderate-income workers. At the time, this estimate was widely criticized as too low, given the prices the allowances would likely bring at auction. (In December 2009, Senators Maria Cantwell [D–Wash.) and Susan Collins [R–Me.] introduced what they call a "cap-and-dividend" bill, which would auction its allowances, return 75 percent of the proceeds to U.S. citizens and legal residents per capita, and use the remaining 25 percent for clean-energy R&D. I discuss this bill more fully later.)

In March 2009, Peter Orszag, Obama's budget director, told a House committee, "If you didn't auction the permits, it would represent the largest corporate welfare program that has ever been enacted in the United States." But after the House committee approved legislation giving away more than 80 percent of the allowances, President Obama called the bill "a historic leap." It certainly was an historic giveaway. By February 2010, when he released his second budget, the terms *cap and trade* and *emission allowances* had disappeared. President Obama seemed to have conceded that coal and other businesses were going to capture most of the value of any permits, at least for the decades ahead.

The House bill attempted to place restrictions on some of its gifts—for example, by requiring certain electric companies to pass them along to consumers in lower prices—but enforcing those restrictions would be a nightmare. Regional variations are certain to emerge depending on how different states regulate electricity prices. And if electricity prices fail to rise to reflect the costs of emitting carbon dioxide, the goal of increasing the costs of such emissions will be undermined. If, instead, recipients of free allowances simply charge whatever prices the market will bear for their products, their shareholders will be the beneficiaries of the gifts of allowances—except, of course, to the extent that the recipients had to pay for obtaining the allowances through campaign contributions and lobbying expenses.

If all allowances were auctioned, a distribution of the cash proceeds by Congress would be clear and very visible to the public. But when permits with identical cash value are "freely allocated," the transfer is complex, mysterious, and opaque to the public. As a retired MIT economics professor told the Senate, the consequences are "very well hidden." Nothing Congress likes better.

In June 2009, Greenpeace announced its opposition to the House version of cap-and-trade legislation, claiming that the "giveaways and preferences in the [legislation] will actually spur a new generation of nuclear and coal-fired power plants to the detriment of real energy solutions." On the same day, Friends of the Earth said that the legislation "will lock us into a system that rewards polluters with massive giveaways and can be gamed by Wall Street." A group calling itself the Community Coalition for Environmental Justice said late in 2009 that this bill "is worse than no climate change bill at all." The Climate Justice Leadership Forum, representing 28 organizations across the country, said that cap-and-trade legislation is "designed to benefit corporate interests—not communities or the climate." These groups called for a carbon tax instead. Most other environmental organizations, however, including the Natural Resources Defense Council and the Sierra Club, continued to support the cap-and-trade legislation. If you thought that handouts of free allowances would induce the most visible environmental advocates to shift their support from cap and trade to a carbon tax, you would have been wrong. They view this giveaway as a necessary price to get Congress to enact a cap on greenhouse gas emissions, emphasizing that whether the allowances are auctioned or given away is "*only* a distributional issue" that does not affect the effectiveness of the cap going forward.

Environmental groups have also supported the use of offsets—for example, the ability of greenhouse gas emitters to pay others not to cut down forests in lieu of reducing their own emissions. The Waxman–Markey legislation allows these offsets to be measured on a subnational basis rather than on a national basis, thereby increasing the risk of "leakage," such as occurs when a timber company is paid not to cut down one area of a forest and simply moves its logging operations to another rain forest not far away. If the targets were measured nationally, the timber company would have to move out of Brazil altogether, for example, increasing the likelihood that trees would actually be saved. In a broadside attack on the role that many environmental groups have played in supporting the cap-and-trade legislation, Johann Hari, writing in *The Nation*, harshly criticizes the alliance of these groups with business interests in supporting subnational offsets. "You would expect the major conservation groups to be railing against this absurd system and demanding an alternative based on real science," he says. After reporting environmental groups' response that "in 'political reality,' this is the only way to raise the cash for the rainforests," Hari concludes: "But this is a strange kind of compromise—since it doesn't actually work." He also claims that the Nature Conservancy and Conservation International in particular have their own financial stakes in the ability to administer subnational rather than national requirements for offsets. Hari quotes James Hansen as saying, "I find the behavior of most environmental NGOs [nongovernmental organizations] to be shocking."

If these enviro–business coalitions are a shock or even just a surprise to you, they should not be. We have seen such coalitions before: recall, for example, the alliance between eastern coal and environmental interests in the 1977 amendments to the Clean Air Act, discussed in chapter 5. What has been genuinely surprising and in great contrast to the experience of the 1970s is the president's relative passivity in allowing Congress to call the shots in this legislation. The White House announced its broad goals for reductions in carbon dioxide emissions and then essentially stepped aside. The failure of Jimmy Carter's detailed energy plan developed in secret by James Schlesinger in 1977 and the subsequent failure of Hillary Clinton's health care proposals developed in a similar manner in the 1990s may have made the Obama administration a bit skittish about trying to present and enact a detailed plan. Even so, however, Obama's White House might have exercised more leadership concerning this energy legislation.

Given the huge stakes involved in the distribution of allowances and other aspects of the energy law, it came as no surprise when the USCAP coalition began to splinter as the legislation made its way through

Congress. In February 2010, the oil and gas giants BP (formerly British Petroleum) and ConocoPhillips as well as the large equipment manufacturer Caterpillar announced that they were leaving USCAP. Each company was worried about its specific interests. These oil and gas companies clearly were angry that the coal industry had done much better in the legislation than they did. Red Cavaney, a Conoco vice president, was candid. Although insisting that enacting climate change legislation is "really important," he said, "It's about what's in the bill, not just about getting a bill." "We need to spend time addressing the issues that impact our shareholders and consumers." To be sure, the devils reside in the details.

USCAP, of course, is not the only important business coalition in this fray. Another large business coalition, for example, that has been active in the climate change legislative battle attempting to shape the details of cap-and-trade legislation is Business for Innovative Climate and Energy Policy—BICEP, for short—a lobbying group founded by Levi Strauss, Nike, Starbucks, Sun Microsystems, and Timberland and consisting of more than 150 businesses. Financial institutions and institutional investors also have lobbied for climate change legislation. The Investor Network on Climate Risk, a group of 52 institutional investors, asset managers, and state treasurers, representing $2.3 trillion in managed funds, wrote senators requesting that they produce a clear federal policy on global warming so that businesses would have some certainty when they are making capital investment decisions. This group includes Deutsche Asset Management, AIG Investments, investment officials from 13 states, and investment banks such as Goldman Sachs and JPMorgan Chase, who are smacking their lips at the money that might be made from a new trillion-dollar market in tradable carbon emissions allowances and related derivatives. The key issues for them are how tightly the market will be regulated and by whom. (Many energy-sector businesses, in contrast, regard a ceiling and a floor on allowance prices as crucial protection against unlimited speculation and volatility.) Financial institutions' enthusiasm for cap and trade produced a modest political backlash; Senator Bryon Dorgan (D–N.D.), for example, proposed limiting or blocking the access of financial companies to the carbon allowance–trading market.

Farmers also have protected their self-interests, with nearly 80 businesses and agricultural interest groups lobbying. Farmers would receive billions in exemptions and subsidies from the energy bills. They have also won an industry-specific concession that would help them take better advantage of emission offsets. In order to secure enough votes for passage in the House of Representatives, farm-state members insisted that the validity of offset credits for U.S. agriculture or forestry projects be evaluated by the Department of Agriculture instead of by the EPA, which would oversee the rest of the offset program. Because the Agriculture Department exists principally to funnel money to farmers, a hefty dose of skepticism about the likely permanence and credibility of projects yielding those offset credits seems warranted.

The mega-agricultural businesses who refine ethanol from corn and their interest group Growth Energy, of course, have also been active, and as they have under other legislation since the 1970s, the corn folks would fare quite well under climate change legislation. Given all the publicity about ethanol's abysmal record, including how much fossil fuel energy is consumed in its production and its impact in raising corn prices, some biofuel promoters are searching for other sources. Nearly 20 companies and organizations hoping to become wealthier from producing, refining, or promoting biofuels from wood or algae also have had their hands out—and they, too, would receive. These groups include technology firms seeking to make biofuels from new sources: companies such as Algenol Biofuels and PetroAlgae, promoting algae as the energy source of the future.

Mandates and incentives for producing ethanol from cellulose, like those contained in the climate change bills, however, may have effects similar to our ethanol experience: energy inefficiency in the production process and increases in the price of wood, thereby increases in construction costs. These incentives might even stimulate companies around the world to cut down trees, which will release carbon into the atmosphere. Some lessons Congress refuses to learn.

To be sure, in addition to the various business groups who see their best interests served by the enactment of a bill favorable to their positions, other members of the business community have taken the more traditional approach to environmental legislation: skepticism or outright opposition. The Business Roundtable, for example, claims to be open-minded but has been lukewarm, at best, toward either a carbon tax or a cap-and-trade regime, although it certainly has no objections to subsidies. In its key publication on climate change, the Roundtable stated its support for "an open and constructive dialogue about the principles that should shape climate policy and the pros and cons of various strategies." It said that a price for carbon emissions starting in 2012 is likely "to result in significantly lower U.S. economic growth in coming decades," adding that if Congress were instead to adopt the Roundtable's recommendations for subsidizing technology development and deployment, the nation would achieve nearly twice the reduction of greenhouse gases at roughly half the economic costs. The Roundtable also emphasized the importance of securing participation by our major trading partners in order to mitigate the problem of "emissions leakage."

The National Association of Manufacturers has vigorously opposed any climate change legislation. Jointly with the American Council for Capital Formation, it projected a 1.8 to 2.4 percent reduction in GDP by 2030 and a 5.3 to 6.5 percent decline in industrial production by 2030 if carbon allowance prices reach $47.50 to $61.24 per ton. (The CBO estimated that under the House bill the carbon emissions allowance price in 2019 would be $26 per ton.) The association also projected substantial declines in household income from climate change legislation.

The U.S. Chamber of Commerce purports to support climate change legislation, but it became the shrillest business community voice opposing the cap-and-trade bills. The chamber insists that cap-and-trade legislation will provoke a trade war with India and China. It also promised to sue the EPA if that agency goes forward with regulation of greenhouse gas emissions. Chamber vice president William Kovacs announced that the chamber would push for a modern-day "Scopes Monkey Trial" on whether climate change actually endangers human health, but chamber president Tom Donohue later expressed regret for those comments (even though the chamber is still contesting the EPA's findings in court). In the fall of 2009, the chamber began to fracture. Four major companies (Exelon, PG&E, PNM Resources, and Apple) resigned their memberships due to disagreements over the chamber's climate change stance. A fifth company (Nike) left its position on the chamber's board of directors. Some members suggested that Donohue was driving the chamber's legislative agenda to serve his own financial self-interest—as if no one else in this saga was behaving that way.

The Chamber of Commerce, of course, is only one of many entities spending prodigious sums of money lobbying Congress in an effort to pass, stop, or shape the details of legislation to limit greenhouse gas emissions. Corporations, environmental and other nonprofit organizations, agricultural interests, unions, and some individuals have opened their wallets. According to the Center for Public Integrity, in 2009 at least 1,150 companies and organizations were actively lobbying Congress just prior to the House passage of the Waxman–Markey cap-and-trade legislation. Energy and manufacturing companies and their advocacy groups accounted for about 200 of the total, power companies and utilities another 130. We will never know how much money they have spent.

The Lobbying Disclosure Act requires only limited disclosure of spending. It does not, for example, require lobbyists to disclose amounts spent organizing and executing grassroots campaigns. The law requires greater disclosure of lobbying on tax returns, but the IRS is prohibited from publicizing this information, so these amounts are rarely made public. The American Coalition for Clean Coal Electricity, in a notable exception, said it reported to the IRS nearly $10 million on lobbying in 2008; by contrast, the lobbying expenditures it reported that year under the Lobbying Disclosure Act were less than $1 million. Less than one-third of the organizations lobbying on climate change publicly report these kinds of numbers.

Although environmental organizations have spent unprecedented sums pushing climate change legislation—by November 2009, one group of organizations had spent a total of $39 million—their expenditures have been dwarfed by business outlays. This is one reason environmentalists have been so anxious to garner business support and have joined with key businesses in coalitions, even when some have held their noses at the costs such support entails.

Although unions have lobbied in favor of climate change legislation and generally prodded their members to support it, they have been concerned primarily with its international trade aspects. The unions, as always, want to protect their members' jobs and benefits. To make sure that climate change legislation does not advantage imports, unions have urged giving the president authority to impose tariffs on goods from countries that fail to adopt their own versions of cap and trade or to tax carbon. A number of unions, such as the Communications Workers of America, the Service Employees International Union, and the Laborer's International Union of North America, joined with certain environmental groups, including the Sierra Club and the Natural Resources Defense Council, in what they called the "BlueGreen Alliance" to make sure that such trade protections are included in any legislation. They have been. And the unions were instrumental in helping the Waxman–Markey bill pass the House by the close vote of 219 to 212.

All the while, efforts on all sides to create the appearance of broad-based public support for policies they favor have become ever more sophisticated, in some instances even crossing the boundaries of propriety. All groups try to mobilize their members, manipulate the media, generate letter-writing and email campaigns, advertise on behalf of their positions, and produce public rallies. The American Petroleum Institute, for example, urged—in a letter released publicly by Greenpeace—its member companies to send employees, vendors, and contractors to opposition rallies it funded around the country. Such efforts, often called "astroturfing" (because they represent false "grassroots" campaigns), sometimes mislead people about the nature of the soliciting organization's financial sponsors and falsely convey to members of Congress the view that local sentiment about a pending piece of legislation is intense and spontaneous.

Astroturfing has played a prominent role in the climate change debate, with schemes ranging from humorous to outrageously deceptive. For example, the Federation for American Coal, Energy, and Security (FACES, for short) purported to be a grassroots campaign composed of "people from all walks of life who are joining forces to educate lawmakers and the general public about the importance of coal and coal mining to our local and national economies and to our nation's energy security." Shortly after the federation launched its Web site, bloggers noticed that the photographs of ordinary Americans populating the site were purchased from a stock photograph service (iStockphotos) and that the Web site was hosted by a Washington lobbying outfit.

One particularly egregious astroturfing incident involved a forged letter scheme originating from Bonner & Associates, a lobbying firm retained by the American Coalition for Clean Coal Electricity. Leading up to the House vote on the Waxman–Markey bill, a Bonner & Associates employee forged 13 letters to three different members of Congress, purporting to represent the views of minority, elderly, and community organizations located in the members' districts. "[I]n one case, the fired employee actually made up not only a fictitious individual but also a fictitious local chapter of a legitimate national organization," Jack Bonner admitted. "Subsequently, we learned that in another instance he forged the name of a legitimate officer in an organization in a letter sent to Congress." When Congressman Edward Markey (D–Mass.), chairman of the Select Committee on Energy Independence and Global Warming, convened a hearing to investigate the forgeries, Bonner focused his testimony on the isolated actions of this "former temporary employee." In this case, astroturfing had run amok. The American Coalition for Clean Coal Electricity said it was contemplating legal action, and the Sierra Club urged the Justice Department to bring criminal charges against Bonner & Associates for wire fraud.

Nobody explained why the affected members of Congress were not contacted prior to the final House vote on the climate change bill, despite the fact that Bonner & Associates detected the fraud three days before the vote. Chairman Markey was not happy. "The Coal Coalition was willing to pay millions to peddle a point of view, but they were unwilling to spend a few cents to call the U.S. Capitol and clear the air," he said. Two of the three representatives receiving the forged letters voted against the bill, but at least one said that the letters "did not have undue influence" on her decision.

In any event, as the legislation moved through the House, the supporting coalitions had made the essential elements of any legislation clear: (1) a slow phase in of any emissions limits; (2) free distribution of cap-and-trade allowances to coal and other fossil fuel interests and electric utilities; (3) omission of the automobile industry from the bill (presumably to be handled separately through CAFE or a tax or fee on petroleum); (4) exemption of the farm sector from emissions limits, along with opportunities for potentially large amounts of new revenues for farmers from selling offsets, plus generous subsidies for biofuels; (5) protection of domestic manufacturers and their employees through potential trade adjustments; (6) some efforts to limit speculative volatility in the market for allowances; (7) protections against large price increases for consumers and specific rebates for low-income families; (8) mandates and large subsidies for renewable fuels and for energy efficiency; and, of course, (9) billions in technology subsidies, especially for carbon capture and storage, all purportedly to stimulate "green" jobs.

With the president, the congressional leadership, important committee chairs in the House and Senate, key business and agricultural interests, organized labor, and the environmental movement pushing for cap-and-trade legislation, along with polls showing that a majority of the public is either indifferent or in support, all of the requisite pieces seemed to have come into place for cap-and-trade legislation to succeed in Congress. But by the end of 2009, it had become clear that there were not 60 votes in the Senate for the broad Waxman–Markey cap-and-trade legislation, so a number of senators moved to reshape the legislation.

Although Senator John Kerry (D–Mass.) had cosponsored with Barbara Boxer (D–Calif.) a similar cap-and-trade bill earlier that year, in December 2009, just before the international climate change meetings in Copenhagen, he, along with Joe Lieberman (I–Conn.) and Lindsay Graham (R–S.C.) released a "framework" for a revised energy bill. They seemed to believe that by strengthening trade protections for U.S. manufacturers, providing benefits and easing burdens for nuclear power, and encouraging domestic drilling for oil and natural gas, they might pick up enough Senate votes for enactment. During the months that followed, while the details of this legislation were being hammered out, Senator Graham dropped out of the group (offering a variety of reasons, including disputes with the White House and Democratic Senate majority leader Harry Reid [D–Nev.] over immigration.) In May 2010, Senators Kerry and Lieberman finally released a bill, called the American Power Act, designed to reduce greenhouse gas emissions to levels quite close to those anticipated under the House bill, but with important differences.

Like the House bill, the statutory changes that Senators Kerry and Lieberman proposed add up to more than a thousand pages, so I don't recite the details here. One major difference with the House bill is that the Kerry–Lieberman bill divides its carbon-pricing mechanism into three parts: (1) using a cap-and-trade system for utilities (with permits provided free for the first four years and auctioned after that); (2) subjecting large manufacturers to a cap-and-trade system beginning in the 2016; and (3) creating a new system requiring fossil fuel producers and importers to purchase allowances at a fixed price (which looks much like a carbon tax dressed up in allowance garb). The senators also include provisions to limit volatility in allowance prices, to protect domestic manufacturers, and to encourage nuclear power and domestic oil and natural-gas production. Subsidies are provided, for example, to "clean coal" technologies, renewable energy, and new vehicle technologies. Farmers, of course, would be treated very well through both exemptions and offset provisions. Unlike the House bill, the Kerry–Lieberman bill would preclude the states from enacting any cap-and-trade legislation of their own.

Meanwhile, in December 2009 Senators Maria Cantwell (D–Wash.) and Susan Collins (R–Me.) introduced a third alternative, known as the CLEAR Act (CLEAR stands for "Carbon Limits and Energy for America's Renewal"), which they called a "cap-and-dividend" approach. This legislation would create a process for selling monthly "carbon shares" to fossil fuel importers and producers. Their legislation provides no trading of those shares, contains no provisions for offsets, returns three-quarters of its proceeds per capita to American citizens and legal residents, and spends most of the remaining 25 percent on energy R&D. This bill resembles a carbon tax and rebate system, but the dreaded word *tax* appears nowhere. The senators emphasize that their legislation takes only 39 pages of statutory language, saying, "Instead of a behemoth bill deigned to conceal backroom deals and giveaways, our framework is a straight path that all Americans can follow."

Even with all of these efforts, none of these proposals was able to garner the necessary 60 votes in the Senate in 2010, so no broad climate change legislation was enacted in time for President Obama to sign before the 111th Congress expired. As the next chapter describes, however, the prospects for a narrower kind of energy legislation picked up dramatically following the massive BP oil disaster in the Gulf of Mexico. As we have seen, Congress becomes far more likely to act following a crisis. Before the BP catastrophe, the most that proponents for change could claim is that a potential disaster looms over the coming decades and that potential EPA regulation will occur if Congress fails to act. This oil blowout moved our nation's energy policy front and center again.

A political wild card, however, for climate change legislation—unprecedented in peacetime for America—is the global reach of the greenhouse gas problem and the necessity for a worldwide solution if we are actually to be effective in addressing it. The absence of global government with power to act drastically complicates matters for Congress and the president. The politics of domestic cap-and-trade legislation has borne a striking resemblance to a large group of pigs racing to feed at a public trough. Now imagine the scramble when the stakes become multiplied worldwide.

The EPA has estimated that the international offsets in the House bill alone would send $1.4 billion a year abroad from American consumers and businesses. Developing countries are anxious for those and additional transfers from richer nations. The richer countries put most emissions into the atmosphere in the past, but mainly because of the growth of the so-called BRIC countries (Brazil, Russia, India, and China) developed countries now account for less than half of total current emissions. (Chapter 10 provided a glimpse into the concerns of China and India, two nations critical to any successful international greenhouse gas effort.) Sounding for all the world just like Jim Rogers requesting compensation for Duke Energy to make the shift away from coal, Saudi Arabia says that large, oil-consuming nations in Europe, Asia, and North America should compensate it if they move away from an oil-based economy. Mexico wants countries to transfer $10 billion into its Green Fund, based on the various nations' levels of GDP and emissions. The so-called G-77, a group of developing countries, staged a walkout of the climate change talks in Copenhagen to protest the richer countries' limited willingness to fund the costs of avoiding or mitigating climate change in the poorer countries. When Europe offered a $10 billion fund, the Sudanese chairman of the G-77 described this offer as "very strange" and suggested that a trillion dollars or so would be appropriate. The World Bank has estimated that about $475 billion a year needs to be spent in developing countries on investments to reduce greenhouse gas emissions. Hands are out, not just those of American businesses, but those of countries around the world. Where to find this money is a challenge that current energy legislation has not begun to face.

Meanwhile, in the Maldives, Indian Ocean islands that rise an average of only about 7 feet above sea level, the president and his cabinet, each accompanied by a diving instructor and a military aide, held a half-hour meeting about 15 feet under water to sign a document calling on all nations to cut their carbon emissions. The Maldives' president said this stunt was to "let the world know what is happening and what will happen to the Maldives if climate change is not checked." He does not care who pays for it.

The U.S. founders designed our political system to limit action by the federal government. In the energy domain, however, their design has not stopped Congress from enacting massive new laws, nor will it in the future. However, a series of changes to the way our political process works that began in the 1970s—from the alteration of the power structures of Congress to the rise of sophisticated and targeted lobbying, accompanied by the debilitating escalation of the importance of money to political outcomes—has severely limited our ability to pass *effective* legislation. The sad fact is that most of the energy bills making their way through our tortured legislative process don't engender confidence that we will address our energy challenges more effectively today than we did through the thousands of pages of legislation enacted in the decade when energy first came to dominate our national attention.

As we have seen throughout this book, unexpected crises have the potential to prod politicians to act and sometimes to engender dramatic changes. In the spring of 2010, just such an energy crisis erupted—one that was more real, more dramatic, and seemed potentially more damaging than any we had seen before.

14 Disaster in the Gulf

On the evening of June 15, 2010, Barack Obama delivered his first Oval Office address to the nation. Surrounded by flags and pictures of his family, President Obama began, "On April 20th, an explosion ripped through [the] BP Deepwater Horizon drilling rig, about 40 miles off the coast of Louisiana. Eleven workers lost their lives. Seventeen others were injured. And soon, nearly a mile beneath the surface of the ocean, oil began spewing into the water. Because there has never been a leak this size at this depth, stopping it has tested the limits of human technology." The president promised to "fight this spill"—which he called "the worst environmental disaster American has ever faced"—with "everything we've got for as long as it takes."

In his 17-minute address, Obama touched all the usual energy policy bases. Speaking nearly two months after the BP explosion in the Gulf of Mexico, the president made it clear that he viewed BP and other oil companies as villains—BP for the catastrophic spill, oil companies generally for showering regulators with "gifts and favors" and "writing their own regulations," and, of course, "oil industry lobbyists." He insisted that BP "will pay for the impact this spill has had on the region." The president said that he had replaced the head of the agency in charge of issuing permits and regulating deepwater drilling, the Minerals Management Service. He announced a restoration plan for the gulf and its neighboring states and a national commission to study the causes of the disaster and to make recommendations. And Obama confirmed an earlier announcement that he had instituted a six-month moratorium on deepwater drilling.

No doubt the most difficult of these tasks will be restoring the gulf region. Some experts have estimated that the costs of restoration there may run as high as $100 billion. And no one knows how long it will take.

Indeed, as President Obama spoke, no one even knew for certain when BP's oil would stop pouring into the gulf's waters or just how far onto the nation's land it might reach. Two things we can be sure of are that not all of the costs will be paid by BP and that restoration will not happen quickly. This was the first major offshore drilling spill since Santa Barbara in 1969, and that was literally a drop in the bucket compared to this one, which spewed nearly 5 million barrels of oil into the gulf. As the old proverb goes, sickness comes in on horseback but goes out on foot. No wonder Obama closed his address with a prayer.

President Obama's speech was roundly criticized—and not just by Republicans. Some observers complained about all his hand gesturing and the aesthetics of the speech setting's background. Many were disappointed by his failure to call for dramatic new action. The greatest disappointment, however, was that President Obama used this catastrophe as just one more occasion to restate his oft-repeated remarks about our nation's "addiction to fossil fuels" and about how the "transition to clean energy" can "create millions of jobs." In a speech prompted by an unprecedented tragedy—a unique occurrence after decades of deepwater drilling—President Obama seized this moment to endorse the cap-and-trade bill that had passed the House of Representatives a year earlier. The president proffered his customary promise "to look at other ideas and approaches from either party." And he told us all—but especially the Senate—that the one thing he would not accept was "inaction." After lamenting our "lack of political courage and candor," however, the president failed to mention that getting rid of our "addiction to fossil fuels" would require us to pay higher prices for those fuels. And, like many presidents before him, he reminded the nation of its successes in World War II and in landing men on the moon, obliquely suggesting —to a nation frustrated from witnessing one failure after another to stop the flow of oil from the floor of the gulf and to keep it from reaching shore—that all we were missing was American grit and ingenuity. (Ironically, the oil industry had proudly proclaimed that its sophisticated technology for deepwater drilling in the gulf rivals that used for space exploration.) Everything had changed, yet nothing had changed.

Despite this oil blowout's unique size, depth, and scope as well as the unprecedented difficulties of stopping it and the heartbreaking scenes of profound economic and ecological distress it had produced in the gulf region, watching President Obama's address seemed, as Yogi Berra is fond of saying, like "déjà vu all over again." On the next night, for those too young to know our energy history or who had forgotten it, Jon Stewart of Comedy Central's *The Daily Show*, who by now just might have succeeded Walter Cronkite as America's most respected news broadcaster, played a sequence of clips from our nation's past eight presidents—from Richard Nixon to Barack Obama: all promising to end our dependence on oil, all offering other energy alternatives, and all setting deadlines for reaching their goals, beginning with Richard Nixon's 1980 timetable and ending with George W. Bush's completion date of 2025.

President Obama attempted to show the nation that he was in control of an event over which he had precious little control. Since the blowout started, television stations had returned to a technique they had perfected during Jimmy Carter's Iranian hostage crisis: every day a crawl at the bottom of the screen counted off the days since the *Deepwater Horizon* explosion. Underwater cameras showed the oil gushing into the gulf. The day Obama spoke was "Disaster in the Gulf—Day 57."

The flow of oil from the bottom of the gulf floor would not be stopped until August 4, 2010, and the well was not finally cemented shut until September 19, five months after the explosion. Early in September BP issued a report blaming the disaster on its own engineers; Transocean, the owner of the Deepwater Horizon drilling rig that had exploded; and Halliburton, the company responsible for cementing the well. Responding to the report, Congressman Markey said, "BP is happy to slice up blame as long as they get the smallest piece." As time passed, it appeared that—despite the oil spill's size—its environmental damage was considerably less than had been feared, although it will be years before the full extent of its economic and environmental destruction will be assessed.

In any event, the day after President Obama's address, at the White House's urging, BP agreed to place $20 billion into an escrow fund to begin compensating some of the region's victims. The president put Kenneth R. Feinberg, who had administered the fund for compensating victims and the families of the September 11 attack, in charge of paying claims. The scope of the damage, however, remained uncertain. Business losses were just one terrible consequence of this explosion.

On June 22, Judge Martin L. C. Feldman of the federal district court in New Orleans enjoined the White House from enforcing its six-month moratorium on drilling wells at a depth greater than 500 feet, handing at least a temporary victory to oil services providers, who had claimed that the moratorium would cause them irreparable economic damage. The judge held that the president's moratorium was arbitrary and that its adverse impact on the region's economy was not justified. "Are all airplanes a danger because one was? Are all tankers like *Exxon Valdez*? All trains? All mines?" the judge asked. "That sort of thinking seems heavy-handed and rather overbearing," he answered his own rhetorical questions. Edward Markey (D–Mass.) said, "This is another bad decision in a disaster riddled with bad decisions by the oil industry." The Obama administration appealed Judge Feldman's injunction and soon thereafter modified but did not reverse its deepwater drilling ban. Judge Feldman and his fellow jurists left no doubt that the courts will continue to play a central role in arbitrating energy and environmental disputes.

The gulf debacle was the largest but not the only fossil fuel disaster in the spring of 2010. Just 15 days before the *Deepwater Horizon* explosion, the coal industry suffered its worst domestic mining disaster in four decades when a methane gas explosion at Massey Energy Company's Upper Big Branch mine in Montcoal, West Virginia, killed 29 miners. Some of the parallels with the explosion in the gulf were eerie. For some time before this explosion, Massey Energy, like BP, had a much worse safety record than its competitors. And blame for this disaster too would be shared by federal regulators. The Labor Department's inspector general found that the Mine Safety and Health Administration had limited the number of mines subject to tough enforcement of safety regulations. Sounding for all the world like the complaints of Interior Secretary Kenneth Salazar about his department's Minerals Management Service, Labor Secretary Hilda Solis described her department's mine safety oversight process as "badly broken" and said that she was working on "regulatory and administrative fixes."

After the miners were known to be dead, the nation's media made relatively little of the coal mine disaster. Despite all the coal mine safety legislation since the 1970s, it was no news that coal mining remains a dangerous occupation.

As we have seen, crises often spur congressional action, but the Senate did not heed President Obama's call and enact cap and trade or any other broad form of energy legislation in 2010 to address either our "addiction to oil" or climate change. Nor was it able that year even to enact legislation limited to addressing regulatory failures, adding some new regulations on offshore drilling, lifting liability limits on oil companies for damages, or mandating and subsidizing more renewable fuels and less energy use. The difficulty in getting the 60 votes necessary to increase prices of fossil fuels through either cap and trade or a carbon tax had certainly not yet disappeared.

The fundamental problem, of course, is that notwithstanding all the new laws Congress has enacted since the oil embargo of 1973, we have still not solved the nation's energy problems. We are still waiting for an answer to the question Jimmy Carter posed more than three decades earlier from the very same office in his "crisis of confidence" speech: "Why have we not been able to get together as a nation to resolve our serious energy problem?" Despite the rise of concerns about climate change since then, this question persists, still germane, still unanswered.

Four decades after energy policy first took center stage in our nation's political discourse, the fundamental difficulties that brought energy onto the policy agenda remain unabated. The United States has 4 percent of the world's population, but we consume 25 percent of the world's oil. In 1970, we imported less than half a billion barrels of oil; by 1980, our imports had increased to nearly 2 billion. Now we import about 3.5 billion barrels annually. We import much of our oil today from our neighbors to the north and the south, but we still depend on OPEC oil to keep our vehicles moving. We import 13 million barrels of oil every day, 5 million of them from OPEC countries.

Thanks to our shift from oil to coal, we now produce the electricity we use from domestic sources, but that production pollutes our atmosphere and threatens to destabilize our planet's climate. Our per capita greenhouse gas emissions are the highest in the world. And as previous chapters have demonstrated, we should not be confident—as Barack Obama seems to be—that enacting a cap-and-trade bill like the one passed in 2009 by the House will either produce energy security for our nation or adequately address the risks of climate change.

Every crisis we have faced—from the OPEC oil embargo to Three Mile Island, from the Iranian Revolution to September 11, from the Santa Barbara oil spill to the BP gulf blowout—has mobilized public opinion and moved Congress to enact new laws. Opportunity knocks often, but our answers have always been inadequate. Politicians continue to lament our plight and promise the coming of clean, inexpensive, domestic energy—just around the next corner.

The fault line for a remarkable amount of economic data concerning the leveling off of our nation's post–World War II economic growth and the flattening of most families' real income growth is 1973. Economists treat that year as the time when our country's long postwar period of broadly distributed growth and prosperity ceased. Since then, middle-class American families have had to struggle—often with little success—to increase their take-home pay faster than inflation. Not coincidentally, 1973 was the year of our first energy crisis, the time when oil prices initially spiked and Americans began routinely to transfer enormous sums of dollars from their pockets into the treasuries of Middle East and other oil-exporting nations—a transfer of wealth that continues today and threatens to persist into the indefinite future.

Following the 1973 Arab oil embargo, the American people were gloomy. The Vietnam disaster and the Watergate debacle contributed to the nation's unease, as did what many perceived as the unraveling of our social solidity, but it would be a mistake to underestimate the effects of the end of the era of cheap energy, the termination of our unconstrained access to cheap domestic oil. When the 1970s ended, the pessimistic book *Zero-Sum Society* by MIT economics professor Lester Thurow became a national best-seller. And in his 1981 Pulitzer Prize winning *Rabbit Is Rich*, the novelist John Updike captured our nation's mood. "Running out of gas," he said, "people out there are getting frantic, they know the great American ride is ending." If we look back to the 1970s, our failure to break free of our bondage to OPEC oil is both surprising and disheartening.

Our nation's new but unmistakable dependence on OPEC's oil spigot didn't just diminish our economic well-being; it also prompted us to shift to an even dirtier domestic fuel for our electricity, transformed our foreign policy, and changed our national security practices. Dollars we exchanged for oil have strengthened countries that oppose us and have helped to fund radical Islamic institutions, including schools, throughout the Middle East. We have bartered our treasure, our national interests, and even the lives of many of our sons and daughters for access to barrels of oil—millions and millions of barrels. Unlike climate change today—with some contesting the importance and even the reality of the challenge—no one denies the costs of our energy dependence, or what eight presidents from Richard

Nixon to Barack Obama have called our "addiction to oil." Any energy policy scorecard for the period since 1973 undoubtedly would show many more errors than hits. It did not have to be this way.

It is child's play to find culprits for our failures. Start with presidential leadership. It let us down. During the critical period, we were led first by a corrupt president, then an accidental one, then an incompetent one. Since Jimmy Carter's failure to get reelected, no president has been willing to ask the nation to sacrifice for the cause of sound energy policy. Dislocations are vaguely hinted at, costs claimed to be trivial.

Congress, becoming ever more beholden to monied interests, has failed to place our national needs over parochial and partisan ones. As we have seen, any energy law enacted by Congress will now be festooned with giveaways and special privileges. Congress is especially bad at addressing long-term problems. Its members are obsessed with reelection; they are busy positioning themselves and obtaining funds for their next electoral campaign. As each member has sought to obtain more opportunity to dole out favors to constituents and contributors, power has dispersed into more committees and subcommittees. In the House, as a counterweight to this diffusion, power has flowed to the Speaker and to party leadership. This process has contributed to greater partisanship, another factor making it more difficult for Congress to fashion legislation benefitting the nation as a whole.

The private sector has served us little better. Auto unions united with the auto industry to fight against increases in gas taxes and requirements for dramatic improvements in fuel efficiency. The big oil companies focused more on strategies for enhancing their short-term profits than for securing our energy security. Coal companies and the UMW treated our new national dilemma as their opportunity and amassed all the political support they could muster to promote coal, never mind its safety and environmental hazards. Business interests became more sophisticated and more successful in their ability to guide legislative decisions. On the other side, environmental leaders were often single-minded, frequently focused on elite NIMBY concerns, became seduced by the small-is-beautiful movement, and embraced the end of economic growth rather than attempting to marry progrowth and environmental requirements. When they did form enviro–business coalitions, these groups frequently supported unfortunate compromises and failed to create any coherent policy basis for moving forward. Environmental organizations brought nuclear power plant construction to a halt, and, despite talking a good game favoring renewable energy, they failed for a very long time to support bringing renewables to scale.

All the while, energy and environmental policies have been buffeted by economic and political forces ostensibly having little or nothing to do with energy. The shift away from presidential power and the diffusion of authority within Congress were largely consequences of Watergate and Vietnam. The rise and spread of the antitax movement gained strength from the stagflation of the 1970s, and the movement's central role in Republican politics since then has hamstrung our ability to require that fossil fuels bear their full costs through taxes. Only tobacco policy has been able to overcome this obstacle.

A very generous accounting of 1970s energy policy might mark decontrol of oil prices and closing the cumbersome price-control bureaucracy for allocating petroleum as successes, but that process was more like shooting oneself in the foot and slowly allowing the wound to heal. On the other hand, the slow, painful decontrol of natural-gas prices and the dismantling of the archaic system for regulating that industry's prices, which had been in place since the 1930s, were a genuine achievement. We have been fortunate that changes in natural-gas pricing and deregulation, along with advances in technology, have led to substantial increases in domestic reserves and production and greatly enhanced utilities' ability to produce electricity economically—and more cleanly with half the greenhouse gas emissions of coal—from natural gas. Substituting natural gas for coal as a source of our electric power would clearly be an important step forward.

Of course, if we had been genuinely serious about decreasing petroleum consumption after 1973, we would not only have immediately shed our controls on oil prices but also have begun to phase in substantial taxes on gasoline and other petroleum products. We know that high gasoline prices change people's behavior. When a gallon of gas cost $4 a gallon in the summer of 2008, the used car market was glutted by people wanting to shed their SUVs. You could not turn on your television without seeing commercials promoting fuel-efficient cars. The fact that higher gasoline prices decrease its consumption is incontrovertible. If Jimmy Carter had been successful in getting his nickel a gallon gas tax increase enacted and if such an increase had occurred in the years following, our nation's automobile fleet would look very different today. But for a decade following the oil embargo in 1973, Congress refused to permit our oil prices even to rise to their world market levels, much less to impose taxes that would raise petroleum prices. Since then the antitax movement has turned such taxes into political poisons. Efforts in the 1990s to tax energy consumption cratered. As a result, growing oil consumption and our reliance on imports continue. Whatever one thinks about climate change, taxing petroleum fuels would have been sound policy in the 1970s, and it still would be today. Economists have estimated that phasing in a gasoline tax of just 25 cents a gallon would have saved as much oil as the CAFE fuel-efficiency standards for automobiles at one-third of CAFE's costs to the economy.

The 1975 enactment of CAFE did help to increase our energy efficiency, even though, as chapter 3 described, when enacted these standards did little more than ratify changes already under way due to market pressures, and they encouraged Japanese automakers to introduce larger, less fuel-efficient vehicles into the American market. CAFE was both less effective and more costly than it should have been. (Chapter 12 urged potential cost-effective structural improvements to CAFE.) Nevertheless, despite CAFE's shortcomings, including the light-truck loophole that proved large enough to drive millions of SUVs through and CAFE's failure to do anything to curtail the number of miles people drive, its enactment did help increase vehicle fuel efficiency. Recent enhancements in the CAFE requirements—along with threats of more to come—may help further reduce the petroleum consumption of our nation's vehicles going forward, but they will not curtail the number of miles we drive. It is an unfortunate and sad fact that CAFE, although far from the best policy we might have had, probably ranks among the best that was enacted.

Our shift from oil- to coal-fired electricity in the 1970s helped us reduce our dependence on oil imports but not without massive costs. In 1973, about 17 percent of our nation's electricity was generated from oil; now just more than one percent is. The good news is that because nearly all of the coal and natural gas used in generating our nation's electricity is from domestic sources, producing electric power is now essentially a domestic enterprise. The bad news is that coal is our dirtiest fuel. Electricity accounts for about 40 percent of U.S. carbon dioxide emissions, and, given concerns about climate change, our heavy reliance on coal-fired electricity is now just another problem to solve.

We did decrease somewhat our oil imports by creating a heavily subsidized ethanol industry that now supplies about 7 percent of the fuel our automobiles burn, but this wasteful and costly pork-barrel program can hardly be scored a success. Its costs have substantially outweighed its benefits—except to the corn farmers and ethanol refiners for whom it has been a financial bonanza. Things might have been worse. We should be grateful for the failure of Jimmy Carter's efforts to fuel our cars and trucks with synfuels from coal. If he had been successful, we would have wasted even more money to power our cars from a much dirtier fuel. More generally, as chapter 11 has detailed, our government's decisions about what, whom, and which technologies to subsidize have been excessively influenced by pork-barrel politics and have not served us well.

What we need now is for scientists and engineers rather than politicians to make the spending and subsidy decisions regarding the taxpayer dollars that are spent on energy technologies. We benefit greatly when energy technology decisions are made through a process that more closely resembles the National Institutes of Health or the National Academy of Science than a congressional committee. Ronald Reagan was correct in his insistence that bringing new technologies to scale requires not just public money, but also the commitments of private capital and private enterprise. The successful initial public offering of the high-end electric car manufacturer, Tesla, in the summer of 2010 was encouraging in this regard, but such innovations still must pass muster in the marketplace. Although government support can serve as a catalyst, developing a new energy economy depends critically on private sector-participation and investments.

As chapter 4 recounted, the 1970s experienced a boom then a bust in construction of nuclear power plants. Our most recent nuclear power plant began generating electricity in 1996, more than two decades after its construction commenced in 1973. No new nuclear plant construction has begun in this country since the Three Mile Island incident. (In 2010, however, President Obama traveled to north Georgia to promise more than $8 billion in loan guarantees for two new plants proposed to be constructed adjacent to two older plants located there.) In 2010, 104 nuclear plants generated nearly 20 percent of our nation's electricity, accounting for nearly three-quarters of the electric production from sources that emit no greenhouse gases. (Hydroelectric power produced the bulk of the remainder.) By contrast, France gets more than three-quarters of its electricity from nuclear plants.

Despite our failure to address the problem of disposing of nuclear waste, we would no doubt be much better positioned today in terms of emissions of greenhouse gases if we had built more nuclear power plants. A forward look, however, tells us that the future of nuclear power remains uncertain, despite the renewed interest around the world in increasing nuclear power capacity. Most environmentalists still object: the Sierra Club's Carl Pope, for example, said that the money President Obama allocated to nuclear power would be better spent retrofitting buildings and lowering energy consumption. A comprehensive MIT study, *The Future of Nuclear Power*, cautions us about embracing nuclear power in response to our worries about climate change. Investments in nuclear power plants remain especially risky due to their capital intensity, and in those countries where such construction is under way, including China, India, and Russia, large government subsidies have proven necessary.

Although the safety and reliability of nuclear power plants have significantly improved, the potential costs of an accident remain staggering. The greatest danger, of course, is the use of nuclear fuel as a weapon—and this prospect is indeed terrifying. We have been able to secure our own fuel, however, so this threat appears to apply more to new nuclear power facilities elsewhere than to new plants in the United States. New nuclear power plants will almost certainly have to be part of a solution to reduce carbon emissions in electricity production. The question is, How big a part? More than 30 new plants are now being planned.

Skepticism and concerns about greater reliance on nuclear power explain why so much hope these days is placed on alternative renewable sources of energy, especially wind and solar power. The oil crises of the 1970s, along with the newly prominent national attention to the environment, created a remarkable opportunity for unprecedented, long-lasting advances in solar and wind technology, development, and deployment. However, although environmental organizations talked a good game, they then offered little support for implementing such projects at a scale to provide a substantial share of our nation's electricity. Instead, they regarded solar and wind power as a means for pursuing a more profound transformation of America—one that would move our society away from large economic, commercial, and social structures to a more pristine small-is-beautiful lifestyle. The determination to change fundamentally our nation's values and culture created a backlash that diminished then virtually destroyed the potential for the kinds of breakthrough technological advantages that we might have secured through large-scale production and transmission of energy from these sources. The relatively puny investments in solar and wind energy that we did make beginning in the 1970s provided our nation with a technological foundation to build on, but one far more flimsy than might have been. And despite today's wind and solar renaissance, we import turbines and components from Asia and Europe, where manufacturing them is and may always be cheaper. Nevertheless, more wind and solar energy must come on line in the years ahead.

Fortunately, we have substantially enhanced our energy efficiency. We now use less than half the energy to produce each real dollar of economic output than we did in 1970. Greater energy efficiency has offset about three-quarters of the increase in total energy demand that would have occurred had our nation continued to use energy with the efficiency of the 1970s. Higher energy prices, improvements in technology, private investments, government mandates, and our shift from a manufacturing to a service economy have played significant roles in developing this greater efficiency. Major improvements have occurred across the board—in industry, commercial and residential buildings, appliances, electronics, and transportation—and abundant opportunities for further advances remain. A 2009 report by McKinsey & Co. estimated that roughly $500 billion of investments in enhancing energy efficiency would save $1.2 trillion in costs by 2020 and would cut our nation's projected energy use by nearly one-quarter. The investments required, however, would be four to five times our current level, and to achieve this kind of savings we would have to overcome substantial economic, political, and structural barriers. The great challenge is to reduce our energy consumption substantially without reducing our quality of life.

Our nation's struggles to fashion successful energy policy since the early 1970s help illuminate the path forward. Even though there is no magic rabbit to pull from a hat, our failures in that decade demonstrate much of what we need to do now to succeed. Phasing in a substantial fee or tax on petroleum products and on carbon would create financial incentives necessary both to decrease our consumption of energy from fossil fuels and to stimulate investments in nonfossil alternative fuels. These effects would in turn help move us toward greater energy security and reduced carbon emissions. To ensure that addressing environmental concerns by increasing energy prices does not harm the economy, the revenues from taxing fossil fuels should be used to reduce other more economically harmful taxes and to help limit our unsustainable level of federal debt and deficits, or be returned to the American people per capita or based on income. In any event, some of these funds would have to be used to protect low- and moderate-income families from the burdens of increasing energy prices. Giving the EPA the ability to adjust the tax rate over time would allow the amount of tax to change if we fall short of our energy security, carbon-curbing, or temperature-limiting goals or if climate science going forward demonstrates either the need to accelerate our energy transformation or the ability to proceed more slowly.

Some significant portion of these tax revenues should also be used to help fund crucial technological developments, but, as I have urged, new institutional arrangements and commitments are critical to assure that R&D funds are directed to a wide range of projects with the greatest potential payoffs, not those with the greatest political pull. The promise of a technology-led transformation of how we fuel our homes and our trucks and cars has never been brighter.

In principle, auctioning allowances in a cap-and-trade system can produce a similar level of revenue that might be used for similar purposes, but our political system has demonstrated in its cap-and-trade bills that this outcome is not likely in practice. Relying on a carbon tax and petroleum taxes for appropriate price signals would introduce greater certainty with less volatility and lower administrative costs and regulatory requirements than cap and trade. A carbon or petroleum tax would also be more advantageous to an oil-importing and consuming nation, such as the United States, than cap and trade and would limit the moves available to oil-exporting nations to avoid its burdens. But no matter how sensible, such energy taxes have been one of the few policy options we have failed to adopt for four decades now.

If one looked back from the end of the 1990s, it must have seemed that the energy crises of the 1970s were a historical anomaly: a unique period of turmoil and upheaval, just a troubling detour from normalcy. By a decade later, however, it had become clear that the 1980s and 1990s, when oil prices were beguilingly low and energy policy had largely disappeared from our national political discourse, were the real anomaly. The fundamental forces that produced the crises of the 1970s—our dependence on foreign oil and the refusal of our government to insist that Americans pay the true costs of their energy—have not only remained present ever since but in fact grew worse. Only by understanding our history, and the government's responses, can we hope to address our present challenges. History, of course, has not ended. But energy as we knew it for most of our history has.

Early in this new millennium, the threat from climate change has moved energy policy back to the forefront of our nation's agenda, and the *Deepwater Horizon* disaster in the Gulf of Mexico has again made vivid the costs of our reliance on petroleum to fuel our transportation. But looking back at our experience over the past four decades does not engender strong faith in our ability to achieve the necessary transformations. Much of the legislation and regulations created in this period has been costly and ineffective, in some cases even counterproductive, and the difficulty of effective action has gotten only greater with the passage of time. Moving forward, our government will most likely deploy all of the techniques from that earlier era: subsidies, mandates, regulations, caps and trades, perhaps even taxes and fees. Thousands of pages of new laws and regulations will undoubtedly come onto our books.

Climate change skeptics may slow or even halt one bill or another, but over time they will find that fighting the risks of an existential threat—now backed by a widely shared moral perspective, especially among the young—with only empirical challenges on their side is a losing venture, absent some dramatic turnabout in climate scientists' assessments. Indeed, one explanation for the failure of all our previous efforts to achieve energy security is that—despite Jimmy Carter's best efforts to convince us that we were engaged in the "moral equivalent of war"—the American people never believed that achieving energy security furthered any moral cause beyond our own self-interest. And we always remained confident in our ability through wealth or force to obtain the oil we needed, so we nibbled around the edges of sound policies. The challenge, of course, is to do much better this time.

The massive expulsion of oil into the Gulf of Mexico and onto the marshes and land adjacent to it beginning in the spring of 2010, along with the Massey Coal mine disaster, have served as a harsh reminder of the risks associated with energy production. Decades of experience demonstrate that we have failed to manage those risks—whether they are physical danger, potential environmental damage, or threats to national security. We are always surprised, shocked even, when things go wrong. We frequently overreact in our policy responses, as we did, for example, after Three Mile Island. We always then retreat to our faith in science and look to new technology for the answers to our problems, answers to our prayers. But the one certain thing about science is its uncertain future, its ongoing process of change. We have failed miserably at managing the intersection of our politics and our science.

Decades after the Arab oil embargo, we have come to view instability and even war in the Middle East as inevitable—like coal mine disasters. We take it for granted that we will go on paying their costs in lives and in treasure. When the risks seem far away in time or place, our attention wanes. We become a nation of Scarlett O'Haras: we'll think about that tomorrow.

As in the 1970s, the barriers to sensible policy today are largely political—and they are potent. People want to blame today's inadequate government responses on the bitter partisanship that now exists, but, as we have seen, our government's capacity to respond to fundamental long-term challenges has always been limited. And, unfortunately, it has diminished over the past 40 years. Now, long after our first oil crisis, the combination of threats from high and volatile oil prices that largely benefit OPEC oil producers, from oil and coal mine disasters, and from climate change offer our nation a new chance to obtain greater energy security and a cleaner, safer planet. We are currently spending hundreds of billions of dollars in an effort to transform the way we produce our electricity and free our transportation sector from dependence on imported oil. Such fundamental changes will not occur quickly, if they occur at all. The big question is where and how we will be getting our energy in midcentury, 40 years hence. The obstacles to change—as they were in the 1970s—are great, but the potential benefits are even larger. The fear, of course, is that knowing our history will not be enough to prevent us from repeating it.

Key Energy Data

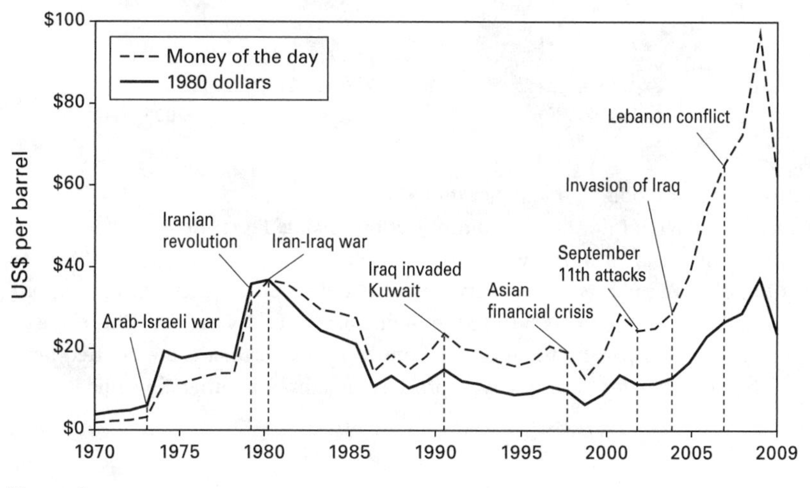

Figure 1
Crude Oil Prices
Source: BP, *BP Statistical Review of World Energy 2010* (June), available at http://www.bp.com/statisticalreview (accessed on October 8, 2010). Figures reflect annual crude oil prices, smoothing the sometimes large price fluctuations that occur within years. Crude oil prices peaked at $145.31 per barrel on July 3, 2008.

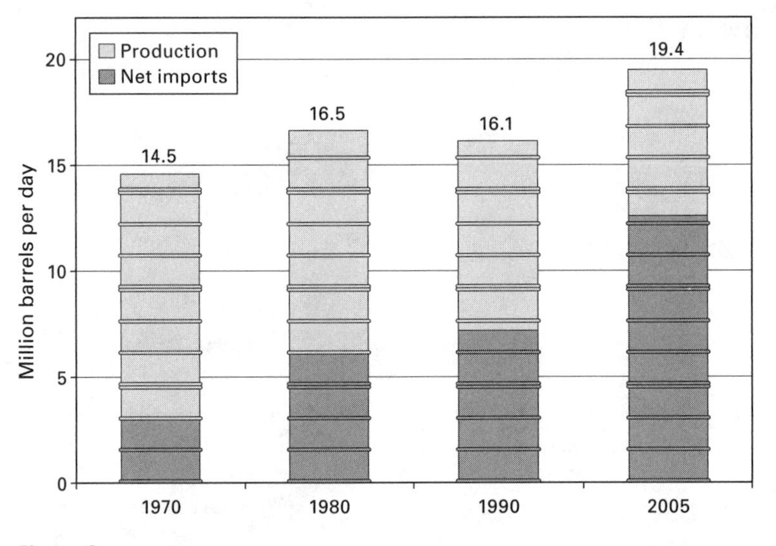

Figure 2
U.S. Petroleum Production and Net Imports
Source: U.S. Energy Information Administration, *Annual Energy Review 2009* (August 2010), available at http://www.eia.gov/aer (accessed on September 6, 2010). *Note*: Net imports are imports less exports. Total petroleum production and net imports decreased from a peak in 2005 to 16.9 million barrels per day in 2009. This decrease is likely driven in part by the economic downturn. Adjustments are omitted here. They include changes in oil reserves, processing gains, and other adjustments.

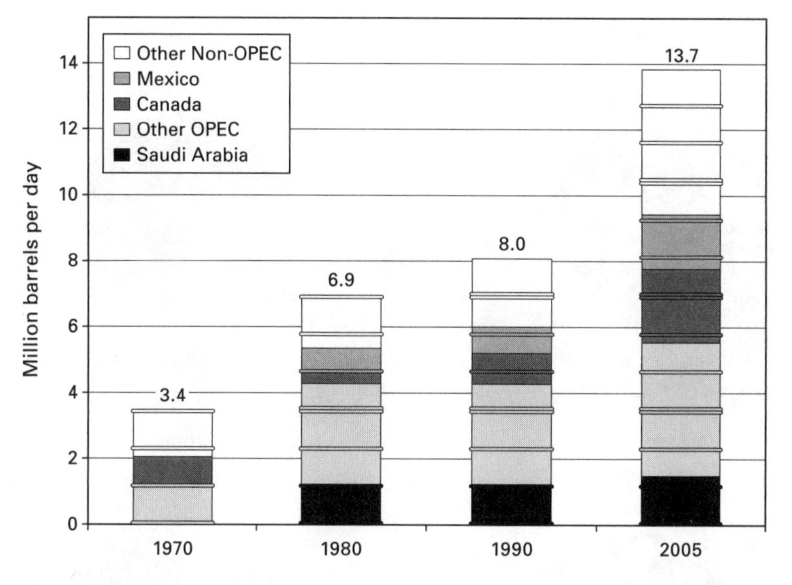

Figure 3

U.S. Petroleum Imports by Country of Origin

Source: U.S. Energy Information Administration, *Annual Energy Review 2009* (August 2010), available at http://www.eia.gov/aer (accessed on September 6, 2010). *Note*: OPEC is the Organization of the Petroleum Exporting Countries, a cartel of 12 oil-exporting countries. Non-OPEC represents countries that export oil but are not members of the OPEC cartel. Petroleum imports decreased from their peak in 2005 to 11.7 million barrels per day in 2009. This decrease is likely driven in part by the economic downturn.

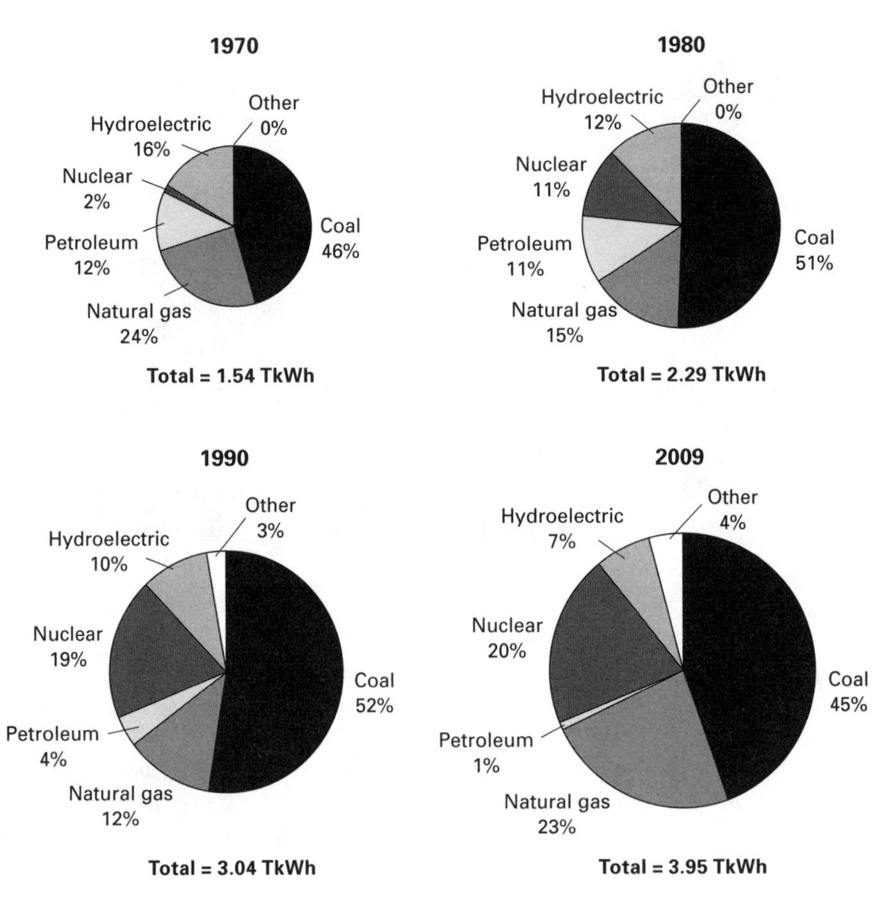

Figure 4

U.S. Net Electricity Generation by Energy Source

Source: U.S. Energy Information Administration, *Annual Energy Review 2009* (August 2010), available at http://www.eia.gov/aer (accessed on September 6, 2010). *Note*: Pie charts are scaled to approximate relative annual totals. Numbers shows are in trillion kilowatt-hours. Other includes wind, biomass, geothermal, and solar.

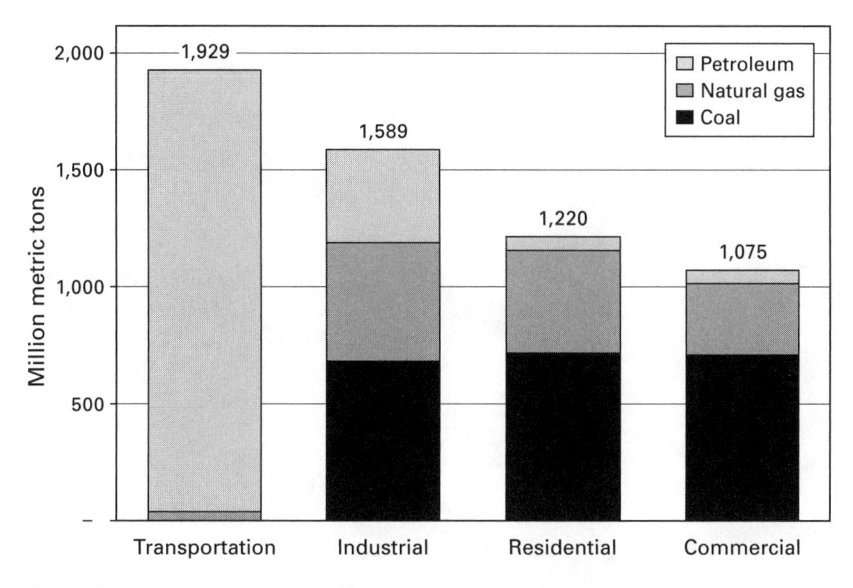

Figure 5
Carbon Dioxide Emissions by Sector, 2008
Source: U.S. Energy Information Administration, *Annual Energy Outlook 2010* (April 2010), available at http://www.eia.gov/oiaf/aeo (accessed on September 6, 2010).

Chronology

August 1953	U.S. and U.K. engineer CIA-led coup in Iran to ensure continued flow of inexpensive oil.
September 1960	OPEC is formed but remains largely ineffective for the next decade.
September 1962	Rachel Carson's *Silent Spring* is published.
1965	Environmentalists begin legal challenges to Consolidated Edison's Storm King Mountain power plant in New York.
1966	PG&E purchases land for Diablo Canyon, California, nuclear plants.
1967	The Environmental Defense Fund is founded.
January 20, 1969	Richard Nixon is inaugurated as the 37th President of the United States.
January 28, 1969	Santa Barbara oil spill begins.
June 22, 1969	The Cuyahoga River in Cleveland, Ohio, erupts in flame.
July 20, 1969	Neil Armstrong and Buzz Aldrin land on the moon.
September 1, 1969	Muammar al-Qaddafi takes over Libya in a coup and initiates a squeeze of Western oil companies.
January 1, 1970	The Natural Resources Defense Council is founded; Richard Nixon signs the National Environmental Policy Act (NEPA) into law.
April 22, 1970	The first Earth Day is held.

December 2, 1970	The Environmental Protection Agency (EPA) is founded.
December 31, 1970	The Clean Air Act of 1970 is enacted.
1971	The Sierra Club Legal Defense Fund (now Earthjustice) is founded.
March 1971	Texas oil reaches peak production.
July 4, 1971	Nixon delivers the "first comprehensive energy message" to Congress.
August 15, 1971	Nixon announces his New Economic Policy, which includes wage and price controls and a delinking of the dollar to gold, prompting a devaluation of the dollar.
December 1971	Last British troops leave the Persian Gulf.
1971–1972	Coal mining in the western United States begins to expand dramatically.
December 22, 1972	Arnold Miller is certified as president of the United Mine Workers after defeating incumbent Tony Boyle in an election.
March 1973	Last American troops leave Vietnam.
October 6, 1973	Syria and Egypt attack Israel, initiating war in the Middle East.
October 20, 1973	OPEC announces a total embargo of oil shipments to the United States and certain other supporters of Israel; in "Saturday Night Massacre," Richard Nixon fires Watergate Special Prosecutor Archibald Cox and accepts resignations of Attorney General Elliot Richardson and Deputy Attorney General William Ruckelshaus.
November 7, 1973	Richard Nixon delivers his "Project Independence" speech.
March 16, 1974	OPEC lifts its embargo, but oil prices remain high.
August 9, 1974	Richard Nixon resigns from the presidency; Gerald Ford is inaugurated as the 38th President of the United States.

January 15, 1975	Gerald Ford delivers his energy proposals at the State of the Union Address.
April 30, 1975	Saigon is captured by North Vietnam; American embassy is evacuated.
December 1975	Gerald Ford signs the Energy Policy and Conservation Act, which, among other provisions, establishes the Strategic Petroleum Reserve and enacts the Corporate Average Fuel Economy (CAFE) standards.
June 1976	The Teton Dam in Idaho collapses.
January 20, 1977	Jimmy Carter is inaugurated as the 39th President of the United States.
February 2, 1977	Jimmy Carter signs the Emergency Natural Gas Act and announces that he will seek a cabinet-level energy department.
April 1977	Jimmy Carter submits his comprehensive energy plan to Congress.
June 20, 1977	Jimmy Carter installs solar panels on White House roof.
August 3, 1977	Jimmy Carter signs the Surface Mining Control and Reclamation Act.
October 1, 1977	The Energy Department opens.
November 1977	Jimmy Carter vetoes bill authorizing continued funding for Clinch River, Tennessee, nuclear breeder reactor; reactor is closed without producing any electricity.
December 1977–March 1978	The United Mine Workers strike.
November 1978	National Energy Act of 1978 is passed, comprising the Public Utility Regulatory Policies Act, the Energy Tax Act, the National Energy Conservation Policy Act, the Power Plant and Industrial Fuel Use Act, and the Natural Gas Policy Act.
May 3, 1978	First (and only) "Sun Day" to promote solar energy is held.

June 6, 1978	California enacts Proposition 13, spurring a national anti-tax movement.
March 28, 1979	Three Mile Island nuclear plant in Pennsylvania shuts down.
April 1, 1979	Iranian Revolution ends with the establishment of an Islamic Republic in Iran.
July 15, 1979	Jimmy Carter delivers his "crisis of confidence" speech.
July 17, 1979	Jimmy Carter asks his entire cabinet to resign and accepts resignation of five, including James Schlesinger, the first Secretary of Energy.
November 4, 1979	The staff of the American embassy in Iran is taken hostage.
April 1980	The U.S. attempt to rescue the hostages in Iran ends in disaster.
September 1980	Saddam Hussein's Iraq attacks Iran.
December 27, 1980	The Soviet Union invades Afghanistan.
January 21, 1981	Ronald Reagan is inaugurated as the 40th President of the United States; the hostages in Iran are released 444 days after their capture.
January 28, 1981	Ronald Reagan announces he will lift all federal controls on oil and gasoline prices.
September 1981	Price controls on oil are lifted.
September 1981	Hundreds of protestors are arrested at a Diablo Canyon nuclear demonstration.
October 15, 1981	Egyptian President Anwar Sadat is assassinated.
May 1982	Ronald Reagan introduces legislation (never passed by Congress) to dissolve the Department of Energy.
November 12, 1984	Diablo Canyon nuclear plants come online.
January 1985	Natural gas prices are decontrolled.

1986	Federal tax credits for renewable energy are allowed to expire; Ronald Reagan removes solar panels from White House roof.
August 20, 1988	Formal cease-fire in Iran-Iraq war.
November 1988	The United Nations establishes the Intergovernmental Panel on Climate Change, a group of 2,500 scientists from more than 130 countries to study and report on climate change.
January 20, 1989	George H. W. Bush is inaugurated as the 41st President of the United States.
November 1989	The Berlin Wall falls.
August 1990–February 1991	The Persian Gulf War sparks a temporary spike in oil prices.
November 1990	George H. W. Bush announces the end of Cold War as the Soviet Union collapses.
1992	AM General begins marketing a civilian version of the M998 High Mobility Multipurpose Wheeled Vehicle (Hum-Vee) as the Hummer. In 1998 General Motors purchased the brand name and began selling three versions of Hummers. On May 24, 2010, manufacture of its last Hummer was completed.
October 24, 1992	The Energy Policy Act of 1992 is enacted.
January 21, 1993	Bill Clinton is inaugurated as the 42nd President of the United States.
November 8, 1994	Republicans capture a majority of the House of Representatives for the first time since 1954.
May 27, 1996	Bar Watts Nuclear Generating Station, the most recent nuclear power plant to open its doors, comes online after more than two decades of construction.
December 11, 1997	The Kyoto Protocol to the UN Framework Convention on Climate Change is adopted.
January 20, 2001	George W. Bush is inaugurated as the 43rd President of the United States.

March 2001	George W. Bush withdraws the Kyoto Protocol from congressional consideration.
September 11, 2001	Terrorists attack the World Trade Center and Pentagon.
March 19, 2003	The United States attacks Iraq.
August 2005	Hurricane Katrina hits the Gulf of Mexico.
May 24, 2006	Al Gore's documentary, *An Inconvenient Truth,* is released.
February 25, 2007	*An Inconvenient Truth* wins the Academy Award for Best Documentary Feature.
April 2, 2007	The Supreme Court decides *Massachusetts v. EPA,* holding that the Clean Air Act requires regulation of greenhouse gas emissions.
December 10, 2007	Al Gore and the Intergovernmental Panel on Climate Change are awarded the Nobel Peace Prize.
December 19, 2007	George W. Bush signs the Energy Independence and Security Act.
July 2008	Oil prices peak at $148 a barrel; gasoline prices in the U.S. exceed $4.00 a gallon.
July 8, 2008	T. Boone Pickens announces his "Pickens Plan" to combat climate change.
August 4, 2008	Presidential candidate Barack Obama lays out energy policy proposals at a speech in Lansing, Michigan.
January 20, 2009	Barack Obama is inaugurated as the 44th President of the United States.
February 2009	Stimulus bill passes, including $60B in "green energy" spending.
April 22, 2009	Barack Obama delivers energy address on fortieth Earth Day.
June 26, 2009	The House of Representatives passes the Waxman-Markey cap and trade bill.
December 10, 2009	Barack Obama accepts the Nobel Peace Prize.

December 2009	The U.N. Framework Convention on Climate Change holds summit in Copenhagen and produces the "Copenhagen Accord."
April 20, 2010	BP's Deepwater Horizon oil drilling rig in Gulf of Mexico explodes and oil begins gushing into the gulf.
August 4, 2010	The BP oil leak is plugged, and it is finally sealed on September 19, 2010.
October 5, 2010	Energy Secretary Steven Chu announces that Barack Obama will have solar panels installed on the White House roof.
November 2, 2010	Republicans gain six Senate seats and more than 60 in the House of Representatives in the midterm elections, once again taking majority control of the House.

Bibliographic Essay*

Prologue

The description of oil's "journey" from well to automobile was based on conversations with oil company personnel and a number of documents describing and analyzing different stages of the oil supply chain.

In general, see Lisa Margonelli, *Oil on the Brain: Petroleum's Long, Strange Trip to Your Tank* (Louisville, KY: Broadway Press, 2008), which provides a nice overview of the whole supply chain; Martin Middlewood, "How Oil Industry Supply Chains Drive Gas Pump Prices," *CRM Buyer*, July 5, 2004, http://www.linuxinsider.com/story/34919.html, which discusses the business objectives influencing how oil companies structure their supply chains; Paul Y. Cheng, Mansi Singhal, Saurabh Bharat, Cheng Khoo, and Lucy Haskins, *Global Oil Choke Points: How Vulnerable Is the Global Oil Market?* (New York: Lehman Brothers, 2008), http://www.deepgreencrystals.com/images/GlobalOilChokePoints.pdf, which provides an overview of potential supply disruptions at each stage of the journey; Dara O'Rourke and Sarah Connolly, "Just Oil? The Distribution of Environmental and Social Impacts of Oil Production and Consumption," *Annual Review of Environment and Resources* 28 (2003): 587–617, discussing the environmental, social, and health impacts of the oil extraction, transportation, and refining process as well as the regulations relating to each; and Alejandro Granado, Chairman, President, and CEO, CITGO Petroleum Co., "Speech at the Global Refining Strategies Summit 2007: A New Era for Refining," September 10–11, 2007, http://www.citgo.com/WebOther/speeches/PresentationGranado070910GlobalRefiningSummit.pdf, providing an overview of the economics and market structure of global oil refining.

For the sources of U.S. oil, see Energy Information Administration, "U.S. Imports by Country of Origin," available at http://tonto.eia.doe.gov/dnav/pet/pet_move_impcus_a2_nus_ep00_im0_mbbl_m.htm. On the export of Saudi oil from the Ghawar and Ras Tanura, in particular, see Peter Maas, *Crude World: The Violent Twilight of Oil* (New York: Knopf, 2009), 9–25. Maas is also the source of the insight that, barring spills or accidents, "no human eyes" see the oil during the entire course of its journey (15). On oil shipments from Saudi Arabia and other countries to the United States, see Alaric Nightingale, "Ship Owners Forced to Pay to Carry Middle

* Unless otherwise noted, all Web addresses cited in this bibliographic essay were last visited in August or September 2010.

East Oil," *Bloomberg,* May 8, 2009, http://www.bloomberg.com/apps/news?pid=newsarchive&sid=ahGIAd6WPfeo&refer=commodities.

There has been much global newspaper coverage revealing the dangers posed by pirates to shipping of oil. See, for example, "Somali Pirates Wreak Havoc along Key Shipping Route," *Xinhua*, November 19, 2008, http://news.xinhuanet.com/english/2008-11/19/content_10382529.htm; Jen DeGregorio, "Tanker Seized by Pirates Bound for Louisiana Offshore Oil Port," *New Orleans Times-Picayune*, November 30, 2009, http://www.nola.com/business/index.ssf/2009/11/tanker_seized_by_pirates_bound.html; Dunstan Prial, "Pirates Becoming a Big Problem for Oil Supply," *Fox Business*, November 18, 2008, http://www.foxbusiness.com/story/markets/commodities/pirates-big-problem-oil-supply. See also Mark Mazzetti and Sharon Otterman, "U.S. Captain Is Hostage of Pirates; Navy Ship Arrives," *New York Times*, April 8, 2009, http://www.nytimes.com/2009/04/09/world/africa/09pirates.html?ref=richard_phillips.

For discussions of the oil supply chain within the United States, the best source is Marathon Petroleum Company's interactive online tool, "The Time It Takes," available at http://www.mapllc.com/the_time_it_takes. For more information on the Louisiana Offshore Oil Port (LOOP), in particular, see Energy Information Administration, State Energy Profiles, Louisiana, http://tonto.eia.doe.gov/state/state_energy_profiles.cfm?sid=LA (describing the LOOP as the "only port in the United States capable of accommodating deepdraft tankers"). See also Katherine Schmidt, "Less Foreign Oil Coming through Louisiana Offshore Oil Port," *Houmatoday.com*, Dec. 20, 2009, http://www.houmatoday.com/article/20091220/ARTICLES/912209979; Jen DeGregorio, "Louisiana Offshore Oil Port Is Well Positioned Because of Dependence on Foreign Energy," *New Orleans Times-Picayune*, May 24, 2009, http://blog.nola.com/tpmoney/2009/05/louisiana_offshore_oil_port_is.html. For further discussion of the logistics and regulation of pipelines within the United States once oil is offloaded, see Cheryl J. Trench, *How Pipelines Make the Oil Market Work—Their Networks, Operation and Regulation* (Dallas: Allegro Energy Group, 2001).

Chapter 1

For an introduction to U.S. economic policy in the 1970s, a good place to start is chapters 5 and 6 of Herbert Stein's *Presidential Economics: The Making of Economic Policy from Roosevelt to Clinton* (Washington, DC: American Enterprise Institute for Public Policy Research, 1994). These chapters provide an excellent overview of the overall economic climate that shaped—and was shaped by—the New Economic Policy, the specific decision making that brought about the policy, and Ford and Carter's subsequent struggles to deal with its ramifications. As chairman of the Council of Economic Advisors under both Nixon and Ford, Stein provides the valuable retrospective perspective of an economist who was actively engaged with these issues at the time. The quotes from Stein are on pages 168 and 191. For a more detailed overview of Nixon's economic policies, see Allen J. Matusow, *Nixon's Economy: Booms, Busts, Dollars, and Votes* (Lawrence: University Press of Kansas, 1998). Of particular note are chapters 4, 5 and 6 (pp. 84–181), which discuss how and why Nixon moved away from a policy of economic "gradualism" to the strict controls of the New Economic Policy, including internal administration debates leading up to the Camp David meeting. The Shultz quote about the difficulty of lifting price controls can be found on page 151.

More color on the internal administration debates and politics at Camp David and surrounding the New Economic Policy can be found in retrospective accounts written by several Nixon administration insiders. Although one might question these accounts' objectivity, they do provide unique and valuable insight. The best is William Saffire's *Before the Fall: An Inside View of the Pre-Watergate White House* (New York: Doubleday, 1975), 497–528. Saffire's retrospective is particularly notable for its emphasis on John Connolly's unique influence over Nixon and their focus on politics rather than on economic policy in shaping the New Economic Policy. Along with Bob Haldeman in *The Haldeman Diaries: Inside the Nixon White House* (New York: G. P. Putnam's Sons, 1994), 340–350, Saffire also provides the best account of the strategy and formation behind the specific articulation of the policy in Nixon's August 15, 1971, speech, including various drafts and iterations. The text of Nixon's final, delivered speech can be found at http://www.ena.lu/speech-richard-nixon-15-august-1971-020002714.html. For contemporaneous journalistic coverage, see, for example, "Battle of the Economy: Stand Pat or Do Something," *Time*, August 16, 1971.

These unilateral decisions about international trade and monetary arrangements, which had profound effects, were interestingly taken without any input from either Nixon's secretary of state William Rogers, Nixon's former law partner in New York who played only a bit role in setting the administration's foreign policy, or his influential national security adviser Henry Kissinger, who at the time was preoccupied with negotiations to end the Vietnam War and was preparing for Nixon's forthcoming visit to China the following February. Nixon had invited neither Rogers nor Kissinger to attend the Camp David meetings the previous weekend.

For an excellent analysis of the New Economic Policy's specific impact on oil prices and energy, see David Hammes and Douglas Wills, "Black Gold: The End of Breton Woods and the Oil-Price Shocks of the 1970s," *Independent Review* 9, no. 4 (Spring 2005): 501–511 (which also discusses the international diplomacy in the aftermath of Nixon's speech and the important impact of the declining value of the dollar). See also Daniel A. Yergin and Joseph Stanislaw, *The Commanding Heights: The Battle for the World Economy* (New York: Touchstone, 2002), 64 (for complex oil price tiers and controls), and B. V. Yarbrough and R. M. Yarbrough, *The World Economy: Trade and Finance*, 3rd ed. (Orlando, FL: Dryden Press, 1994), 64 (for data on U.S. trade deficits in the 1970s). For the 118-page Cost of Living Council report that comprehensively described the four phases of rules and regulations of the oil price controls and the consequences described here, see U.S. Office of Economic Stabilization, Department of the Treasury, *Historical Working Papers on the Economic Stabilization Program, August 15 1971 to April 30, 1974* (Washington, DC: U.S. Government Printing Office, 1974), 1223–1340.

Several articles characterized Nixon's New Economic Policy as "closing the gold window." See "Closing the Gold Window," *Chicago Tribune*, June 24, 1973, A4; "U.S. Economists Applaud Nixon's Severing the Gold-Dollar Link," *Christian Science Monitor*, August 17, 1971, 3; and "Closing the Gold Window Freed U.S.," Shultz Says," *Wall Street Journal*, August 14, 1972, 3.

A broader discussion of sources on the global oil market and U.S. oil policy can be found in the bibliographic essay for chapter 2. Those sources also informed the brief discussion of the global oil market in this chapter. Specifically see Carl Solberg, *Oil Power* (New York: Mason/Charter, 1976), 223 n. 5 (for the Kissinger quote about energy surplus); John M. Blair, *The Control of Oil* (New York: Pantheon Books, 1976), 213–214 (data on Middle Eastern oil prices in the 1960s and 1970s); James Onley, "Britain's Informal Empire in the Gulf, 1820–1971," *Journal of Social Affairs* 22, no. 87 (Fall 2005): 29–43 (Britain's decision to leave the gulf); Henry Kissinger, *Years of Upheaval* (Boston: Little Brown, 1982), 854–858 (shah's visit to Washington).

Chapter 2

The discussion of the global oil market is indebted foremost to Daniel Yergin's *The Prize: The Epic Quest for Oil, Money, and Power* (New York: Free Press, 1991). Yergin's magisterial book is the definitive work and shaped the overall thinking behind this chapter in countless ways. It was also a source for a number of specific facts and discussions in the chapter, including, for example, the origins of the oil market in the early twentieth century (106–110); the oil politics of the United States and the Middle East in the 1950s and 1960s (523–567); the Iran coup and the U.S. relationship with Iran more generally (450–507); the disputes between the oil companies and OPEC in the late 1960s and early 1970s, including the details of Libya's battle with the companies (563–587); the facts of the oil crisis (614–616); and Nixon's November 1973 energy proposals (617–618).

On the history of the global market oil market, Matthew Yeoman's *Oil: A Concise Guide to the Most Important Product on Earth* (New York: New Press, 2005) provides a succinct introduction to the subject. Leonard Mosley's *Power Play: Oil in the Middle East* (New York: Random House, 1973) provides good background on the origins of the Middle East oil market and OPEC's failure to organize effectively in its early years. Carl Solberg's *Oil Power* (New York: Mason/Charter, 1976), John M. Blair's *The Control of Oil* (New York: Pantheon Books, 1976), and Robert Engler's *The Politics of Oil: A Study of Private Power and Democratic Directions* (Chicago: University of Chicago Press, 1967) give excellent background on the rise and fall of the Western oil companies, including their control over oil prices and disregard for OPEC in its early years (see especially Solberg, 191, for the quote about their disregard). James Bamberg's *British Petroleum and Global Oil (1950–1975)* (Cambridge, UK: Cambridge Univ. Press, 2000) offers an in-depth account of the structural evolution of the oil market in the 1950s, 1960s, and 1970s, and Shukri Ghanem's *OPEC: The Rise and Fall of an Exclusive Club* (London: Routledge & Kegan Paul, 1986) gives excellent background on OPEC's origins. A collection of published primary sources on the same subject (including the quoted text of the 1945 Anglo-American Agreement on Petroleum) can be found in A. L. P. Burdett, ed., *OPEC: Origins & Strategy 1947–1973* (Chippenham, UK: Archive, 2004).

For a history of the 1973 embargo itself, Benjamin Zimmer's "Empire of Omission: The Failure of American Middle East Policy 1970–1973," senior honors thesis, Harvard University, 2007, provides an excellent discussion of how and why the Nixon administration's Middle East policy failed to anticipate or prevent the embargo. Archival research that Zimmer conducted for this paper (mostly at the Nixon Presidential Materials Collection at the National Archives in College Park, Maryland) served as the source for much of the discussion and quotes related to Nixon's Middle East policy, including coordination with Saudi Arabia (quotes from pp. 32, 33 of Zimmer), the OECD Oil Committee working group report on consumer nation coordination (quote from p. 35 of Zimmer), NSSM 174 (quotes from pp. 36–39 of Zimmer), Henry Kissinger's statements at the Washington Special Action Group (p. 4 of Zimmer), the Kuwait press reaction to American support for Israel (quotes from pp. 91–92 of Zimmer), Kissinger's October 16 meeting with Arab foreign ministers (quotes from pp. 92–93 of Zimmer), and the administration's consideration and rejection of military intervention in the Middle East (quotes from pp. 94–96 of Zimmer).

For further discussion of the foreign-policy components of the oil crisis, see William Bundy, *A Tangled Web: The Making of Foreign Policy in the Nixon Presidency* (New York: Farrar, Straus and Giroux, 1998), 428–472; William B. Quandt, *Decade of*

Decisions: American Policy toward the Arab–Israeli Conflict, 1967–1976 (Berkeley and Los Angeles: University of California Press, 1977); Henry Kissinger, *Years of Upheaval* (Boston: Little, Brown, 1982), 450–544, 854–895 (see page 865 for the Kissinger quote about the 1971 Tripoli and Tehran agreements); Rachel Bronson, *Thicker Than Oil: America's Uneasy Partnership with Saudi Arabia* (New York: Oxford University Press, 2006); and Michael B. Oren, *Power, Faith, and Fantasy: America in the Middle East, 1776 to the Present* (New York: W. W. Norton, 2007), 528–538. Background on earlier American and British policy toward the Middle East is contained in Oren's book as well as in William Roger Louis, *Ends of British Imperialism: The Scramble for Empire, Suez, and Decolonization* (London: I. B. Tauris, 2006), 727–787, and Stephen Kinzer, *All the Shah's Men: An American Coup and the Roots of Middle East Terror* (Hoboken, NJ: Wiley, 2003).

On the economics of the oil crisis, one of the best accounts and the source of much of the data on oil prices and imports used in this chapter is Matthew R. Simmons, *An Energy White Paper* (Simmons & Company International, 2000). Further sources of data on the oil market include Don E. Kash and Robert W Rycroft, *U.S. Energy Policy: Crisis and Complacency* (Norman: University of Oklahoma Press, 1984), 58; Rustum Ali, *Saudi Arabia and Oil Diplomacy* (New York: Praeger, 1976), 32–39; Joseph P. Kalt, *The Economics and Politics of Oil Price Regulation: Federal Policy in the Post-Embargo Era* (Cambridge, MA: MIT Press, 1981) (provides detailed economic analysis of the impact of federal price controls and oil regulation in the decade following the embargo); and Allen J. Matusow, *Nixon's Economy: Booms, Busts, Dollars, and Votes* (Lawrence: University Press of Kansas, 1998), 241–275. Matusow's excellent chapter on oil also served as an important source for the Nixon administration's response to the oil crisis, including the internal administration chaos, the role of Watergate, and quotes from Nixon and members of his administration.

In addition to books and academic papers, newspaper and magazine articles from the late 1960s and early 1970s provided important sources of information for this chapter. For example, newspaper and magazine sources from the early 1970s gave important examples of largely ignored warnings of U.S. oil vulnerability in advance of the crisis. Among the most prominent of these warnings were James E. Akins's article "The Oil Crisis: This Time the Wolf Is Here," *Foreign Affairs* 51 (April 1973), 462, and the cover of the April 2, 1973, issue of *Time*, which read, "The Arab World: Oil, Power, Violence." For more background, see also "The Political Power of Mideast Oil," *Time*, October 19, 1970, 81; "Oil: Looking for a Fair Sheik," *Time*, February 1, 1971, 74; "Oil: Power to the Producers," *Time*, March 1, 1971, 74; Clyde H. Farnsworth, "Most World Currency Dealing Is Halted," *New York Times*, August 17, 1971, 1; "Nixon's Economics: President's Move Met by Praise and Confusion in U.S., Caution Abroad," *Wall Street Journal*, August 17, 1971, 1; Thomas E. Mullaney, "President's Package Adds Imponderables to Economic View," *New York Times*, September 5, 1971, F1; Clyde H. Farnsworth, "The Common Market Is Torn by Dollar Crisis," *New York Times*, September 12, 1971, F1; John M. Lee, "Group of 10 Agreement on Need for Realigning Currencies Is Cited," *New York Times*, September 18, 1971, 37; George W. Ball, "The Danger of a Wave of Competitive Protectionism: Import Surcharge as a Bargaining Device," *Washington Post*, September 22, 1971, A22; Rowland Evans and Robert Novak, "Nixon's Mideast Squeeze," *New York Times*, October 1, 1971, A27; Edwin L. Dale, "I.M.F. Asks Fixing of Money Rates, Cut in Trade Bars," *New York Times*, October 2, 1971, 1; Gerd Wilcke, "Export Benefit Seen by Connally," *New York Times*, October 9, 1971, 17; Hans J. Stueck, "Bonn Asks Surtax Data," *New York Times*, October 19, 1971, 61; Anthony Lewis, "In Reply to Nixon, Threats of Trade War," *New York Times*, November 7, 1971, E2; Robert Metz, "Market Place: Depressed Oils, The Bright Side," *New York Times*,

December 1, 1971, 70; "The World: Oil and Amity," *Time*, April 24, 1972, 38; "Oil: Arab Victory," *Time*, October 16, 1972, http://www.time.com/time/magazine/article/0,9171,906610,00.html; Donald L. Barlett and James B. Steele, "Oil, the Created Crisis: Oil Firms Sell Abroad, U.S. Pays," *Philadelphia Inquirer*, July 22, 1973, http://www.barlettandsteele.com/journalism/inq_oil_1.php; "Oil: Libya's 100-Percenter," *Time*, May 28, 1973, 70, http://www.time.com/time/magazine/article/0,9171,903988,00.html; "Middle East: Policeman of the Persian Gulf," *Time*, August 6, 1973, http://www.time.com/time/magazine/article/0,9171,903988,00.html; "Middle East: The Arabs' Final Weapon," *Time*, September 17, 1973, http://www.time.com/time/magazine/article/0,9171,907884,00.html; "Nationalization: Counterattack in Libya," *Time*, October 8, 1973, http://www.time.com/time/magazine/article/0,9171,908015,00.html; and Gerard Clarke, "Time Essay: What Went Wrong," *Time*, December 10, 1973, http://www.time.com/time/magazine/article/0,9171,908282,00.html.

Newspaper and magazine sources also served as an important source for the chapter's discussion of the impact of the embargo in the months following the October 1973 embargo. Articles from *Time* magazine, in particular, served as the source for a number of quotes and facts in the latter section of this chapter. See "Man from Colorado in the Hot Seat," *Time*, July 9, 1973, 52 (John Love quote); "Middle East: An Israeli Blitz v. Arab Summitry," *Time*, September 24, 1973, http://www.time.com/time/magazine/article/0,9171,907932,00.html; "Now, a Change in Wasteful Habits," *Time*, October 29, 1973, 57 (Michael Ameen and Melvin Laird quotes); "Diplomacy: Winding Up War, Working toward Peace," *Time*, November 5, 1973, http://www.time.com/time/magazine/article/0,9171,908101-3,00.html; "Energy: The Arabs' New Oil Squeeze: Dimouts, Slowdowns, Chills," *Time*, November 19, 1973, http://www.time.com/time/magazine/article/0,9171,944743,00.html; "Oil: Stepping on the Gas to Meet a Threat," *Time*, November 26, 1973, http://www.time.com/time/magazine/article/0,9171,908158,00.html; "Energy: The Emissary from Arabia," *Time*, December 17, 1973, http://www.time.com/time/magazine/article/0,9171,908323,00.html; "The Fuel Crisis Begins to Hurt," *Time*, December 17, 1973, http://www.time.com/time/magazine/article/0,9171,908321,00.html (discussion of angry motorist threatening to kill service station attendant); "Policy: Striking Back at the Chill," *Time*, December 24, 1973, http://www.time.com/time/magazine/article/0,9171,943494,00.html ("monumental blooper" quote); "Supply: From Output to Squeeze to Price Embargo," *Time*, January 7, 1974, 36; "Policy: A Global Deal on Prices?" *Time*, January 14, 1974, 15 (Kissinger "worldwide depression" quote); "Policy: The Whirlwind Confronts the Skeptics," *Time*, January 21, 1974, 22; "Policy: No Shortage of Skepticism," *Time*, January 28, 1974, 30; "Policy: Oil Profits Under Fire," *Time*, February 4, 1974, 32 (Scoop Jackson hearings); "Energy: Payoff for Terror on the Road," *Time*, February 18, 1974, 36 (trucker strike); "Shortages: Gas Fever: Happiness is a Full Tank," *Time*, February 18, 1974, http://www.time.com/time/magazine/article/0,9171,942763,00.html (Don Jacobson quote); "Rationing Spotty Local Starts," February 25, 1974, 26 (gas rationing); "Policy: After the Veto," *Time*, March 25, 1974, 29; "Attitudes: Return of the Heavy Foot," *Time*, April 15, 1974; "Oil: How Much Will Prices Drop?" *Time*, April 22, 1974, http://www.time.com/time/magazine/article/0,9171,943622,00.html (predictions of aggressive price drops in aftermath of embargo).

The *New York Times*, *Washington Post*, *Wall Street Journal*, and other city newspapers similarly trace these events. See, for example, "Jersey Schools to Cut Heat to Help Meet Oil Shortages," *New York Times*, October 22, 1973, 13; "Heating: American Families Are Seeking Alternative Ways to Keep Houses Warm," *New York Times*,

November 11, 1973, 52; "Chilling Prospect: Cities, Towns, Schools, Hospitals Fear Worst Worst As Fuel Crisis Looms," *Wall Street Journal*, November 12, 1973, 1; David Bird, "Rochdale Village Facing Cutoff of All Fuels by Thanksgiving," *New York Times*, November 14, 1973, 14; "Energy Crunch Grips Campuses," *Hartford Courant*, November 19, 1973, 32; Thomas O'Toole, "Nixon to Ban Sunday Sale of Gasoline," *Washington Post*, November 25, 1973, A1; "Nixon Orders Home Oil Rationing," *Hartford Courant*, November 26, 1973, 1A; "Heating Oil Rules Outlined," *Chicago Tribune*, November 27, 1973, 2; "Schools Will Extend Holiday by 3 Days to Save Heating Oil," *New York Times*, December 6, 1973, 100; Linda Newton Jones, "Fights Erupt for Fuel," *Washington Post*, December 25, 1973, C1; "Home Heat Cut 6 Degrees," *Chicago Tribune*, December 29, 1973, 1; Laurie Johnston, "Policeman Faces Own Crises Keeping Peace at Pumps," *New York Times*, January 3, 1974, 25; "Oregonians Fight for Gas, Hatfield Says," *Los Angeles Times*, January 11, 1974, 18; David Bird, "Federal Aides Assure Beame on Home-Heating Oil," *New York Times*, January 12, 1974, 66; Robert D. McFadden, "Lack of Gasoline 'Worst Ever' Here," *New York Times*, February 3, 1974, 1; Robert A. Wright and Martin Waldron, "Oil Is a National Problem, but Seriousness Varies," *New York Times*, February 3, 1974, 176; "Rationing of Gasoline Seen Almost Inevitable by Petroleum Institute," *Wall Street Journal*, February 5, 1974, 10; Les Gapay, "U.S. Will Risk Growing Gasoline Shortage to Avert a Crisis in Heating, Residual Oils," *Wall Street Journal*, February 7, 1974, 7; "Empty Tanks," *Wall Street Journal*, February 8, 1974, 1; Laurie Johnston, "Gas Station Throws in Towel, Is 'Serving Everybody' in Line," *New York Times*, February 13, 1974, 29; "Gasoline Aplenty: 52 Cities Have No Lines at the Pumps," *Los Angeles Times*, February 13, 1974, A15; "Gasoline and Allocation," *Washington Post*, February 18, 1974, A18; and Louise Cook, "Violence by Motorists Erupts at Gas Stations," *Washington Post*, February 21, 1974, A5.

Several sources provided important background for this chapter's discussion of the Ford administration's economic and energy policies. James Everett Katz's *Congress and National Energy Policy* (New Brunswick, NJ: Transaction, 1984) provides excellent background on the political challenges behind the failure of Ford's deregulation attempts (57–78). See also Pietro S. Nivola, *The Politics of Energy Conservation* (Washington, DC: Brookings Institution Press, 1986), 24; Yanek Mieczkowski, *Gerald Ford and the Challenges of the 1970s* (Lexington: University Press of Kentucky, 2005), 257 (Ford quote looking back on his decision not to veto the extension of oil price controls), 212 (Ikard quote about energy crisis), 221–222 (Burns and Greenspan quotes about crisis); William Simon, *A Time for Truth* (New York: McGraw-Hill, 1978), 79, 81 (for Simon on Ford's decision not to veto the 1975 energy legislation). For Alan Greenspan's subsequent commentary on Ford's policies, see Alan Greenspan, *The Age of Turbulence: Adventures in a New World* (New York: Penguin Press, 2007), 54–76. For the 1954 Supreme Court ruling on federal regulation of natural gas that set the context for Ford's attempted deregulation, see *Phillips Petroleum Co. v. Wisconsin*, 374 U.S. 672 (1954).

Chapter 3

Among the many books written on the origins of the U.S. environmental movement, two have been most influential in this chapter: Robert Gottleib's *Forcing the Spring* (Washington, DC: Island Press, 1993) and Samuel P. Hayes's *Beauty, Health, and Permanence: Environmental Politics in the United States, 1955–1985* (Cambridge, UK: Cambridge University Press, 1987). Each of these books provides a comprehensive descriptive and analytical overview of the movement's origins, growth, and impact, and both generally informed all aspects of this chapter.

In particular, *Forcing the Spring* served as a source for the origins and events of Earth Day (105–114; Hickel quote, 112: *Time* quote, 113), including an interview with Denis Hayes and the analysis that "in the months following Earth Day, a new environmentalism [began] to take shape" (113); the legacy of Rachel Carson's *Silent Spring* (81–86; quotes on page 86); and the growth of grassroots political lobbying movements, including further discussion of the antitoxin movement here and the Diablo Canyon nuclear demonstrations in chapter 4 (182). "Nation: Summons to a New Cause," *Time*, February 2, 1970, http://www.time.com/time/magazine/article/0,9171,878145-2,00.html, discusses the enthusiasm of the day (including the "no class or color distinctions" quote).

Beauty, Health, and Permanence served in particular as a source for the discussion of the environmental movement's uniquely successful political activism in the realms of legislation, administrative agency regulations, and litigation (458–490). Hayes discusses the rise and success of environmental lobbyists in the 1960s, described by one Congressman as the "most effective lobbying action in Washington since the Civil Rights movement" (465). Hayes especially highlights environmental lobbyists' success in leveraging and growing administrative agencies such as the EPA, understanding that these agencies did not merely "carr[y] out the law," but also "made new decisions of vast importance" (470). Hayes also discusses the antitoxin and nuclear movements (171–206) and the environmental movement's push for energy conservation that "came to be almost a religion" (230).

In addition to Hayes and Gottlieb, a number of sources informed the different issues discussed in the chapter. On the history of the development of environmental law, Richard J. Lazarus's *The Making of Environmental Law* (Chicago: Chicago University Press, 2004) is excellent. His chapters 4 and 5 were especially helpful. For example, see page 77 for the Nixon quote about environmentalists "going crazy" and page 84, which notes that David Sive has often been called the "father" of environmental law. See also Joseph L. Sax, *Defending the Environment* (New York: Bonzai Books, 1970). On the transition in the U.S. environmental movement from "conservation" to "environmentalism" in general, see Robert Cameron Mitchell, Angela G. Mertig, and Riley E. Dunlap, "Twenty Years of Environmental Mobilization: Trends among National Environmental Organizations" in *American Environmentalism: The U.S. Environmental Movement, 1970–1990*, ed. Riley E. Dunlap and Angela G. Mertig (New York: Taylor and Francis, 1992), 11–38 (it is also the source of data regarding the magnitude of "environmental impact statements" by federal agencies and the Environmental Defense Fund's role in inducing these statements); Stacy J. Silveira, "The American Environmental Movement: Surviving through Diversity," *Boston College Environmental Affairs Law Review* (Winter 2001): 497–532 (details of transition of environmental movement from an elite conservationist movement to a mainstream environmental movement, including the roles of *Silent Spring*, the Santa Barbara oil spill, and Earth Day); and Philip Shabecof, *A Fierce Green Fire: The American Environmental Movement* (New York: Hill and Wang, 1993), 118–119 (source for Sierra Club executive director Michael McCloskey's quote about this transition).

For more details on Earth Day 1970, Gaylord Nelson, the Wisconsin senator who helped bring it about, has written a number of informative retrospective accounts. His "Earth Day '70: What It Meant," *EPA Journal* (April 1980), http://www.epa.gov/history/topics/earthday/02.htm, provides a basic overview of the day's impact, and his article "How the First Earth Day Came About," *EnviroLink*, http://earthday

.envirolink.org/history.html, provides a firsthand account of its origins and organization. Many of Nelson's other Earth Day materials can be found in the digital collection of the Wisconsin Historical Society, http://www.wisconsinhistory.org/libraryarchives/collections/digital.asp. Denis Hayes, Earth Day's national coordinator, also offers a firsthand account of how he became involved in the event in "Earth Day," in *Encyclopedia of World Environmental History* (New York: Routledge, 2003), 355–356. For additional sources on Hayes's role, see "Earth Day Sponsors to Stay Together," *New York Times*, April 22, 1970, 35 (on Hayes's refusal to meet with Erlichman) and "After the Teach-In," *Science News*, April 4, 1970, 341–342. For more general commentary on Earth Day and its importance within the environmental movement, see Jesse H. Ausubel, David G. Victor, and Iddo K. Wernik, "The Environment since 1970," *Consequences: The Nature & Implications of Environmental Change* 1, no. 3 (1995) (arguing "1970 marked the birth of the modern environmental movement, symbolized by the first observance of 'earth day' in April of that year"), http://www.gcrio.org/CONSEQUENCES/fall95/envir.html, and John Steele Gordon, "The American Environment," *American Heritage*, June 30, 1993, http://www.americanheritage.com/articles/magazine/ah/1993/6/1993_6_30.shtml. For a skeptical view of the causes and legal aftermath of the Cuyahoga fire, see Jonathan H. Adler, "Fables of the Cuyahoga: Reconstructing a History of Environmental Protection," *Fordham Environmental Law Review* 14 (2002): 89–146.

For further discussion of the Santa Barbara oil spill and the other environmental disasters that helped inspire Earth Day and the rise of the environmental movement more generally, contemporaneous news media coverage provides an excellent source of information. For the Santa Barbara oil spill, the entire chronology of events is best captured in the *Los Angeles Times*, which provided regular and detailed coverage throughout 1969. For a sampling, see, Dick Main and Dial Torgerson, "Shifting Winds Push Oil Slick onto Santa Barbara's Beaches," *Los Angeles Times*, February 5, 1969, A1; Robert L. Jackson, "Oil Firms Shielded in Spillage Cases, Senate Aides Say," *Los Angeles Times*, February 17, 1969, A1. Two decades later the *Los Angeles Times* also published a useful retrospective. See Miles Corwin, "1969 Santa Barbara Oil Spill," *Los Angeles Times*, January 28, 1989, I23. For a more academic discussion, see K. C. Clarke and Jeffrey J. Hemphill, "The Santa Barbara Oil Spill: A Retrospective," in *Yearbook of the Association of Pacific Coast Geographers*, vol. 64, ed. Darrick Danta (Honolulu: University of Hawaii Press, 2002), 157–162 (quoting Richard Nixon as saying, "The Santa Barbara incident has frankly touched the conscience of the American people"). Sources on disasters beyond the Santa Barbara spill include Lynn G. Llewellyn and Claire Peiser, *National Environmental Policy Act (NEPA) and the Environmental Movement: A Brief History* (Washington, DC: Technical Analysis Division, Environmental Protection Agency, 1973) (Torrey Canyon oil spill) and the August 1, 1969, edition of *Time* (Cuyahoga River disaster).

On the importance of Rachel Carson's *Silent Spring*, the *New York Times*'s obituary of Carson provides a good overview. Frank Graham Jr.'s *Since Silent Spring* (New York: Houghton Mifflin, 1970) details the book's impact, including details on the DDT suit against Suffolk County inspired by the book (251–258). Victor B. Scheffer's *The Shaping of Environmentalism in America* (Seattle: University of Washington Press, 1991) also discusses the DDT suit as well as the role of *Silent Spring* in bringing about the creation of the Environmental Defense Fund and the general proliferation of environmental lawsuits in the late 1960s and 1970s (135–138).

For more discussion on the legislative, administrative, and litigation politics of the environmental movement, including explanations of the movement's distinctive successes, a good starting point is E. Donald Elliott, Bruce Ackerman, and John

C. Millians, "Toward a Theory of Statutory Evolution: The Federalization of Environmental Law," *Journal of Law, Economics, and Organization* (Fall 1985): 313–340. This article provides an overview of the growth of environmental statutes in the 1970s with a focus on their implications for theoretical understanding of statutory lawmaking and administrative law. It also discusses the role of environmental regulations in establishing patterns for federal legislation more generally. See also James E. Krier and Edmund Ursin, *Pollution and Policy: A Case Essay on California and Federal Experience with Motor Vehicle Air Pollution, 1940–1975* (Berkeley and Los Angeles: University of California Press, 1977); Charles O. Jones, *The Policies and Politics of Pollution Control* (Pittsburg: University of Pittsburg Press, 1975); John E. Bonine and Thomas O. McGarity, *The Law of Environmental Protection: Cases, Legislation, Policies* (St. Paul, MN: West, 1984); Michael S. Greve, "Environmentalism and the Rule of Law: Administrative Law and Movement Politics in West Germany and the United States," Ph.D. diss., Cornell University, May 1987; David Vogel, *Trading Up: Consumer and Environmental Regulation in a Global Economy* (Cambridge, MA: Harvard University Press, 1995); and Frank R. Baumgartner and Bryan D. Jones, *Agendas and Instability in American Politics* (Chicago: University of Chicago Press, 1993), 185–189 (source of data around growth of environmental groups in the 1960s and 70s). Lazarus, *The Making of Environmental Law*, p. 70, lists eighteen major environmental protection statutes enacted in the 1970s, and pages 83, 84, and 118 provide data concerning increases in the memberships of environmental organizations. The *New York Times* also provided contemporaneous coverage. See, for example, Gladwin Itell, "Environmental Impact Statements, Practically a Revolution," *New York Times*, December 5, 1976, 208 (source for "revolutionary effects" of new NEPA law).

For more details on environmental litigation in particular, see Lettie M. Wenner, *The Environmental Decade in Court* (Bloomington: Indiana University Press, 1982) (account of litigation in the period, including the data on the number of cases produced under NEPA and the Clean Air Act); Carol Harlow and Richard Rawlings, *Pressure through Law* (London: Routledge, 1992); Evan J. Ringquist, "A Question of Justice: Equity in Environmental Litigation, 1974–1991," *Journal of Politics* 60, no. 4 (November 1998), 1148 (analyzes role of judicial decisions in leading to the fact that "minorities and the poor are exposed to disproportionately high levels of environmental risk," concluding that "judicial contributions to inequities in the distribution of environmental risk are negligible"). For the article relating the importance and impact of the Storm King Mountain litigation, see William Tucker, "Environmentalism and the Leisure Class," *Harper's* (December 1977): 50–75. The key decision by the Court of Appeals is *Scenic Hudson Preservation Conference v. FPC*, 354 F. 2d 608 (2d Cir. 1965). The *New York Times* also covered the details of the crisis in copious detail with many articles. See Dana Adams Schmidt, "Morton Says Court Rulings Have Delayed Alaska Pipeline Action," *New York Times*, February 11, 1972, 38.

For the published works on the future and conservation mentioned in this chapter, see Paul Ehrlich, *Population Bomb* (New York: Buccaneer Books, 1968); Club of Rome, *The Limits of Growth* (New York: Universe Books, 1972); Stewart Brand, *Whole Earth Catalog* (Menlo Park, CA: Portola Institute, 1968); Amory B. Lovins, "Energy Strategy: The Road Not Taken?" *Foreign Affairs* 55, no. 1 (1976): 65–96; *The Unfinished Agenda: The Citizen's Policy Guide to Environmental Issues*, ed. Gerald O. Barney (New York: Crowell, 1977).

For analysis of the history and impact of CAFE, the best source is Daniel Luskin's unpublished manuscript "Conservation by Accident: The Surprising History of CAFE" (Yale University, August 2008). Luskin details and analyzes the origins of CAFE and its legislative history, including the political interest-group compromises underlying the bill; the limits of its actual impact and value; the reconciliation of critics and supporters' arguments in discussions of the extent to which increased oil prices coinciding with the bill account for disputes over its impact; and the role of auto industry in the lead-up to CAFE, including all of the quotes except for the speech of GM's chairman Thomas Murphy cited in this chapter. Murphy's speech, "The Empty Promise of the Planner: Private Enterprise or National Economic Planning?", was delivered before the Greater Detroit Chamber of Commerce on June, 5, 1975, and published in *Vital Speeches of the Day* 41, no. 19: 591–594. For Daniel Yergin's take on CAFE, see *The Prize*, p. 718. Among the "other analysts" who question CAFE's effectiveness are Jerry L. Mashaw and David L. Harfst, *The Struggle for Auto Safety* (Cambridge, MA: Harvard University Press, 1990), 235–236. See also Michael Moritz and Barrett Seaman, *Going for Broke: The Chrysler Story* (Garden City, NY: Doubleday, 1981) (discussing changes in the U.S. auto-manufacturing industry before CAFE and America's continued "love affair" with the auto in its aftermath, including the quote from Iacocca [4]). And for a contemporaneous critique of CAFE, see Elizabeth Drew "The Energy Bazaar," *The New Yorker* (June 21, 1975): 35–72. For an argument in support of CAFE, see Wenonah Hauter, testimony on CAFE standards before the Transportation and Related Agencies Appropriations Subcommittee, U.S. House of Representatives, February 10, 2000, https://www.citizen.org/cmep/energy_enviro_nuclear/electricity/Oil_and_Gas/articles.cfm?ID=6365.

For the quote from Nixon's January 1 speech on the signing of NEPA, see Rick Perlstein, *Nixonland: The Rise of a President and the Fracturing of America* (New York: Scribner, 2008), 460. Nixon's 1970 State of the Union Address is available online at http://www.infoplease.com/t/hist/state-of-the-union/183.html; Jimmy Carter's statements on conservation are available in his July 15, 1979, address, http://www.pbs.org/wgbh/amex/carter/filmmore/ps_crisis.html; and for Ralph Nader's testimony to Congress on energy prices, see Pietro S. Nivola, *The Politics of Energy Conservation* (Washington, DC: Brookings Institution Press, 1986), 55.

Chapter 4

For background on the Three Mile Island accident, see Daniel J. Kevles, *Nuclear Power in the United States: The Rise and Fall of a Hazardous Technology*, working paper (New Haven, CT: Yale University, 2006); "A Nuclear Nightmare," *Time*, April 9, 1979, 24–26; Peter A. Bradford, written testimony, *Three Mile Island: Thirty Years of Lessons Learned: Hearing before the Subcommittee on Clean Air and Nuclear Safety, Committee on Environment and Public Works, U.S. Senate*, 111th Cong., 1st sess., March 24, 2009; Harold R. Denton, former director of Office of Nuclear Reactor Regulation, written testimony, *Three Mile Island—Looking Back on Thirty Years of Lessons Learned: Hearing before the Subcommittee on Clean Air and Nuclear Safety, Committee on Environment and Public Works, U.S. Senate*, 111th Cong., 1st sess., March 24, 2009; Marvin S. Fertel, President and Chief Executive Officer, Nuclear Energy Institute, written testimony to the Subcommittee on Clean Air and Nuclear Safety, Committee on Environment and Public Works, U.S. Senate, 111th Cong., 1st sess., March 24, 2009; Dale E. Klein, Chairman, U.S. Nuclear Regulatory Commission, written testimony to the Committee on Environment and Public Works, Subcommittee on Clean Air and Nuclear Safety, U.S. Senate, 111th Cong., 1st sess., March 24, 2009; Dick Thornburgh, former governor of Pennsylvania, Counsel, K&L Gates LLP, written testimony, *Three Mile Island—Looking Back on Thirty Years of Lessons Learned*, hearing, 111th Cong., 1st sess., March 24, 2009.

Before the accident at Three Mile Island, there was a widespread belief that nuclear power would become America's primary source of electricity. For this perspective, see, for example, Alvin M. Weinberg, "Unlimited Energy Is Not Just a Dream," *New York Times*, January 6, 1969, 145; H. G. Wells, *The World Set Free: A Story of Mankind* (New York: E. P. Dutton, 1914).

Momentum for nuclear power grew after Congress enacted the Atomic Energy Act in 1946. See Kevles, *Nuclear Power in the United States*, 2, 8. For prevailing notions during the 1960s that electricity from nuclear power would be low cost and widespread, see Kevles, *Nuclear Power in the United States*, 15.

For information on the rise and fall of nuclear power plants between the mid-1960s and late 1970s, see Harold P. Green and Alan Rosenthal, *Government of the Atom: The Integration of Powers* (New York: Atherton Press, 1963); John L. Campbell, *Collapse of an Industry: Nuclear Power and the Contradictions of U.S. Policy* (Ithaca, N.Y.: Cornell University Press, 1988), 23, 33, 73 (on the Oyster Creek plant, on the peak years of nuclear plant manufacturing and sales in the early 1970s, and on the sharp decline in nuclear plant sales in the late 1970s); S. D. Thomas, *The Realities of Nuclear Power: International Economic and Regulatory Experience* (Cambridge, UK: Cambridge University Press, 1988), 79, 107–108 (on the sharp decline of nuclear plant sales in the late 1970s, on lack of standardized nuclear plant designs, and on rising costs of nuclear plant construction throughout the 1970s); J. Samuel Walker, *Containing the Atom: Nuclear Regulation in a Changing Environment, 1963–1971* (Berkeley and Los Angeles: University of California Press, 1992); Robert J. Duffy, *Nuclear Politics in America: A History and Theory of Government Regulation* (Lawrence: University Press of Kansas, 1997), 103 (providing statistics on nuclear plant construction and AEC forecasts of nuclear plant construction); and Kevles, *Nuclear Power in the United States*, 16, 23 (on the peak years of nuclear plant construction in the early 1970s).

In 1957, JCAE pushed through Congress the Price–Anderson Act, named for Congressman Melvin Price and Senator Clinton Anderson, limiting the liability of private parties for nuclear accidents and providing for government-funded insurance. The political support for developing nuclear power was so strong then that this legislation passed the House by a voice vote and the Senate without any debate. As late as 1975, the JCAE remained powerful enough to secure a ten-year extension of the Price–Anderson legislation, even though the original law was not scheduled to expire until 1977.

For information on congressional efforts to reduce AEC authority over nuclear matters and on AEC investment in reactor technology and power plant construction in the early 1970s, see Duffy, *Nuclear Politics in America*, 103, 105. Peter Bradford, who served as both a federal and a state nuclear regulator, has said that the construction of power plants in the early 1970s "was as if the airline industry had gone from Kitty Hawk to jumbo jets in 15 years." Peter A. Bradford, written testimony, *Three Mile Island*, March 24, 2009.

For background on nuclear energy policy under the Nixon administration, including the appointment of James Schlesinger, see Duffy, *Nuclear Politics in America*, 108–124; Leslie H. Gelb, "In a World of 'Peanuts' and MAD: Schlesinger," *New York Times*, August 4, 1974, 198; President Nixon's speech to the Annual Meeting of the Atomic Industrial Forum and the American Nuclear Society, October 20, 1971, quoted in Duffy, *Nuclear Politics in America*, 108–111; "Environment: Changes in Dixyland," *Time*, November 5, 1973, 102.

Private parties regularly sought to impose additional environmental responsibilities on the government in federal courts. These efforts resulted in several government losses and some victories. For the D.C. Circuit's 1971 decision finding the

AEC's regulatory regime to be in violation of NEPA, see *Calvert Cliffs' Coordinating Committee, Inc. v. Atomic Energy Commission*, 449 F2d. 1109 (D.C. Cir. 1971). For information on the AEC's regulatory response to the *Calvert Cliffs* decision, see Duffy, *Nuclear Politics in America*, 93. But for the Supreme Court's 1978 reversal of a lower court's decision to impose additional procedural requirements on the NRC's licensing process, see *Vermont Yankee Nuclear Power Corp. v. Natural Resources Defense Council*, 435 U.S. 519 (1978).

Legal and environmental controversy swirled around efforts to build and operate nuclear power plants in California. For background on the Diablo Canyon saga, see Noel Greenwood, "Period of Great Nuclear Expansion," *Los Angeles Times*, February 26, 1970, 3; Philip Fradkin, "Conservationists Gear to Block Nuclear Plant," *Los Angeles Times*, April 4, 1971, H12; "Nuclear Plant Line Delayed," *Los Angeles Times*, December 2, 1971, G18; "The Environment," *Los Angeles Times*, April 25, 1972, B2; "Conservationists Lose on Diablo Atom Plant," *Los Angeles Times*, June 9, 1972, B2; George Gladney, "PG&E Reaches Midway Mark on Disputed $600 Million Plant," *Los Angeles Times*, September 4, 1971, E11; "Potential Water Supply Danger Reported," *Los Angeles Times*, June 1, 1973, A2; George Lowe, "Bus Tour Takes in Big Nuclear Plant," *Los Angeles Times*, August 5, 1973, F13; "Licenses Sought for 2 A-Power Plants," *Los Angeles Times*, October 17, 1973, B5; Lee Dye, "Study of Quake Potential in Sea Near Nuclear Plant under Way," *Los Angeles Times*, December 1, 1973, A1; Lee Dye, "Diablo Canyon A-Plant Work Won't Be Halted," *Los Angeles Times*, January 11, 1974, 3C; Charles Hillinger, "A Really Sweet Deal Turns Sour," *Los Angeles Times*, June 16, 1974, F1; "Ban Sought on Nuclear Storage at Plant," *Los Angeles Times*, May 30, 1975, B2; "Satellite to Scan Ozone Layer for Damage," *Los Angeles Times*, November 19, 1975, B12A; "A-Fuel Hearing Delay Rejected," *Los Angeles Times*, December 4, 1975, D4; and Art Seidenbaum, "Anguish of the Atoms," *Los Angeles Times*, October 29, 1975, C1.

For background on the public concern over the risk of earthquake at Diablo Canyon, see "Quake Fault May Delay License for Nuclear Plant," *Los Angeles Times*, January 15, 1976, B3. The NRC's chief of seismology and geology added that the fault system "certainly is a matter of concern to us. We believe the fault is active." At this time, the costs of the plants were estimated to be $900–985 million; the first was 90 percent complete, the second 60 percent. PG&E said that the plants could withstand what it estimated to be the largest potential shock. But an article in *Science Magazine* in December 1975 by the chairman of UCLA's geology department ("San Simeon-Hosgri Fault System, Coastal California: Economic and Environmental Implications") showed that the fault was both longer and younger than anyone originally thought. Scientists with the U.S. geological survey believed it might produce an earthquake up to 7.5 on the Richter scale. According to PG&E, the plant was designed to absorb a 6.75-magnitude earthquake directly below it. See Larry Pryor, "Fate of Atom Plant in Doubt," *Los Angeles Times*, January 18, 1976, 3. As New Year 1976 approached, PG&E delivered the first truckload of nuclear fuel to the plant.

For a discussion of the myriad political obstacles to the operation of the Diablo Canyon plants, see Paul Houston and Paul E. Steiger, "State's Need to Push Nuclear Energy Told," *Los Angeles Times*, April 24, 1977, A1. In January 1976, the California State Senate held hearings concerning the plants' safety, as did the state's Coastline Commission and its Energy Resources and Conservation Commission. Jerry Gillam, "Probe of PG&E Action on Quake Peril Sought," *Los Angeles Times*, January 28, 1976, B3; Jerry Gillam, "Atom Safety Bills Endorsed by State Energy Chairman," *Los Angeles Times*, April 28, 1976, OC10. In March 1977, facing a shortage of electric power, the San Luis Obispo County supervisors voted (by a three-to-two vote) to ask the NRC to speed up its operating licenses for Diablo Canyon. A month later James Schlesinger, then serving as Jimmy Carter's energy czar, predicted that the state of California would have to build several more new nuclear power plants to meet its energy needs for the rest of the century. Houston and Steiger, "State's Need to Push Nuclear Energy Told."

For additional background on the political and social forces against the opening of the Diablo Canyon plants, see "Cutoff of Atom Fuel to Plant Site Sought," *Los Angeles Times*, April 2, 1976, OCA2; "Strike Halts Work on A-Power Plant," *Los Angeles Times*, April 23, 1976, B2; "Scientists Study Possibility of Link in 3 Offshore Faults," *Los Angeles Times*, April 25, 1976, C1; George Alexander, "Quakes and A-Plants: How Great Is the Threat?" *Los Angeles Times*, May 30, 1976, B1; Jerry Gillam, "Plan Ordered on Atom Plant Safety," *Los Angeles Times*, September 16, 1976, C5; Paul E. Steiger, "Officials Push Diablo Plants Despite Major Quake Fault," *Los Angeles Times*, June 29, 1977, 1; Ellen Hume, "Interim License for Diablo Plant Possible," *Los Angeles Times*, July 1, 1977, B4; and "35 A-Plant Workers Suffer from Fumes," *Los Angeles Times*, June 30, 1978, OC2.

At one point, scientists seemed finally to agree that the Diablo Canyon plants were safe. Bryce Nelson, "Diablo Canyon A-Plants Safe, Panel Declares," *Los Angeles Times*, July 12, 1978, A13; "Firm Says Diablo A-Plant Could Be Running in April," *Los Angeles Times*, December 5, 1978, 27. Nonetheless, the Diablo Canyon plants continued to face opposition. "20 Nuclear Demonstrators to Be Tried," *Los Angeles Times*, October 11, 1978, OC2; "20 Antinuclear Demonstrators Get Fines, Probation," *Los Angeles Times*, December 24, 1978, A10; J. Cordner Gibson, "Myths about Diablo Canyon Nuclear Plant," *Los Angeles Times*, March 10, 1979, B4; Kenneth Reich, "State to Review Procedures at Nuclear Plants," *Los Angeles Times*, March 31, 1979, OCA1; Claude Walbert, "San Luis Obispo's Antinuclear Activists," *Los Angeles Times*, April 1, 1979, H1; Myrna Oliver, "Antinuclear Alliance Finds Some Help in the 'Bag Lady,'" *Los Angeles Times*, June 7, 1979, A18 (including the "alter government policy" quote); Myrna Oliver, "18,000 Attend L.A. Rally Against Nuclear Power," June 11, 1979, A4 (including the quotes from Ralph Nader); John R. Phillips and Steve Kristovich, "The Diablo Canyon Plant: More Regulatory Acrobatics at the NRC," *Los Angeles Times*, June 24, 1979, E1; and "See the Governor Grovel," *Los Angeles Times*, July 3, 1979, C4.

In April 1980, Governor Jerry Brown spent some of his office funds to hire a lawyer to urge the NRC to require PG&E to convert its Diablo Canyon plants to a nonnuclear fuel. PG&E said that would cost more than building new plants. Soon thereafter the California Assembly passed a bill requiring the state's Energy Commission to engage in a year-long study of what such a conversion would cost. For background on these events and their context, see Cathleen Decker, "County Chamber Urges Development of Nuclear Power, Criticizes Brown Stance," *Los Angeles Times*, July 31, 1979, OCA5; Gayle Fisher, "The Sierra Club: A Force of Nature," *Los Angeles Times*, October 11, 1979, OCA11; "Pro-nuclear Rally: Applause for A-Power," *Los Angeles Times*, October 14, 1979, A3; Robert Gillette, "3 Mile Panel Urges Action by Congress," *Los Angeles Times*, November 1, 1979, B7; Ronald L. Soble, "Spending Initiative Endorsed by Brown," *Los Angeles Times*, November 2, 1979, B3; "300 Stay Off Jobs at Diablo A-Plant," *Los Angeles Times*, November 9, 1979, OC2; Jerry Gilliam, "Brown Asks Halt to Review of A-Plant," *Los Angeles Times*, December 8, 1979, A30; "7,000 Join Pennsylvania Nuclear Protest," *Los Angeles Times*, March 30, 1980, OC2; "Brown Hires Lawyer to Fight A-Plant," *Los Angeles Times*, April 20, 1980, OC2; Peter De Wetter, "Brown's Legal Move against Power Plant," *Los Angeles Times*, April 25,

1980, D6; Graham L. Jones, "Quake Triggers New Wave of Anti-nuclear Protests," *Los Angeles Times*, May 30, 1980, B28; "Testing at Diablo Canyon Proposed," *Los Angeles Times*, July 17, 1980, OC2; "PUC Stands by Certification of A-Plant," *Los Angeles Times*, July 29, 1980, A3; "Energy and Environment: Voiding of Pesticide Order Asked," *Los Angeles Times*, October 21, 1980, SD2; George Alexander, "Participants Endure Debate on Diablo Safety," *Los Angeles Times*, October 27, 1980, 3; Lee Dembart, "State Orders New Plans for N-Plants," *Los Angeles Times*, November 11, 1980, SDA1; Robert A. Rosenblatt, "Speedup Urged for Licensing Nuclear Plants," *Los Angeles Times*, March 18, 1981, B1; Ellen Hume, "Diablo Canyon to Be Test Case for Lessons of 3 Mile Island," *Los Angeles Times*, April 10, 1981 (including the "fiddles while oil burns" quote) ; "3,000 Protest at Hearings on A-Plant License," *Los Angeles Times*, May 20, 1981, B25; George Alexander, "Diablo Plant Hearing Disrupted by Protestors," *Los Angeles Times*, May 23, 1981, OCA22; and George Alexander, "U.S. Close to Granting Diablo Canyon License," *Los Angeles Times*, May 25, 1981, B1.

After the NRC appeals board determined the Diablo Canyon plants to be safe, Governor Brown and antinuclear groups promised to file suit in federal court. See "Ruling on Atom Plant Safety Upheld," *Los Angeles Times*, June 17, 1981, 2; John Hurst and Mark Landsbaum, "Foes Gear Up for Diablo Canyon Siege," *Los Angeles Times*, September 11, 1981, A3; "Deputies with Batons Break Diablo Blockade," *Los Angeles Times*, September 17, 1981, A3; "Arrest Count Tops 1,000 at Diablo Canyon," *Los Angeles Times*, September 19, 1981, OCA1; and "Diablo Canyon: Dispel the Doubts," *Los Angeles Times*, September 23, 1981, D6.

In a countermove, supporters of the plant, including a California Assembly member, filed suit in state court against the Abalone Alliance, asking that it reimburse the state for $1 million spent by law enforcement authorities during the most recent twelve days of protests. "Diablo Canyon Nuclear Protestors Sued for $1 Million," *Los Angeles Times*, September 26, 1981, A26. In November 1981, the NRC told PG&E that it still needed to make three additional safety assessments of the $2.3 billion Diablo plants: David Treadwell, "PG&E Told to Make 3 Design Studies at Diablo Canyon Plant," *Los Angeles Times*, November 4, 1981, B3; and Tom Redburn, "How PG&E 'Shot Itself in the Foot,'" *Los Angeles Times*, March 16, 1982, B1.

For more information about these events, see Robert A. Rosenblatt, "Court Delays Diablo Plan Fuel Loading," *Los Angeles Times*, November 12, 1983, C1; Robert A. Rosenblatt, "Uranium Loading Begins at Diablo," *Los Angeles Times*, November 16, 1983, OC3; and Lee Dembart, "Diablo Plant Finally Starts Up," *Los Angeles Times*, April 20, 1984, 1.

For more information on opposition faced by utilities in other states, including the reactor in Seabrook, New Hampshire, see Duffy, *Nuclear Politics in America*, 175.

For background on the federal government's initial financial and political support for and ultimate withdrawal from fast breeder reactor technology, see "Breeder Reactor Built in Soviet," *New York Times*, January 4, 1969, 55; Alvin M. Weinberg, "Unlimited Energy Is Not Just a Dream," *New York Times*, January 6, 1969, 145; "Bonn to Step Up Reactor Program," *New York Times*, February 23, 1969, 24; John L. Hess, "France to Build Atom Plants Based on U.S. Design," *New York Times*, November 15, 1969, 14; Anthony Ripley, "Nixon Aides Weigh Reshaping A.E.C. for a Wider Role," *New York Times*, June 12, 1970, 1; Walter Sullivan, "'Clean' Reactors Delayed in Drive for Atom Power," *New York Times*, March 8, 1971, 1; John Noble Wilford, "Scientists' Group Sues A.E.C. for Data on New-Type Reactor," *New York Times*, May 26, 1971, 84; Philip Shabecoff, "Nixon Offers Broad Plan for More 'Clean Energy,'" *New York Times*, June 5, 1971, 1; Gene Smith, "U.S. Effort Urged on a New

Reactor," *New York Times*, June 8, 1971, 49; Gene Smith, "Utilities Stress Reactor Project," *New York Times*, June 9, 1971, 65; Edward Cowan, "Soviet Ends Construction of Word's First Large Commercial Fast Breeder Nuclear Reactor," *New York Times*, January 5, 1972, 11; Richard D. Lyons, "A.E.C. to Build a Reactor Creating Power and Fuel," *New York Times*, January 15, 1972, 1 (including the "our best hope" quote); "Court Rejects Plea for U.S. to Clarify Perils of Reactors," *New York Times*, March 25, 1972, 62; Edward Cowan, "Scientists Oppose Breeder Reactor," *New York Times*, April 26, 1972, 7; Walter Sullivan, "British Are Pressing for the Completion of Atomic Breeder Reactor," *New York Times*, August 1, 1972; Edward Cowan, "Group to Be Led by Westinghouse," *New York Times*, November 23, 1972, 57; and Gene Smith, "Leadership by U.S. Seen for Future," *New York Times*, December 5, 1972, 69.

The drama surrounding the rise of the breeder reactor in the United States included a decision by the D.C. Circuit Court of Appeals holding that the AEC was bound to adhere to NEPA's requirements. The AEC predictably produced a five-volume environmental impact statement two years later, claiming not only that the breeder reactor was environmentally safe, but also that it was the only power-generating technology adequate to meet the nation's expanding electricity needs over the next 40 years or so. A scientists' group soon thereafter denounced that report as "frivolous and shallow," and the EPA concluded that it was "inadequate," giving it its lowest possible rating. In 1974, the AEC estimated that the costs of the breeder reactor had increased by more than $1 billion and in one of its last acts before going out of business announced that commercial breeder reactors would not come online until the 1990s, a decade later than it had originally estimated. By 1975, Congress was debating vigorously whether the effort to produce the breeder—which some estimated had now cost about $10.6 billion—should be halted altogether. For information on these events and developments, see Richard D. Lyons, "Nuclear Reactor Delayed by Court," *New York Times*, June 13, 1973, 15; "AEC to File Impact Report on Reactor Plan," *New York Times*, June 18, 1973, A2; Gene Smith, "G.E. Calls Nuclear Power Answer to Crisis," *New York Times*, November 8, 1973, 71; Victor K. McElheny, "'Breeder' Reactor Plan Facing Delays," *New York Times*, December 19, 1973, 89; "New Sources of Energy Sought in Expanded Funding for A.E.C.," *New York Times*, February 5, 1975, 22; "Breeder Reactor Expected by 1987," *New York Times*, March 15, 1974, 38; "Scientists Criticize A.E.C. Assessment of Reactor Hazard," *New York Times*, April 24, 1974, 24; Edward Cowan, "A.E.C. Is Criticized on 'Breeder' Plan," *New York Times*, May 7, 1974, 21; "Atom Breeder Plant up $1-Billion in Cost since 1972 Estimate," *New York Times*, September 21, 1974, 57; "A.E.C. Rejects Plan on Breeder Reactor," *New York Times*, November 26, 1974, 35; David Burnham, "E.P.A. Doubts Data on Atomic Mishap," *New York Times*, December 3, 1974, 23; Victor K. McElheny, "A.E.C. Stresses Reactor's Value," *New York Times*, December 29, 1974, 25; Anthony Ripley, "Reactor Schedule Delayed to 1990's," *New York Times*, January 18, 1975, 27; David Burnham, "Congress Faces 3 Key Decisions on Nuclear Reactors," *New York Times*, March 30, 1975, 26; Reginald Stuart, "Nuclear Power Campaign Is On," *New York Times*, May 26, 1975, 21; Edward Cowan, "Energy Development Plan Offers Priorities for U.S.," *New York Times*, July 1, 1975, 63; and David Burnham, "Safety of Breeder Reactors Questioned," *New York Times*, February 16, 1976, 30.

The Government Accounting Office estimated that at least 33 congressional committees and 29 executive-branch agencies shared responsibility for the breeder and related energy policies. One of these agencies, the Energy Research and Development Administration, had been planning for 124 breeder reactors at a cost of $150 billion (in 1974 dollars). See David Burnham, "G.A.O. Says Ford Plutonium View Perils Future of Breeder Reactor," *New York Times*, December 1, 1976, 18.

For information about the disfavored treatment of breeder reactors under the Carter administration, see Philip Shabecoff, "Udall Would Go Slow on Nuclear Energy," *New York Times*, February 1, 1977, 37; David Burnham, "More Funds Sought for Reactor That Carter Opposed in Campaign," *New York Times*, February 7, 1977, 14; Edward Cowan, "Carter Seeks to Cut $200 Million from Breeder Reactor Program," *New York Times*, February 23, 1977, 1; David Burnham, "Ford Panel Urges Major Policy Shift on Nuclear Power," *New York Times*, March 22, 1977, 54; Edward Cowan, "U.S. Ready to Cease Push for Plutonium as Fuel in Reactors," *New York Times*, April 7, 1977, 69; Edward Cowan, "Carter Bids Other Nations Join U.S. in Curbing Plutonium," *New York Times*, April 8, 1977, 70; Edward Cowan, "Carter's Opposition to Tennessee Breeder Reactor Is Reasserted before House Panel Vote," *New York Times*, May 24, 1977, 20 (including the "highly expensive" quote from Schlesinger); Steven Rattner, "Senate Snubs Carter in Backing Reactor," *New York Times*, July 12, 1977, 9; Martin Tolchin, "In Carter Compromise, Conferees Drop Funds for Breeder Reactor but Approve Nine Water Projects," *New York Times*, July 21, 1977, 15; Martin Tolchin, "House, in Rebuff to Carter, Votes Funds for Clinch River Reactor," *New York Times*, September 21, 1977, 78; "Senate Votes $80 Million for Clinch River Reactor That President Opposes," *New York Times*, November 2, 1977, 17; "President Uses First Veto to Bar Nuclear Reactor," *New York Times*, November 6, 1977, 1; and "Carter Signs Appropriations Bill; Will Close Clinch River Project," *New York Times*, March 8, 1978, 10.

For discussion of President Obama's decision not to go forward with storing nuclear waste at Yucca Mountain and alternative storage in the United States, see Rebecca Smith, "Atomic Waste Gets 'Temporary' Home," *Wall Street Journal*, June 2, 2010, A1. The NRC's three-judge panel's decision that the Obama administration did not have legal authority to do so is reported in Siobhan Hughes and Rebecca Smith, "Panel Blocks Move to Scrap at Yucca Site," *Wall Street Journal*, June 30, 2010, A6. For public-health concerns related to water leaks at existing nuclear power plants, see "Funding Cut for U.S. Nuclear Waste Dump," *Nature* 458, no. 30 (April 2009): 1086; and Mathew J. Wald, "At the Indian Point Nuclear Plant, a Pipe Leak Raises Concerns," *New York Times*, May 2, 2009, 15 (including the "systemic failure of the licensee" quote).

For background on the prevalence of nuclear power in countries besides the United States, see Jose Goldenberg, "Nuclear Energy in Developing Countries," *Daedalus* 138, no. 4 (Fall 2009): 71–80.

Chapter 5

Dan Rottenberg, *In the Kingdom of Coal: An American Family and the Rock That Changed the World* (New York: Routledge, 2003), examines the coal industry's contentious history in the United States over the course of the twentieth century by focusing on two families—one composed of coal magnates, the other of miners—that together serve as a microcosm illuminating the labor sagas endemic to the extraction of coal. In *Coal: A Human History* (New York: Penguin, 2003), Barbara Freese chronicles the role that coal has played in human civilization from the earliest days of its use through contemporary debates on global warming. Freese's engaging narrative, focusing on Great Britain, the United States, and China, emphasizes the challenges and pitfalls that reliance on coal has brought to these societies. Chapters 6 and 7 in her book address the impact of coal on American labor, environmental, and political developments since the late nineteenth century. The "campaign of misleading publicity" quote can be found on p. 170. Jeff Goodell's *Big Coal: The Dirty Secret behind America's Energy Future* (New York: Mariner Books, 2007) addresses the power and influence of the American coal industry, past and present. Underlying Goodell's wide-ranging survey is the argument that the industry's harmful effects on safety, health, politics, and the climate make coal ill suited for the nation's energy needs. Yanek Mieczkowski, *Gerald Ford and the Challenges of the 1970s* (Lexington, KY: University Press of Kentucky, 2005) details the Ford administration's response to the energy crisis, including Gerald Ford's description of Coal as "our ace in the hole" (267).

Time magazine provided extensive coverage of the 1977–1978 UMW strike. The early days of the strike, including *Time*'s prediction of "a long stoppage," are reported in "The Coal Miners Walk Out," *Time*, December 12, 1977, and "But Life Can be Cruel," *Time*, December 19, 1977, which describes the state of the unionized miner workforce. "Darkness in the Coal Country," *Time*, February 13, 1978, and "Entering the Doomsday Area," *Time*, February 27, 1978, chronicle the effects of the energy squeeze caused by the strike, including the depletion of coal reserves, the violent interruption of coal deliveries from the Midwest, the impact on local communities, and the threatened effect on the automobile industry and the U.S. economy. The latter article discusses Arnold Miller and the controversy within UMW ranks surrounding his leadership.

More information about Miller is provided in "Black-Lung Hillbilly in a Big Job," *Time*, November 25, 1974 (including the "hillbilly" quote), and Rottenberg, *In the Kingdom of Coal*, 210 (including the "soft-spoken" quote). Rottenberg also describes Yablonski's quest for leadership of the UMW and Boyle's orchestration of his murder (203–205, 219). For John L. Lewis's tenure as president of the UMW, see Freese, *Coal*, 158–160. The transition from the Lewis to the Boyle presidency and the subsequent corruption in the UMW are described in "The New Militancy: A Cry for More," *Time*, November 25, 1974, and Miller's election as president is discussed in Rottenberg, *In the Kingdom of Coal*, 212, which includes the *New York Times* quote.

On the spread of wildcat strikes in the early 1970s and the resulting loss of workdays, see President's Commission on Coal, *Recommendations and Summary Findings* (Washington, DC.: U.S. Government Printing Office, March 1980), 13–14. "To Work," *Time*, March 20, 1978, describes the importance of safety concerns as an impetus for wildcat strikes, including the miner's "when I kiss my wife goodbye" quote. The month-long 1974 UMW strike is reported in "The New Militancy: A Cry for More," *Time*, November 25, 1974. Goodell, *Big Coal*, 60–61, explains the close relationship between prominent mining accidents and subsequent safety legislation, noting that "when it comes to enacting and enforcing safety laws against Big Coal, the only good lobbyists are dead miners" and discussing the 1969 legislation. The 1977 act is described in Richard Bonskowski, William D. Watson, and Fred Freme, *Coal Production in the United States—An Historical Overview* (Washington, DC: Energy Information Administration, October 2006), 9, http://www.eia.doe.gov/cneaf/coal/page/coal_production_review.pdf.

On the Buffalo Creek Hollow flood and its aftermath, see "Disaster in the Hollow," *Time*, March 13, 1972, and "After the Deluge," *Time*, October 9, 1972. The latter article includes the Pittson representative's "act of God" quote and reveals the earlier warning to state officials and the prior recognition of danger by residents. "Out of the Hole with Coal," *Time*, January 28, 1974, provides figures on the closure of eastern mines due to increased operating expenses and decline in worker productivity by 1974, as well as the Miller speech quote.

For increased coal-production costs and decreased worker productivity resulting from the 1969 safety legislation, see F. William Brownell, "Energy Independence—The Return to Coal: Constraints on Production and Utilization of Our Most Abundant Natural Energy Resource," *Saint Mary's Law Journal* 11 (1979–1980), 677, 693–694.

For an extensive examination of the controversy over leasing policies relating to federal coal reserves in the West, see Robert H. Nelson, *The Making of Federal Coal Policy* (Durham, NC: Duke University Press, 1983), chap. 3. The Powder River basin and its promising coal supply are described in Calvin Kentfield, "New Showdown in the West," *New York Times*, January 28, 1973, which also contains predictions of the population influx that western mining would bring and provides the "back to the olden days" observation by a resident. Douglas E. Kneeland, "To Ranch in West or Strip for Coal: A Difficult Choice," *New York Times*, February 18, 1974, likewise projects half a million new residents in the basin, describes the effect of the influx on Forsyth, Montana, and includes the quote from the Forsyth mayor and the "Let the Bastards Freeze in the Dark" bumper sticker. For concerns about the miner influx in Gillette, Wyoming, see "Town Scarred by Oil Boom Waits Apprehensively for Miners," *New York Times*, April 11, 1974, 37. Rottenberg, *In the Kingdom of Coal*, pp. 220–221, 224–226, recounts the Crow tribe's leasing of mining rights to Westmoreland Resources and the company's development of new rail lines.

Plans for the Burlington Northern Railroad are reported in Dave Earley, "$6.4 Million Rail Spur Planned for Coal Haul," *Billings Gazette*, July 8, 1971, 1. Further details are provided in "Burlington to Ask Approval for Building Coal Carrier," *New York Times*, October 9, 1972, 48, and "Burlington Northern Outlay," *New York Times*, December 3, 1974, 62. Nelson, *The Making of Federal Coal Policy*, 4–5, explains the impetus for the shift to western coal production in the 1970s. Freese, *Coal*, p. 178–180, also describes the shift to low-sulfur western coal and mechanized surface mining, noting Wyoming's current status as the top coal-producing state. The *Time* quote on leasing rights is from "Out of the Hole with Coal," *Time*, January 28, 1974.

The Court of Appeals and Supreme Court strip-mining decisions are *Sierra Club v. Morton*, 514 F.2d 856 (D.C. Cir. 1975), and *Kleppe v. Sierra Club*, 427 U.S. 390 (1976). Media coverage of the Sierra Club's lawsuit to halt strip mining in the Powder River basin can be found in Ben A. Franklin, "Court Extends Strip Mining Injunction," *New York Times*, June 22, 1975, 40 (describing the Court of Appeals decision); Grace Lichtenstein, "Town's Boom Halted by a Strip Mine Ban," *New York Times*, November 20, 1975, 1, 59 (surveying the impact on Gillette, Wyoming, of the court's injunction temporarily halting mining); and Ben A. Franklin, "High Court Ruling to Ease Strip Mining of U.S. Coal," *New York Times*, June 29, 1976, 14 (describing the Supreme Court's reversal of the appellate decision and lifting of the injunction).

The economic and environmental consequences of Appalachian strip mining that had occurred by 1971 are detailed in "The Price of Strip Mining," *Time*, March 22, 1971. The article also describes strip mining's higher productivity as well as the accidents and lung disease common in traditional mining. For the legislative history culminating in passage of SMCRA, see Richard O. Miller, "The Surface Mining Control and Reclamation Act: Policy Structure, Policy Choices, and the Legacy of Legislation," in *Moving the Earth: Cooperative Federalism and Implementation of the Surface Mining Act*, ed. Uday Desai (Westport, CT: Greenwood Press, 1993), 17–30. On the purposes and the state-centered structure of the SMCRA, see James M. McElfish Jr. and Ann E. Beier, *Environmental Regulation of Coal Mining: SMCRA's Second Decade* (Washington, DC: Environmental Law Institute, 1990), 20–22; page 42 describes the "no more stringent" maneuver utilized by several states. The Office of

Surface Mining's rush to approve state program applications before the onset of the Reagan administration is described in David Howard Davis, "Energy on Federal Lands," in *Western Public Lands and Environmental Politics*, ed. Charles Davis (Boulder, CO: Westview Press, 1993), 122, 134. SMCRA's mandate to restore the land's "approximate original contour" can be found in Section 515(b)(3) of the act, codified at 30 U.S.C. §1265(b)(3). On SMCRA's reclamation bond system, see McElfish and Beier, *Environmental Regulation of Coal Mining*, chap. 5. The OSM narrowed the circumstances requiring federal intervention in OSM, DOI, Directive REG-8 (June 20, 1996), http://www.osmre.gov/guidance/directives/archive/directive844.pdf. For the $5.5 billion and $1 billion expenditure figures, see Robert L. Bamberger, Congressional Research Service, *Abandoned Mine Land Reauthorization: Selected Issues* (Washington, DC: Congressional Research Service, March 8, 2005), 2, http://ncseonline.org/NLE/CRSreports/05mar/RL32373.pdf.

On the status of the UMW strike in February 1978 and President Carter's intervention to facilitate negotiations (including the miner's "no gun can make me work faster than I want to" quote), see "Entering the Doomsday Area," *Time*, February 27, 1978. Carter's invocation of the Taft–Hartley Act, his pronouncement that "the law will be enforced," and the miners' refusal to obey the injunction are detailed in "To Work," *Time*, March 20, 1978. The article describes the miners' economic suffering amid the strike, the increased production of nonunion western coal during this period, and the contract concessions offered in exchange for the power to end wildcat strikes. Rottenberg, *In the Kingdom of Coal*, pp. 241–242, describes the White House–sponsored contract negotiations, the administration's failure to enforce the back-to-work order, the end of the strike, and how "the critical issue of wildcat strikes that had so vexed both sides remained unresolved." Judge Robinson's refusal to extend the back-to-work order are reported in "Once Again, a Coal Agreement," *Time*, March 27, 1978. The political damage to Carter for his perceived mismanagement of the coal strike is discussed in "At Last, Peace in the Coalfields," *Time*, April 3, 1978. "But Life Can be Cruel" *Time*, December 19, 1977, recounts the miners' resort to hunting, stockpiling of coal and canned goods, difficulty in paying for Christmas presents, and loss of health insurance. "Darkness in the Coal Country," *Time*, February 13, 1978, reports the halting of pension plan checks, the inability to pay hospital bills, and other impacts on miners and their families (including Hamrick's "they're gonna stick it out" quote). The "principles are nice" quote is from "At Last, Peace in the Coalfields," which also provides the details of the third, successful contract proposal. "Once Again, a Coal Agreement" reports the coal operators' loss of man-days of work in 1977, their desire for the right to punish wildcat strike leaders, and how this question was ultimately left to an arbitration board.

President Carter's call for increased coal production and decreased reliance on oil and natural gas is reported in "The Coal Miners Walk Out," *Time*, December 12, 1977. Freese, *Coal*, notes the repeated doubling of sulfur dioxide emissions over several decades and the emergence of awareness about acid rain, including the coal industry representative's "campaign of misleading publicity" quote and Congress's 1990 adoption of the Acid Rain Program (168–170). "Nixon's Second Round," *Time*, February 22, 1971, describes Nixon's proposed "sulfur tax" and includes his statements on the dangers of sulfur dioxide. "Nixon's Third Round," *Time*, February 21, 1972, provides more background on the sulfur tax, including Nixon's "costs of pollution should be included in the price" quote and the lack of support his proposal received in Congress.

Bruce Ackerman and William T. Hassler, *Clean Coal, Dirty Air* (New Haven, CT: Yale University Press, 1981), discusses the role of eastern coal interests in collaboration with environmental groups in the 1977 clean air legislation. See pages 31–32 for the "bizarre coalition" quote, page 27 for the "mutually incompatible . . . hopelessly incoherent" quote, and page 45 for the "clean air symbols" quote. For information on the Coal Policy Project, see Tom Alexander, "Détente for Coal?" *Environment: Science and Policy for Sustainable Development* 20, no. 5 (June 1978) (including the "industrial lions" quote).

Freese, *Coal*, discusses harmful emissions from coal-fired power plants, including the surprisingly low costs of reducing sulfur dioxide emissions since 1990, the development of "scrubbers," and the availability of low-sulfur coal in the western states (170–171); the impact of sulfur dioxide emissions on air visibility (171); the contribution of coal power plants to nitrogen oxides (causing smog) and high mercury levels in fish (172–173); the study estimating annual U.S. deaths due to power plant emissions (175); as well as contemporary indications that the 1990 limits on sulfur emissions may be inadequate (178).

On establishment of the OCR, see Linda R. Cohen and Roger G. Noll, "Synthetic Fuels from Coal," in Linda R. Cohen and Roger G. Noll, with Jeffry N. Banks, Susan A. Edelman, and William M. Pegram, *The Technology Pork Barrel* (Washington, DC: Brookings Institution, 1991), 270. A description of President Ford's 1975 State of the Union and the establishment of the synfuels task force can be found on pp. 281–284. For Nixon's Project Independence funding of the OCR, see Michael Crow, Barry Bozeman, Walter Meyer, and Ralph Shangraw Jr., *Synthetic Fuel Technology Development in the United States: A Retrospective Assessment* (New York: Praeger, 1988), 69. Germany's prewar development of coal liquefaction is recounted in Cohen and Noll, "Synthetic Fuels from Coal," 265, and in Goodell, *Big Coal*, 218–219. For America's World War II and postwar efforts to develop synthetic fuels, see Crow et al., *Synthetic Fuel Technology Development*, 115–117. On the Bureau of Mines mission, see U.S. Department of Energy (DOE), *The Early Days of Coal Research*, available at http://www.fe.doe.gov/aboutus/history/syntheticfuels_history.html. On Project Paperclip, see Anthony N. Stranges, "The US Bureau of Mines's Synthetic Fuel Programme, 1920–1950s: German Connections and American Advances," *Annals of Science* 54 (1997), 29–30. For Nixon's request for one-third private funding for new synfuel plants, see Cohen and Noll, "Synthetic Fuels from Coal," 273. On the failed Catlettsburg project, see Crow et al., *Synthetic Fuel Technology Development*, 74–77.

President Carter's energy plan and the history of the Synthetic Fuels Corporation are described in Cohen and Noll, "Synthetic Fuels from Coal," 288–289. Carter's first proposal to subsidize production of synthetic fuels is reported in "Carter at the Crossroads," *Time*, July 23, 1979, and congressional approval (with modifications) of his second proposal, including the creation of the Synthetic Fuels Corporation, is reported in "Synfuel Success," *Time*, June 2, 1980. Predictions of a coal-powered automobile available by the early 1990s are recounted in John Holusha, "G.M. Displays Car Fueled with Coal Dust," *New York Times*, June 4, 1981, D4. On industry's withdrawal from the SFC and its reorganization within Reagan's Clean Coal Program, see Crow et al., *Synthetic Fuel Technology Development*, 81, 123. Carter's outline of his energy proposals and discussion of past stages of energy innovation can be found in Jimmy Carter, "The President's Proposed Energy Policy," televised speech, April 18, 1977, http://www.pbs.org/wgbh/amex/carter/filmmore/ps_energy.html (including the "moral equivalent of war" quote).

Goodell, *Big Coal*, xiii, reveals that half of U.S. electricity derives from coal. Freese, *Coal*, discusses changes in the nature of the coal industry since the 1970s (179–180) and the shortcomings of carbon sequestration as a solution to carbon dioxide emissions (237–239). Rentech's proposal to convert coal into synthetic transportation fuel is reported in Kathleen Schalch, "Administration Backs Making Liquid Fuel from Coal," *National Public Radio*, July 31, 2007, http://www.npr.org/templates/story/story.php?storyId=12314966. The closing quote from President Carter is from "The President's Proposed Energy Policy," the April 1977 televised speech on energy policy cited in the previous paragraph.

Chapter 6

For background on the natural-gas industry in the United States, see I. C. Bupp and Frank Schuller, "Natural Gas: How to Slice a Shrinking Pie," in *Energy Future: The Report of the Harvard Business School Energy Project*, ed. Roger Stobaugh and Daniel Yergin (New York: Random House, 1979), 56–78; W. M. Burnett and S. D. Ban, "Changing Prospects for Natural Gas in the United States," *Science* 244 (April 21, 1989): 305; Arlon R. Tussing and David Hatcher, "Federal Regulation of Interstate Gas Sales, 1939–1980," in *The Natural Gas Industry: Evolution, Structure, and Economics*, ed. Arlon Tussing and Bob Tippee (Tulsa, OK.: PennWell, 1995), 139; Arlon R. Tussing, Bob Tippee, and Ronald D. Ripple, "Overview of the Natural Gas Industry," in *The Natural Gas Industry*, ed. Tussing and Tippee, 15; Vaclav Smil, *Energy at the Crossroads: Global Perspectives and Uncertainties* (Cambridge, MA: MIT Press, 2003); Joe Barnes, Mark H. Hayes, Amy M. Jaffe, and David G. Victor, "Introduction to the Study," in *Natural Gas and Geopolitics: From 1970 to 2040*, ed. David G. Victor, Amy M. Jaffe, and Mark H. Hayes (Cambridge, UK: Cambridge University Press, 2006), 3–26; and *BP Statistical Review of World Energy* (London: BP, 2008), 23, 27. For general history on the natural gas industry in the early 1970s, see also Patricia E. Starratt, *The Natural Gas Shortage and the Congress* (Washington, DC: American Enterprise Institute for Public Policy Research, 1975); and Victor, Jaffe, and Hayes, eds., *Natural Gas and Geopolitics*.

For background on the 1972–1978 winters when natural-gas shortages first occurred, see "Energy: The Frigid Nightmare," *Time*, December 25, 1972; "Environment: Who Shut the Heat Off?" *Time*, February 12, 1973, 43; Paul Hodge, "Fuel Shortage Threat Creates Shortage of Firewood, Stoves," *Washington Post*, November 7, 1973, C1; "Energy Crunch Grips Campuses," *Hartford Courant*, November 19, 1973, 32; "Heating Oil Rules Outlined," *Chicago Tribune*, November 27, 1973, 2; "Home Heat Cut 6 Degrees," *Chicago Tribune*, December 29, 1973, 1; "From Output Squeeze to 'Price Embargo,'" *Time*, January 7, 1974, 36; Robert A. Wright and Martin Waldron, "Oil Is a National Problem, but Seriousness Varies," *New York Times*, February 3, 1974, 2; Les Gapay, "U.S. Will Risk Growing Gasoline Shortage to Avert a Crisis in Heating, Residual Oils," *Wall Street Journal*, February 7, 1974, 7; Kevin P. Phillips and Albert E. Sindlinger, "Americans Becoming Angrier in Fuel Woes," *Hartford Courant*, March 4, 1974, 4; "Electric Catch-22," *Time*, April 1, 1974, 23; "Leaning on the Consumer," *Time*, August 25, 1975, 57; "Row over Scarce Gas," *Time*, October 13, 1975, 61; "We've Been Asked; How Bad a Natural-Gas Shortage?" *U.S. News & World Report*, December 1, 1975, 63; "The Alaskan Gas Rush," *Time*, December 29, 1975, 60; "Edging Back from the Brink," *Time*, January 12, 1976, 58; and "What a Way to Run a Railroad," *Time*, August 9, 1976, 55.

For information on the natural-gas shortage and record low temperatures in 1976–1977, its effects, and the Carter administration's response, see Steven Rattner, "Less Natural Gas Expected in Winter: Little Effect Seen," *New York Times*, November 18, 1976, 61; Edward Cowan, "F.P.C. Is Exhorted to Raise Amount of Interstate Gas," *New York Times*, December 17, 1976, 81; Steven Rattner, "Cold Raises Fear of a Gas

Shortage," *New York Times*, December 29, 1976, 1; Steven Rattner, "Cold Wave Causes Energy Shortages and Plant Closings," *New York Times*, January 18, 1977, 1; Robert McFadden, "Record –1° Cold Disrupts Travel, Closes Some Schools in Suburbs," *New York Times*, January 18, 1977, 1; Edward Cowan, "Carter Said to Plan Early Action to Combat Shortages of Energy," *New York Times*, January 19, 1977, 67; Steven Rattner, "Cold Deepens Natural Gas Shortage; Continued Supply Cutbacks Expected," *New York Times*, January 19, 1977, D1, D7; Peter Kihss, "Cold Wave Continues to Grip Eastern U.S.," *New York Times*, January 19, 1977, 1; "The Wolf Has Arrived," *Wall Street Journal*, January 19, 1977, 16; Seth S. King, "Midwest Feels Bite of a Frozen Month," *New York Times*, January 20, 1977, 42; Steven Rattner, "Gas Shortages into 1978 Held Probable," *New York Times*, January 21, 1977, 68; Reginald A. Stuart, "Ohio, Citing Energy Shortage Caused by the Cold, Shuts Schools and Curbs Businesses in 24-County Area," *New York Times*, January 21, 1977, 13; Edward Cowan, "President Urges 65° as Top Heat in Homes to Ease Energy Crisis," *New York Times*, January 22, 1977, 1; Reginald Stuart, "Ohio Retracts 24-County Energy Emergency Order," *New York Times*, January 22, 1977, 8; Robert Young, "Carter Plea to U.S. 'Turn Down Thermostats,'" *Chicago Tribune*, January 22, 1977, 1, 5; "Emergency Plan to Allow U.S. to Allocate Natural Gas Is Being Weighed by Carter," *Wall Street Journal*, January 24, 1977, 4; "Georgia's Fuel Crisis Forces 75,000 off Jobs and Economic Loss of $30 Million," *Atlanta Daily World*, January 27, 1977, 1; Robert D. McFadden, "Byrne Issues Order," *New York Times*, January 28, 1977, 42; Steven Rattner, "Ohio Schools Close," *New York Times*, January 28, 1977, A1, D12; Hon. Richard L. Dunham, Federal Power Commission, statement to the Subcommittee on Energy and Power of the Committee on Interstate and Foreign Commerce, U.S. House of Representatives, 93rd Cong., 1st sess., January 28, 1977, 25–26; James R. Schlesinger, appendix to statement to the Subcommittee on Energy and Power of the Committee on Interstate and Foreign Commerce, U.S. House of Representatives, 93rd Cong., 1st sess., January 28, 1977, 9; John Kifner, "Blizzard Tightens Energy Crisis; Central States Are Hardest Hit; Cold Grips East, Gas Is Reduced," *New York Times*, January 29, 1977, 1; James R. Schlesinger, statement to the Subcommittee on Energy and Power of the Committee on Interstate and Foreign Commerce, U.S. House of Representatives, 93rd Cong., 1st sess., January 28, 1977, 9; "Luck Runs Out on Natural Gas," *Time*, January 31, 1977, 35; Peter Kihss, "Shortage in Energy Has Idled Thousands," *New York Times*, February 4, 1977, 1; "Warmer Weather Forecast for East," *Los Angeles Times*, February 8, 1977, B14; Megan Rosenfeld, "Va. Seeks 1.3 Billion Cubic Feet of Natural Gas through March," *Washington Post*, February 11, 1977, C3; "A Surplus of Suspicion," *Time*, February 21, 1977, 60; "After the Chill Comes the Bitter Bill," *Time*, February 28, 1977, 46; and "Congress Clears Emergency Gas Bill Quickly," in *CQ Almanac 1977* (Washington, DC: Congressional Quarterly, 1978), http://library.cqpress.com/cqalmanac/cqal77-1203891.

For the contrasting response in gas-producing states, see Arlon R. Tussing and Connie C. Barlow, *The Natural Gas Industry: Evolution, Structure, and Economics* (Cambridge, MA: Ballinger, 1984), 110, and "When the Gas Stops," *Time* magazine, October 17, 1977, 18.

On the chemical properties of natural gas, see, for example, David Howard Davis, *Energy Politics* (New York: St. Martin's Press, 1978), 132; and I. C. Bupp and Frank Schuller, "Natural Gas: How to Slice a Shrinking Pie," in *Energy Future*, ed. Stobaugh and Yergin, 56.

For a comprehensive analysis of the economics of the natural-gas industry, see Paul W. MacAvoy, *The Natural Gas Market* (New Haven, CT: Yale University Press, 2000). For more information on the regulation of the natural-gas industry in the

1930s, see Davis, *Energy Politics*, 136; and Ralph K. Hurtt, "Natural Gas Regulation under the Holding Company Act," *Law and Contemporary Problems* 19 (1954), 456. For the 1954 Supreme Court case concerning FPC regulation of wellhead prices, see *Phillips Petroleum Co. v. Wisconsin*, 347 U.S. 622 (1954). For information on the Natural Gas Act of 1938 and development of FPC regulation up through the 1950s, see Arlon R. Tussing and David Hatcher, "Federal Regulation of Interstate Gas Sales, 1939–1980," in *The Natural Gas Industry*, ed. Tussing and Tippee, 148–151.

For more information about President Eisenhower's response to legislation concerning regulation of the natural gas industry, see "The Gas Blast," *Time*, February 27, 1956; Dwight D. Eisenhower Memorial Commission, "President Eisenhower Vetoes His Own Bill," available at http://www.eisenhowermemorial.org/stories/Ike-vetoes-own-bill.htm; Davis, *Energy Politics*, 144–145. President Eisenhower said that "private persons representing some energy companies . . . have been seeking to further their own interests by highly questionable activities, [including] efforts that I deem so arrogant and so much in defiance of acceptable standards of propriety to risk creating doubt among the American people concerning the integrity of governmental processes" (quoted in Davis, *Energy Politics*, 145). Oil and gas interests were not chastened. In 1958, when Texas Republicans organized a fund-raising dinner for Joe Martin of Massachusetts, the House Republican leader, they listed Martin's efforts on behalf of Texas oil and gas interests in the publicity surrounding the event. That halted that year's legislative efforts to deregulate natural-gas prices (Davis, *Energy Politics*, 145).

For more information on trends in natural-gas prices and production during the 1960s and 1970s, see I. C. Bupp and Frank Schuller, "Natural Gas: How to Slice a Shrinking Pie," in *Energy Future*, ed. Stobaugh and Yergin, 63, table 3-1; and Tussing and Barlow, *The Natural Gas Industry*, 106. On the difference between interstate and intrastate markets, see I. C. Bupp and Frank Schuller, "Natural Gas: How to Slice a Shrinking Pie," 64. For discussion of the decrease in natural-gas reserves after 1973 and Ray Suave's related remarks, see "Congress Clears Emergency Gas Bill Quickly," in *CQ Almanac 1977*. For the OECD's analysis of the natural-gas shortage in the 1970s, see Organization of Economic Cooperation and Development (OECD), International Energy Agency, *Natural Gas: Prospects to 2000* (Paris: OECD, 1982), 80. On Congress's refusal to deregulate the natural-gas industry amid the oil crisis, see Tussing and Barlow, *The Natural Gas Industry*, 107.

For more information on the FPC's tiered pricing system in 1976, see Davis, *Energy Politics*, 155–156; and "Congress Clears Emergency Gas Bill Quickly," in *CQ Almanac 1977*. *Time* magazine reported different prices: 29.5 cents per thousand cubic feet for pre-1973 gas and 52 cents per thousand cubic feet for "new" gas: "Big Boost for Gas," *Time*, August 9, 1976, 55.

For background on President Carter's Emergency Natural Gas Act of 1977 and the criticism it received, see "Congress Clears Emergency Gas Bill Quickly," in *CQ Almanac 1977*; "Presidential Statement: Carter's Emergency Natural Gas Plan," in *CQ Almanac 1977*at http://library.cqpress.com/cqalmanac/cqal77-863-26256-1200398; "Emergency Gas Sales, 1977 Legislative Chronology," in *Congress and the Nation, 1977–1980*, vol. 5 (Washington, DC: Congressional Quarterly, 1981), http://library.cqpress.com/congress/catn77-0010174660; Edward Cowan, "Carter Energy Bill Goes to Congress; 65° Plea Renewed," *New York Times*, January 27, 1977, 1, 52; Charles Mohr, "Carter, Putting Leadership to Test, Again Asks Saving of Heating Fuel," *New York Times*, January 27, 1977, 1; "Business Newsmakers," *Washington Post*, January 16, 1978, D10; Governor Hugh Carey, statement to the Subcommittee on

Energy and Power of the Committee on Interstate and Foreign Commerce, U.S. House of Representatives, 93rd Cong., 1st sess., January 28, 1977, 50–53; Governor Dolph Briscoe, statement to the Subcommittee on Energy and Power of the Committee on Interstate and Foreign Commerce, House of Representatives, 93rd Cong., 1st sess., January 28, 1977, 49; Edward C. Burks, "Carey and Byrne, in Washington, Ask for Additional Gas Allocations," *New York Times*, January 29, 1977, 26; Edward Cowan, "Cutoff of Gas for Some Homes Draws Near, Schlesinger Warns," *New York Times*, January 29, 1977, 1, 26; Charles Mohr, "Carter Asks Cabinet to Help States Cope," *New York Times*, January 30, 1977, 29; President James Carter, Emergency Natural Gas Act of 1977, Remarks on Signing S. 474 and Related Documents, February 2, 1977, http://www.presidency.ucsb.edu/ws/index.php?pid=7422; Edward Cowan, "President Signs National Gas Bill, Marking First Legislative Victory," *New York Times*, February 3, 1977, 1; "I Have Already Acted on Several of My Promises," *Washington Post*, February 3, 1977, A6; Richard Lyons, "Congress, Carter Approve Gas Emergency Bill," *Washington Post*, February 3, 1977, A6; Jack Nelson, "Carter Appeals to All to Conserve Energy," *Los Angeles Times*, February 3, 1977, B1; "Power Impasse Deepens," *Hartford Courant*, February 6, 1977, 33; Peter Kihss, "White House Says Weather-Crisis Layoffs Totaled 1.8 Million at Height," *New York Times*, February 8, 1977, 14; "Warmer Weather," *Los Angeles Times*, February 8, 1977, B3; "Gas Shift Reports Called 'Misleading' in Tenneco Rebuttal," *New York Times*, March 9, 1977, 73; "Nation: More Gas for Transco," *Washington Post*, March 23, 1977, C1; Peter Goldman, Eleanor Clift, and Thomas M. DeFrank, "Jimmy So Far," *Newsweek*, May 2, 1977, 48; "In Congress," *Washington Post*, August 11, 1977, VA4; and "Fuel Emergency Plan Set by Energy Agency," *Wall Street Journal*, December 2, 1977, 8.

For the article concerning undiscovered natural-gas reserves in the United States, see Nicholas C. Chriss, "Drillers Say Natural Gas Abounds—'Deep Down,'" *Los Angeles Times*, February 3, 1977, B1. For information on the U.S.–Mexico negotiations over natural-gas sales, see Tussing and Barlow, *The Natural Gas Industry*, 149–152.

The best discussion and analysis of Jimmy Carter's energy policy and its development can be found in John C. Barrow, "An Age of Limits: Jimmy Carter and the Quest for a National Energy Policy," in *The Carter Presidency: Policy Choices in the Post–New Deal Era*, ed. Gary M. Fink and Hugh Davis (Lawrence: University Press of Kansas, 1998), 158–178, and in the dissertation on which that chapter is based, John C. Barrow, "An Era of Limits: Jimmy Carter and the Quest for a National Energy Policy," Ph. D. diss., Vanderbilt University, 1996. The Schlesinger quote that the basic constituency for the energy problem is in the future is from the latter at page 60. See also James Everett Katz, *Congress and National Energy Policy* (New Brunswick, NJ: Transaction, 1984), 97–117, on Carter's energy policy; and Bruce Schulman, *The Seventies: The Great Shift in American Culture, Society, and Politics* (Cambridge, MA: Free Press, 2002), 127 (overview of Carter's energy policy goals described in quote by James Schlesinger).

For background information on President Carter's Natural Gas Policy Act of 1978, remarks from several key legislators, and the response to the act, see "Carter's Program: Will It Work?" *Time*, May 2, 1977, 17–22; "On Tiptoe toward the Big Battle Ahead," *Time*, May 9, 1977, 75; "Lobbying the Carter UFO," *Time*, June 20, 1977, 74–75; Steven Rattner, "Schlesinger Visits Jackson to Ask Aid on Natural Gas Bill," *New York Times*, January 7, 1978, 27; Brendan Jones, "People and Business: Charles W. Robinson Is Elected Vice Chairman of Eastman Dillon," *New York Times*, January 10, 1978, 56; "Natural Gas Act Beastly Tangle," *Hartford Courant*, September 1, 1978, 43; "The Fight to Pipe Alaska's Gas," *Time*, September 19, 1977, 80; "Hard Going for

Carter's Plan," *Time*, September 26, 1977, 48; "A Filibuster Ends, but Not the Gas War," *Time*, October 17, 1977, 10; "Where the Carter Plan Stands," *Time*, November 28, 1977, 85; Bill Stall, "And Now, the Next Crisis," *Los Angeles Times*, February 5, 1978, A1; "Energy Bill: The End of an Odyssey," *CQ Almanac 1978* (Washington, DC: Congressional Quarterly, 1979), 12–13 (source of Dingell quotes), 24 (source of Byrd quote), 34–37 (Carter quote on p. 35; Strauss quote on p. 37), http://library.cqpress.com/cqalmanac/cqal78-1236641; Jo Thomas, "Senate Energy Panel Due to Vote on Nominee with Oil Industry Ties," *New York Times*, February 7, 1978, 6; "Scoop v. the Energy Knot," *Time*, February 13, 1978, 19; Morton Mintz, "Supreme Court Upholds Price Rise for Natural Gas," *Washington Post*, February 28, 1978, D11; "A Fast Fix for a Scarce Fuel," *Time*, March 13, 1978, 65; Warren Weaver Jr., "Bubble-Gum Makers Warm Up for Test on Baseball Cards," *New York Times*, April 18, 1978, 53; "California Is Winner in Gas Decision," *Los Angeles Times*, June 1, 1978, E14; "Intrastate Gas Switch Barred," *New York Times*, June 1, 1978, D3; "Top Court Backs Energy Agency on Sales of Gas," *Wall Street Journal*, June 1, 1978, 4; "U.S. Approval Required for Gas Markets Switch," *Washington Post*, June 1, 1978, D16; Morton Mintz, "Broiler Cooperative Ruled under Antitrust Laws," *Washington Post*, June 13, 1978, D10; "U.S. Probes Possible Improper Gas Sale in Intrastate Market," *Wall Street Journal*, June 22, 1978, 10; "U.S. to Investigate Units of Florida Gas and Indiana Standard," *Wall Street Journal*, August 22, 1978, 14; "Justices to Examine Bar on U.S. Efforts to Assure Interstate Sales of Natural Gas," *Wall Street Journal*, October 3, 1978, 6; "High Court to Hear Two Cases Involving Interstate Fights over Energy Resources," *Wall Street Journal*, October 11, 1978, 4; "Supreme Court Cases Today," *Washington Post*, November 28, 1978, A4; "Natural Gas: Sudden Glut," *Time*, January 8, 1979; "Natural Gas Up," *Time*, March 19, 1979, 74; Susan Fraker, John Walcott, Henry W. Hubbard, and Deborah Witherspoon, "Energy: A Year Later," *Newsweek*, April 24, 1978, 36; "OPEC Economics," *Wall Street Journal*, December 19, 1978, 24; and Pietro S. Nivola, "Energy Policy and the Congress: The Politics of the Natural Gas Policy Act of 1978," *Public Policy* 28 (1980), 491–492.

For more information on the implementation, consequences, and aftermath of the Natural Gas Policy Act of 1978, see "Natural Gas Act Beastly Tangle," *Hartford Courant*, September 1, 1978, 43 (including the "better than a zero" quote); James Abourezk, "Is the Natural-Gas Bill Better Than Nothing?" *Washington Post*, September 15, 1978, A15; Davis, *Energy Politics*, 162; Don E. Kash and Robert W. Rycroft, *U.S. Energy Policy: Crisis and Complacency* (Norman: University of Oklahoma Press, 1984), 208; MacAvoy, *The Natural Gas Market*, 15; Tussing and Barlow, *The Natural Gas Industry*, 229; and "Natural Gas," *Time*, January 18, 1983, 55.

On the Alaska natural gas pipeline, see Energy Information Administration, *Annual Energy Outlook 2009 with Projections to 2030* (Washington, DC: Energy Information Administration, 2009), at http://www.eia.doe.gov/oiaf/aeo/gas.html.

For discussion on the embattled history of LNG in the United States, see Michael Seiler, "Protesters Jam Hearing on LNG Facility Site," *Los Angeles Times*, April 12, 1978, B25; Penny Girard, "U.S. Aides Call Law on LNG Sites Invalid," *Los Angeles Times*, May 18, 1978, B3; Bupp and Schuller, "Natural Gas: How to Slice a Shrinking Pie," in *Energy Future*, ed. Stobaugh and Yergin, 69–70; OECD, International Energy Agency, *Natural Gas: Prospects to 2000*, 157; Dan Walters, "Natural Gas: An Old Story," *Sacramento Bee*, March 14, 2001, A3; "A Fast Fix for a Scarce Fuel," *Time*, March 13, 1978, 65; and "Natural Gas," *Time*, January 18, 1983, 63–65.

For more information on the history and development of natural gas vehicles and their associated federal tax credits, see *Energy Policy Act of 2005*, Public Law 109-58, §1341, 119 Stat. 594 (2005); National Renewable Energy Laboratory, U.S.

Department of Energy (DOE), *Natural Gas as a Transportation Fuel: Benefits, Challenges, and Implementation*, NREL/PR-540-41884 (Washington, DC: U.S. DOE, 2007), http://www.nrel.gov/docs/fy07osti/41884.pdf; Natural Gas Vehicles for America, *Federal Incentives for Natural Gas Vehicles* (Washington, DC: Natural Gas Vehicles for America, 2009), http://www.ngvc.org/pdfs/FederalVehicleTaxCredit.pdf; and Brent D. Yacobucci, Congressional Research Service, *Natural Gas Passenger Vehicles: Availability, Cost, and Performance* (Washington, DC: Congressional Research Service, 2008), http://opencrs.com/document/RS22971.

Much has been written about the growing domestic reserves and potential of natural gas. For estimates, see, for example, U.S. Energy Information Administration, *U.S. Crude Oil, Natural Gas, and Natural Gas Liquid Reserves* (Washington, DC: U.S. Energy Information Administration, October 29, 2009), http://www.eia.doe.gov/natural_gas/data_publications/crude_oil_natural_gas_reserves/cr.html, and U.S. Energy Information Administration, *Natural Gas Annual 2008* (Washington, DC: U.S. Energy Information Administration, March 2, 2010), http://www.eia.doe.gov/natural_gas/data_publications/natural_gas_annual/nga.html. Other estimates can be found at http://www.naturalgas.org/overview/resources.asp. The Colorado School of Mines reports the findings of the Potential Gas Committee at *Potential Gas Committee Reports Unprecedented Increases in Magnitude of U.S. Natural Gas Reserve Base* (Golden: Colorado School of Mines, June 18, 2009), http://www.mines.edu/Potential-Gas-Committee-reports-unprecedented-increase-in-magnitude-of-U.S.-natural-gas-resource-base. See also May Meyers Jaffe, "How Shale Gas Is Going to Rock the World," *Wall Street Journal*, May 10, 2010, R1; Jad Mouwed, "Estimate Places Natural Gas Reserves 35% Higher," *New York Times*, June 18, 2009, http://www.nytimes.com/2009/06/18/business/energy-environment/18gas.html.

Chapter 7

An excellent analysis of the effort to develop alternative energy sources is contained in Tomás Carbonell, "Confronting a 'Creeping Crisis': Lessons from the History of Alternative Energy Policy in the United States," May 1, 2008, unpublished manuscript, Yale University. For general background relating to renewable energy policy in the United States in the 1970s, see Amory B. Lovins, "Energy Strategy: The Road Not Taken?" *Foreign Affairs* 55, no. 1 (1976): 65–96; Barry Commoner, *The Politics of Energy* (New York: Knopf, 1979) (including the "economy of scale" quote); A. S. McFarland, *Public Interest Lobbies: Decision Making on Energy* (Washington, DC: American Enterprise Institute, 1976), 45–107; Peter V. Davis, "Selling Saved Energy: A New Role for the Utilities," in *Uncertain Power*, ed. Dorothy S. Zinberg (New York: Pergamon Press, 1983), 182: Modesto A. Maidique, "Solar America," in *Energy Future: The Report of the Harvard Business School Energy Project*, ed. Robert Stobaugh and Daniel Yergin (New York: Random House, 1983), 1238–1271; Richard E. Sclove, "Energy Policy and Democratic Theory," in *Uncertain Power*, ed. Zinberg, 37–65; Robert Stobaugh and Daniel Yergin, "Energy Wars," in *Energy Future*, ed. Stobaugh and Yergin, 1272–1290; Daniel Yergin, "Conservation: The Key Energy Source," in *Energy Future*, ed. Stobaugh and Yergin, 1173–1237; Richard H. K. Vietor and Louis Galambos, *Energy Policy in America since 1945: A Study of Business–Government Relations* (Cambridge, UK: Cambridge University Press, 1987); see also Allen L. Hammond, "The Sun and the Wind Are Soft, Nuclear Power and Coal Are Hard," *New York Times*, August 28, 1977, 148; Anthony J. Parisi, "'Soft' Energy, Hard Choices," *New York Times*, October 16, 1977, 123; Richard Corrigan and Dick

Kirschten, "The Energy Package—What Has Congress Wrought," *National Journal* 10, no. 44 (November 4, 1978), 1760; Denis Hayes, "Untangling America's Solar Energy Policy," *Los Angeles Times*, February 24, 1979, 1; Timothy B. Clark, "After a Decade of Doing Battle, Public Interest Groups Show Their Age," *National Journal* 12, no. 28 (July 12, 1980), 1136; Walter Rosenbaum, *Energy, Politics, and Public Policy* (Washington, DC: Congressional Quarterly Press, 1987); Robert M. Margolis and Daniel M. Kammen, "Evidence of Under-investment in Energy R&D in the United States and the Impact of Federal Policy," *Energy Policy* 27 (1999): 575–584; S. Gouchoe, V. Everette, and R. Haynes, National Renewable Energy Laboratory, *Case Studies on the Effectiveness of State Financial Incentives for Renewable Energy*, NREL/SR-620-32819 (Golden, CO: National Renewable Energy Laboratory, September 2002); Paul Roberts, *The End of Oil* (New York: First Mariner, 2004); Vaclav Smil, *Energy at the Crossroads* (Cambridge, MA: MIT Press, 2003); see also John J. Fialka and Jeffrey Ball, "Addiction Treatment: Bush's Latest Energy Solution, Like It's Forebears, Faces Hurdles; Fuel from 'Cellulosic Ethanol' Is Costly, Hard to Dispense; Broad Political Support; Enthusiasm from Detroit," *Wall Street Journal*, February 2, 2006, A1; Vaclav Smil, "Energy at the Crossroads," background notes for a presentation at the Global Science Forum Conference on Scientific Challenges for Energy Research, Paris, May 17–18, 2006; Gregory F. Nemet and Daniel M. Kammen, "U.S. Energy Research and Development: Declining Investment, Increasing Need, and the Feasibility of Expansion," *Energy Policy* 35 (2007): 746–755; and Terry Tamminen, *Lives per Gallon: The True Cost of Our Oil Addiction* (Washington, DC: Island Press, 2009).

For background information on the rise of solar energy leading up to and during the Carter administration, see Burton G. Malkiel, "What to Do about the End of the World," *New York Times*, January 26, 1975, 309; Michael Harwood, "But Not Soon," *New York Times*, March 16, 1975, SM111; "Highlights of the Energy Plan," *New York Times*, April 14, 1977, 1; "Transcript of Carter's Address to the Nation about Energy Problems," *New York Times*, April 19, 1977, 24; Charles Mohr, "Carter Asks Strict Fuel Saving; Urges 'Moral Equivalent of War' to Bar a 'National Catastrophe,'" *New York Times*, April 19, 1977, 69; Norma Skurka, "Tax Credit Plan Makes Sun Power Bargain at Home," *New York Times*, April 21, 1979, 56; Richard Haitch, "The President's 'Solar Advocate,'" *New York Times*, June 5, 1977, NJ1; Bayard Webster, "Test Indicates Solar Heating Isn't Economical Yet," *New York Times*, June 15, 1977, 81; Kirkpatrick Sale, "Consider the Windmill," *New York Times*, July 24, 1977, BR3; "E. F. Schumacher, 66, Economist Who Believed That 'Small Is Beautiful,'" *New York Times*, September 6, 1977, 42; Marlene Cimons, "The Goal of Sun Day: Solar Power," *Los Angeles Times*, December 1, 1977, A1 (including Munson quotes); Thomas W. Janes, "Homeowners Use Sun Power," *New York Times*, February 26, 1978, NJ13; "Carter Proclaims 'Sun Day,'" *New York Times*, March 27, 1978, 15; Bryce Nelson, "Panel Disputes Pessimistic Forecasts: Solar Energy Given Larger U.S. Role," *Los Angeles Times*, April 13, 1978, B19; "Environmentalists Accuse Carter of Selling Out," *Los Angeles Times*, April 24, 1978, A2; "'Soft Energy' Promoted as U.S. Solution," *Los Angeles Times*, April 26, 1978, D6; Bayard Webster, "Solar-Energy Fetes Due throughout U.S.," *New York Times*, May 2, 1978, 19; Gladwin Hill, "California Sees Itself as Leading Country to Era of Solar Energy," *New York Times*, May 4, 1978, B7; Martin Tolchin, "Carter Orders a Rise for Solar Research," *New York Times*, May 4, 1978, B6 (Including "speeding the use of solar technologies" quote and "equal financial footing" quote); Steven Rattner, "California Pushing Solar Power," *New York Times*, July 31, 1978, D1; Richard Phalon, "Energy Savers Tax Credits," *New York Times*, November 28, 1978, D4; Walter Sullivan, "Solar Energy Held Still Decades Away,"

New York Times, February 1, 1979, A7; Richard Halloran, "Solar Lobby—Striking While the Iron Is Still Hot," *New York Times*, April 29, 1979, E20; Anthony J. Parisi, "Solar Power—The Skies Have Not Yet Opened Up," *New York Times*, April 29, 1979, E20; Ward Sinclair, "Sun Shines Brightly, on Payday, at Solar Energy Center," *Washington Post*, June 9, 1979, A3; Steven Rattner, "President Setting Solar Power Goal," *New York Times*, June 19, 1979, A1; Martin Tolchin, "Carter Welcomes Solar Power," *New York Times*, June 21, 1979 (including the "20 percent" quote), D1; Tom Redburn, "Energy Crisis Spurs Study of Dozens of Alternatives," *Los Angeles Times*, June 25, 1979, B1; "Mr. Carter Flirts with the Sun," *New York Times*, June 26, 1979, A18; "Solar Energy Appointee," *New York Times*, July 27, 1979, D6; "Solar Activist, Denis Hayes, Heads SERI," *Science* 205 (August 10, 1979): 563; "Pentagon Plans Boost for Basic Research," *Science* 205 (August 10, 1979): 566; Paul Horvitz, "Solar Energy Incentives Spur Development," *New York Times*, August 13, 1979, D8; Harrison H. Donnelly, "Alternatives Overshadow Solar Energy," *Los Angeles Times*, October 18, 1979, K8; "$20 Billion Measure for Energy Voted; New Fuels Stressed," *New York Times*, November 10, 1979, 1; Anthony J. Parisi, "Despite U.S. Nudges, Solar Energy Moves Slowly," *New York Times*, December 9, 1979, 1; Anthony J. Parisi, "New Path for Solar Institute," *New York Times*, December 26, 1979, D1; Neal R. Pierce, "Will the 1980s Turn Out to Be 2nd Environmental Decade?" *Los Angeles Times*, April 22, 1980, C5; Steve Lohr, "Duncan Memo Asks Cut in Solar Energy Funds," *New York Times*, May 24, 1980, 29; Steve Lohr, "U.S. Affirms Solar Energy Goal," *New York Times*, May 29, 1980, D3; Tracy Kidder, "The Future of the Photovoltaic Cell," *Atlantic Monthly* (June 1980): 68–74, quoting Barry Commoner on pages 71–73; (also including the "expensive, relatively short-lived" quote at 71); Steve Lohr, "Solar Energy: New Procedure," *New York Times*, August 14, 1980, 2; "U.S. Solar Aide Says He's Being Ousted," *New York Times*, June 23, 1981, A17; "An Acting Director Is Chosen for Solar Institute," *New York Times*, August 4, 1981, A12; Denis Hayes, "Washington Decrees a Solar Eclipse," *New York Times*, August 12, 1981, A27; Diana Shaman, "New Tax Breaks for Solar Systems," *New York Times*, September 27, 1981, LI1; Nancy Traver, "An Advocate Replaced by a Loyalist: Researcher Takes Over Solar Institute," *Los Angeles Times*, November 27, 1981, M14; George W. Goodman, "States Still Cool to Solar Energy," *New York Times*, August 8, 1982, R6; Elizabeth M. Gunn, Steven C. Ballard, and Michael D. Devine, "The Public Utility Regulatory Policies Act: Issues in Federal and State Implementation," *Policy Studies Journal* 13, no. 2 (December 1984): 353–363; Matthew L. Wald, "Planning for Solar Energy While Tax Credits Continue," *New York Times*, May 17, 1984, C10; "Solar Moves toward Photovoltaics," *New York Times*, June 23, 1985, F15; Matthew L. Wald, "Solar Power's Future Unclear as Tax Credit Faces End," *New York Times*, December 30, 1985, A10; and Michael J. Weiss, "Everybody Loves Solar Energy, but . . . ," *New York Times*, September 24, 1989, BW64. For less optimistic views, see American Physical Society, *Principal Conclusions of the American Physical Society Study Group on Solar Photovoltaic Energy Conversion* (New York: American Physical Society, 1979). For the amount of solar energy in the year 2000 as a percentage of the total, see U.S. Energy Information Administration, *Annual Energy Review 2000* (Washington, DC: U.S. Energy Information Administration, August 2001), figure 10.1, http://tonto.eia.doe.gov/ftproot/multifuel/038400.pdf. Denis Hayes, a leading figure in the development of solar power, discusses solar energy in his book *Rays of Hope: The Transition to a Post-Petroleum World* (New York: W. W. Norton & Co., 1977).

For information on the microeconomic and technological challenges facing solar power, see Richard Phalon, "Energy Savers Tax Credits," *New York Times*, November 28, 1978, D4; William M. Pegram, "The Photovoltaic's Commercialization Program," in Linda R. Cohen and Roger G. Noll, with Jeffrey S. Banks, Susan A. Edelman, and William M. Pegram, *The Technology Pork Barrel* (Washington, DC: Brookings Institution, 1991), 321–383; and Beth Daley, "Jimmy Carter's Solar Panels Help Power a Maine College, Then Star in Film," Boston.com, June 20, 2008, at http://www.boston.com/lifestyle/green/greenblog/2008/06/jimmy_carters_solar_panels_hel_1.html.

On the Reagan administration's approach to solar energy, see "U.S. Solar Aide Says He's Being Ousted," *New York Times*, June 23, 1981, A17; "An Acting Director Is Chosen for Solar Research Institute," *New York Times*, August 4, 1981, A12; Denis Hayes, "Washington Decrees a Solar Eclipse," *New York Times*, August 12, 1981, A27; "U.S. Help Called Solar Industry Key," *Los Angeles Times*, September 22, 1981, OC2; Diana Shaman, "New Tax Breaks for Solar Systems," *New York Times*, September 27, 1981, LI1; Michael deCourcy Hinds, "U.S. Assistance Wanes for the Solar Industry," *New York Times*, November 4, 1982, C13; Michael deCourcy Hinds, "Solar Bank Now Open," *New York Times*, November 12, 1982, A24; Matthew L. Wald, "Planning for Solar Energy While Tax Credits Continue," *New York Times*, May 17, 1984, C10; and Matthew Wald, "Solar Power's Future Unclear As Tax Credit Faces End," *New York Times*, December 30, 1985, A10.

For more information about the history of hydropower in the United States, see Senator John F. Kennedy and Vice President Richard M. Nixon, First Joint Radio–Television Broadcast, Monday, September 26, 1960, http://www.jfklibrary.org/Historical+Resources/Archives/Reference+Desk/Speeches/JFK/JFK+Pre-Pres/1960/Senator+John+F.+Kennedy+and+Vice+President+Richard+M.+Nixon+First+Joint+Radio-Television+Broadcast.htm; U.S. General Accounting Office, *Implementation of the National Dam Inspection Act of 1972* (Washington, DC: U.S. General Accounting Office, March 15, 1977), http://www.gao.gov/products/100522; "Environment: Teton: Eyewitness to Disaster," *Time*, June 21, 1976, http://www.time.com/time/magazine/article/0,9171,918216,00.html (for Dale Howard quote); "The Nation: A Dam Breaks in Georgia," *Time*, November 21, 1977, http://www.time.com/time/magazine/article/0,9171,915720-1,00.html; Idaho Public TV, *The Bureau That Changed the West: One That Got Away: Teton Dam: Impressions from Governor Cecil Andrus*, http://idahoptv.org/outdoors/shows/bofr/teton/andrus.html (describing the Pat Oliphant cartoon); Walter Pincus, "When a Campaign Vow Crashes into a Pork Barrel," *Washington Post*, April 1, 1977, A1; C. L. Sanders and V. B. Sauer, U.S. Geological Survey Hydrologic Investigations, *Kelly Barnes Dam Flood of November 6, 1977, near Toccoa, Georgia*, Atlas HA-613 (Washington, DC: U.S. Geological Survey, 1979), http://ga.water.usgs.gov/publications/atlas/ha-613/index.html; Allison M. Conner and James E. Francford, Idaho National Engineering and Environmental Laboratory, *U.S. Hydropower Resource Assessment for Georgia*, report prepared for the U.S. Department of Energy (Washington, D.C.: U.S. Department of Energy, 1988), http://hydropower.inel.gov/resourceassessment/pdfs/states/ga.pdf; William Joe Simonds, "The Bureau of Reclamation and Its Archeology: A Brief History," *Cultural Resource Management* 23, no. 1 (2000), http://crm.cr.nps.gov/archive/23-01/23-01-2.pdf; Idaho National Laboratory, *Hydropower: Hydropower's Historical Progression* (Idaho Falls: Idaho National Laboratory, July 18, 2005), http://hydropower.inel.gov/hydrofacts/historical_progression.shtml; U.S. Department of the Interior, Bureau of Reclamation Power Resources Office, *Hydropower: Managing Water in the West* (Washington, DC: U.S. Department of the Interior, July 2005); Anthony DePalma, "East River Fights Effort to Tap Its Currents for Electricity," *New York Times*, August 13, 2007, B1; Kenneth M. Murchison, *The Snail Darter Case: TVA versus the Endangered Species Act* (Lawrence: University Press of Kansas, 2007); Adrian Guerrero, Benna

Crawford, Victoria Hayes, and Catie Heindel, "*Tennessee Valley Authority v. Hiriam Hill, et al.*: The Endangered Species Act and Dam Construction," student paper, Chicago-Kent College of Law, 2008; Zygmunt J. B. Plater, "Tiny Fish Big Battle," *Tennessee Bar Journal* 44, no. 4 (April 2008), http://www.tba.org/journal_new/index.php/component/content/article/256?ed=4; U.S. Army Corps of Engineers, *Hydropower: Value to the Nation* (Washington, DC: U.S. Army Corps of Engineers, 2009), http://www.vtn.iwr.usace.army.mil/hydro/default.htm; U.S. Department of Energy, "Obama Administration Announces up to $32 Million Initiative to Expand Hydropower," June 30, 2009, http://www.energy.gov/news2009/7555.htm; U.S. Department of the Interior, Bureau of Reclamation, *The History of Hydropower Development in the United States* (Washington, DC: U.S. Department of the Interior, August 12, 2009), http://www.usbr.gov/power/edu/history.html; Pierce Atwood, "Energy Secretary Chu States Support for Expansion of Hydropower," October 1, 2009, http://www.pierceatwood.com/showarticle.asp?Show=892; U.S. Geological Survey, "Hydroelectric Power Water Use," October 19, 2009, http://ga.water.usgs.gov/edu/wuhy.html; Glen Canyon Institute, *History of the River Restoration Movement*, http://www.glencanyon.org/library/riverrestoration.php; and Hydro Research Foundation, "Hydropower Today," available at http://www.hydrofoundation.org/hydropowereducation.html. That two-thirds of electricity generation of renewables in 2008 were hydropower can be found at http://www.eia.doe.gov/cneaf/electricity/epa/epa_sum.html. For historical data on electricity generation by energy source, see U.S. Energy Information Administration, *Annual Energy Review 2008* (Washington, DC: U.S. Energy Information Administration, June 2009), table 8.4b, http://www.eia.gov/emeu/aer/pdf/pages/sec8_18.pdf.

On the history of wind energy in the United States from the 1970s through the 1980s and the challenges faced, see R. Buckminster Fuller, "Energy through Wind Power," *New York Times*, January 17, 1974, 39; Nancy Hicks, "Energy Crisis Impels Many to Study and Erect Windmills as Power Source," *New York Times*, May 20, 1974, 67 (source of Welliver story and quote); "State Finances $30,000 Study on Windmill-Generated Electricity," *New York Times*, July 27, 1974, 64; "Around the World, Alternate Energy Is Sought," *New York Times*, January 26, 1975, 195; "Reinventing the Windmill—And Selling It," *New York Times*, March 16, 1975, F15; Lawrence C. Levy, "L.I. Man Setting Out to Prove That Energy of the Future Is Blowin' in the Wind," *New York Times*, June 22, 1975, 97; "Windmills to Be Studied as a State Energy Source," *New York Times*, July 3, 1975, 61; "A Huge Windmill to Be Built by U.S.," *New York Times*, August 8, 1976, 20; Richard D. Lyons, "U.S. Testing Six New Windmills as a Source of Power for Farms," *New York Times*, February 21, 1977, 45; Grace Lichtenstein, "Role for Wind Energy Seen Far Off," *New York Times*, May 13, 1977, 10; "House Panel Adopts Revised Energy Bill," *New York Times*, July 1, 1977, 6; Kirkpatrick Sale, "Consider the Windmill," *New York Times*, July 24, 1977, BR3; John M. Crewdson and Richard Halloran, "Energy Act Is Signed; Limits Seen," *New York Times*, November 10, 1978, A1; David Johnston, "Utility Regulation: U.S. or State Choice?" *Los Angeles Times*, November 20, 1978, D18; Betsy Brown, "Letting Hot Air Pay Off in Tax Credits," *New York Times*, November 26, 1978, WC8; "20 California Windmills Planned to Supply Power to 1,000 People," *New York Times*, April 22, 1979, 52; Deborah Rankin, "Taxes: Energy Credits Prove Popular," *New York Times*, September 4, 1979, D2; John Noble Wilford, "Winds and Tides Offer Hope on Energy," *New York Times*, December 12, 1979, D22; Frank Farwell, "New Design Spurs a Resurgence of Windmills," *New York Times*, February 3, 1980, NJ2; Frank Farwell, "New Energy: A Burgeoning Business in Windmills," *New York Times*, April 27, 1980, F3; Tom Furlong, "Wind Farm Site Sparks Land Boom," *Los Angeles Times*,

November 17, 1981, SDC1; "Wind-Energy Inquiries Show Marked Increase," *New York Times*, December 3, 1981, C9; Richard O'Reilly, "Wind Energy Plans Weather the Storms," *Los Angeles Times*, February 21, 1982, B1; Robert E. Tomasson, "Harnessing Wind to Generate Power for Homes," *New York Times*, November 14, 1982, R7; Wallace Turner, "Private Investors Selling Wind Power to Utilities," *New York Times*, February 14, 1983, A14; William J. Broad, "Small, Cheap Windwills Outdo High Tech," *New York Times*, August 14, 1984, C1; Sylvia White, "Towers Multiply, and Environment Is Gone with the Wind," *Los Angeles Times*, November 26, 1984, C5; "Wind Power Shifts into High Gear," *New York Times*, June 23, 1985, F15; Larry Fisher, "The Threat to Wind Energy," *New York Times*, October 26, 1985, 33 (source of Stark quote); Robert Reinhold, "The Promise of Wind Awaits a New Energy Crisis," *New York Times*, May 22, 1988, 151 (source of Gray quote and Liebold quote); Thomas A. Starrs, "Legislative Incentives and Energy Technologies: Government's Role in the Development of the California Wind Industry," *Ecology Law Quarterly* 15 (1988), 116. See also Robert W. Richter, *Wind Energy in America: A History* (Norman: University of Oklahoma Press, 1996), 225–226; Richard F. Hirsh, *Power Loss: The Origins of Deregulation and Restructuring in the American Electric Utility System* (Cambridge, MA: MIT Press, 1999), 96–97; Mark Gielecki, Fred Mayes, and Lawrence Prete, "Incentive, Mandates, and Government Programs for Promoting Renewable Energy," in *Renewable Energy 2000: Issues and Trends*, DOE/EIA-0628(2000) (Washington, DC: U.S. Energy Information Administration, 2001), 10–12, http://tonto.eia.doe.gov/FTPROOT/renewables/06282000.pdf (source of Ekland quote); and U.S. Energy Information Administration, *International Electricity Generation*, http://www.eia.doe.gov/emeu/international/electricitygeneration.html. For more recent analysis, see Louise Guey-Lee, "Forces behind Wind Power," in *Renewable Energy 2000*, 73, and U.S. Energy Information Administration, *Annual Energy Review 2008*, table 8.4b.

For discussion of the alternative technologies movement, see Ernst Friedrich Schumacher, *Small Is Beautiful: Economics As If People Mattered* (New York: Harper and Row, 1973); Burton G. Malkiel, "What to Do about the End of the World," *New York Times*, January 26, 1975, 309; Ann Crittendon, "Move Back to Smallness Urged for Manufacturers," *New York Times*, October 22, 1975, 72; James T. Wooten, "A Mined-Out Coal Town Gains New Life as Social Laboratory," *New York Times*, November 12, 1975, 45; Amory B. Lovins, *World Energy Strategies: Facts, Issues, and Options* (New York: Harper and Row, 1975); Amory B. Lovins and John H. Price, *Non-nuclear Futures: The Case for an Ethical Energy Strategy* (New York: Harper Colophon, 1975); Amory B. Lovins, "Energy Strategy: The Road Not Taken?" *Foreign Affairs* 55, no. 1 (1976): 65–96, advocating technologies that "rely on renewable energy flows that are always there whether we use them or not, such as sun and wind and vegetations" (77); Stewart Brand, *Whole Earth Catalog* (Menlo Park, CA: Portola Institute, 1968); *Rain* magazine staff, *Rainbook: Resources for Appropriate Technology* (New York: Schocken Books, 1977); Kirkpatrick Sale, "Consider the Windmill," *New York Times*, July 24, 1977, BR3 (source of "alternative technologies movement" and "manifestations everywhere" quotes; Allen L. Hammond, "The Hard and the Soft Technology of Energy," *New York Times*, August 28, 1977, 148; "E. F. Schumacher, 66, Economist Who Believed That 'Small Is Beautiful,'" *New York Times*, September 6, 1977, 42; "At Conference on Alternative Energy, 'Simple Is Essential,'" *New York Times*, December 6, 1977, 18 (quoting Anthony Maggiore); Amory B. Lovins, *Soft Energy Paths: Toward a Durable Peace* (New York: Harper and Row, 1979); and Eleanor Charles, "Homesteading the Back Yard," *New York Times*, March 23, 1980, WC1.

For background on the history of ethanol as an alternative energy source in the United States, see Walter S. Mossberg, "Can U.S. Reduce Imports with Gasohol?

Some Say Yes, but Officials Are Dubious," *Wall Street Journal*, July 12, 1978, 39; "Administration Would Like to Give Gasohol a Push," *Los Angeles Times*, July 13, 1978, G14; Harry Anderson, "'Gasohol'—New Spark for Old Idea," *Los Angeles Times*, October 7, 1978, A1; Richard D. Lyons, "America Is Looking Twice for Enough Energy to Burn," *New York Times*, February 25, 1979, E18; "Further Ethanol Price Rise," *New York Times*, March 8, 1979, "Gasohol Virtues—Fuel Saver or Fast Sales Pitch on Snake Oil?" *Chicago Tribune*, April 15, 1979, K22; Alexander R. Hammer, "Gasoline Problems Weaken Market," *New York Times*, May 15, 1979, D8; John T. McQuiston, "Gasohol: The Issue Is Practicality," *New York Times*, May 19, 1979, 29; Richard Orr, "U.S.: Don't Plan on Gasohol from Food Crops," *Chicago Tribune*, May 21, 1979, D9; John Gorman, "U.S. Study Urges Tax Policy to Foster Gasohol," *Chicago Tribune*, July 12, 1979, C7; "Case against Gasohol," *Chicago Tribune*, July 23, 1979, D7; Hal Bernton, "The Godfather of Gasohol Is . . . Henry Ford!" *Washington Post*, August 5, 1979, B1; Jane Seaberry, "Gasohol," *Washington Post*, August 26, 1979, K1; John T. McQuiston, "Sales of Gasohol Gaining Support in 3-State Area," *New York Times*, September 17, 1979, B2; Warren Weaver Jr., "Senate Backs Plan on Synthetic Fuels; Victory for Carter," *New York Times*, November 9, 1979, A1; "New York City to Try Gasohol," *New York Times*, November 14, 1979, B4; William Robbins, "Corn and the Economy," *New York Times*, December 2, 1979, F1; William H. Hoge, "Developing Gasohol," *New York Times*, December 3, 1979, 25; and Commoner, *The Politics of Energy* (Zeithamer story and quote on pages 41–44). See also "Carter Aide Says U.S.-Embargoed Corn Will Be Used in New Gasohol Program," *Los Angeles Times*, January 7, 1980, 7; Morton Mintz, "Major New Gasohol Program Planned," *Washington Post*, January 7, 1980, A1; Charles Mohr, "Carter to Announce Gasohol Plan Soon," *New York Times*, January 7, 1980, D3; Richard D. Lyons, "U.S. Struggles for Gasohol Plan," *New York Times*, January 8, 1980, D4; "Exxon Puts Limits on Gasohol Sales," *New York Times*, January 9, 1980, D20; Richard D. Lyons, "U.S. Details Gasohol Program," *New York Times*, January 12, 1980, 27; "Gasohol without Intoxication," *New York Times*, January 14, 1980, A16; Richard D. Lyons, "The Outlook on Fuel: Plentiful but Costly," *New York Times*, January 27, 1980, AUTO3; "Ethanol Price to Rise," *New York Times*, March 5, 1980, D16; Terry Brown, "Gasohol's Effect on Food Supply," *Chicago Tribune*, March 19, 1980, C10; Fred H. Sanderson, "Gasohol from Corn—Good Sense?" *New York Times*, May 3, 1980, 23; "Ethanol Complex Planned," *New York Times*, June 23, 1980, D3; "Department of No Shame," *Wall Street Journal*, July 2, 1980, 14; "Company News: Texaco and CPC Join in Ethanol Plant," *New York Times*, July 17, 1980, D5; Steve Lohr, "Business People: 'Synfuel Man' Planning Large Gasohol Plant," *New York Times*, July 31, 1980, D2; "A Welcome for Ethanol Plant," *New York Times*, August 8, 1980, D1; "Gasohol Gain Reported," *New York Times*, August 23, 1980, 37; John S. Rosenberg, "State's Role in Synthetic Fuel," *New York Times*, October 26, 1980, CN21; Robert D. Hershey, "Gasohol's Market Share Is Meager 1%–2% in U.S.," *New York Times*, October 13, 1980, D2; Clyde H. Farnsworth, "Washington Watch: Farm Lobby's Gasohol Victory," *New York Times*, December 1, 1980, D2; "Ethanol-Plant Guarantees," *New York Times*, December 26, 1980, D5; "Company News: New Ethanol Plant," *New York Times*, February 12, 1981, D4; Beverly Kitching, "Converting Corn into Alcohol for Fuel," *Chicago Tribune*, February 19, 1981, B2; Douglas Martin, "Budget Cuts, Weak Market Hurt Gasohol," *New York Times*, December 14, 1981, A1; Nicholas Madigan, "An Idea Whose Time Is Going," *New York Times*, January 31, 1982, section 12, page 10; "Agricultural Problem Seen," *New York Times*, February 15, 1982, D2; Ernst R. Habicht Jr., "Prepare for the Next Oil Shock," *New York Times*, March 2, 1982, A22; "A History of Hedging Political Bets," *New York Times*, March 25, 1985, D3; Steven Greenhouse,

"Crusading for Grain Exports," *New York Times*, March 25, 1985, D1; "Import Duty on Ethanol," *New York Times*, August 3, 1985, 41; Michael Isikoff, "Ethanol Producer Reaps 54% of U.S. Subsidy," *Washington Post*, January 29, 1987, A14; Carolyn Dimitri and Anne Effland, "Fueling the Automobile: An Economic Exploration of Early Adoption of Gasoline over Ethanol," *Journal of Agricultural and Food Industrial Organization* 5, no. 2 (2007): art. 11, http: www.bepress.com/jafio/vol5/iss2/art11; U.S. Government Accountability Office, *Petroleum and Ethanol Fuels: Tax Incentives and Related GAO Work*, GAO/RCED-00-301R (Washington, DC: U.S. Government Accountability Office, September 25, 2000), http://www.gao.gov/new.items/rc00301r.pdf; Robert Bryce, "The Corn Ethanol Juggernaut," *Yale Environment* 360 (September 15, 2008), http://e360.yale.edu/content/feature.msp?id=2063; and "Ethanol Fuel in the United States," *Wikipedia*, http://en.wikipedia.org/wiki/Ethanol_fuel_in_the_United_States (accessed on June 11, 2009).

For further discussion of ethanol policy, see Jerry Flint, "Fuel for Debate: Gasoline or Alcohol?" *New York Times*, January 29, 1978, AUTO11; Anthony J. Parisi, "Technology: Gasoline–Alcohol Mixture Ignites Dispute," *New York Times*, May 3, 1978, D3; J. P. Smith, "Gasohol Advocates Lobbying Intensely," *Washington Post*, April 23, 1979, A1 (including the "gasohol lobby says" quote); Pasqual A. Donvito, "Gasohol: The Real Issue Is BTU's," *New York Times*, July 13, 1980, F18 (including oil industry study saying plant would use 15 percent more energy than it would produce); Robert D. Hershey, "Blessing or Boondoggle? The $88 Billion Quest for Synthetic Fuels," *New York Times*, September 21, 1980, F1; Earle E. Gavett, Gerald E. Grinnell, and Nancy L. Smith, *Fuel Ethanol and Agriculture: An Economic Assessment*, Agricultural Economic Report no. 562 (Washington, DC: U.S. Department of Agriculture, August 1986), 4 (source of 4.7 percent fewer quote); James Bovard, *Archer Daniels Midland: A Case Study in Corporate Welfare*, Policy Analysis no. 241 (Washington, DC: Cato Institute, September 26, 1995), http:/www.cato.org/pubs/pas/pa-241.html; Stephen Moore, *Push Ethanol off the Dole* (Washington, DC: Cato Institute, July 10, 1997), http://www. cato.org/pub_display.php?pub_id=6123.

For more contemporary discussion, see Robert K. Niven, "Ethanol in Gasoline: Environmental Impacts and Sustainability Review Article," *Renewable and Sustainable Energy Reviews* 9 (2005): 535–555; Roel Hammerschlag, "Ethanol's Energy Return on Investment: A Survey of the Literature 1990–Present," *Environmental Science and Technology* 40, no. 6 (2006): 1744–1750; John J. Fialka, "Politics & Economics: Coalition Pushes Wider Ethanol Use; Group Seeks to Include Proposals on Farm Bill, Lobby for $64.5 Billion," *Wall Street Journal*, February 28, 2007, A6; Joseph E. Carolan, Satish V. Joshi, and Bruce E. Dale, "Technical and Financial Feasibility Analysis of Distributed Bioprocessing Using Regional Biomass Pre-processing Centers," *Journal of Agricultural & Food Industrial Organization* 5, no. 2 (2007): art. 10, http://www.bepress.com/jafio/vol5/iss2/art10; Carolyn Dimitri and Anne Effland, "Fueling the Automobile: An Economic Exploration of Early Adoption of Gasoline over Ethanol," *Journal of Agricultural & Food Industrial Organization* 5, no. 2 (2007): art. 11, http://www.bepress.com/jafio/vol5/iss2/art11; Amani Elobeid and Chad Hart, "Ethanol Expansion in the Food versus Fuel Debate: How Will Developing Countries Fare?" *Journal of Agricultural & Food Industrial Organization* 5, no. 2 (2007): art. 6, http://www.bepress.com/jafio/vol5/iss2/art6; Bruce Gardner, "Fuel Ethanol Subsidies and Farm Price Support," *Journal of Agricultural & Food Industrial Organization* 5, no. 2 (2007): art. 4, http://www.bepress.com/jafio/vol5/iss2/art4; Bruce Gardner and Wallace Tyler, "Explorations in Biofuels Economics, Policy, and History: Introduction to the Special Issue," *Journal of Agricultural & Food Industrial Organization* 5, no. 2 (2007): art. 1, http://www.bepress.com/jafio/vol5/iss2/art1; Angelo

Gurgel, John M. Reilly, and Sergey Paltsev, "Potential Land Use Implications of a Global Biofuels Industry," *Journal of Agricultural & Food Industrial Organization* 5, no. 2 (2007): art. 9, http://www.bepress.com/jafio/vol5/iss2/art9; 4150–57; Tatsuji Koizumi and Keiji Ohga, "Biofuels Policies in Asian Countries: Impact of the Expanded Biofuels Programs on World Agricultural Markets," *Journal of Agricultural & Food Industrial Organization* 5, no. 2 (2007): art. 8, http://www.bepress.com/jafio/vol5/iss2/art8; Ariadna Martinez-Gonzalez, Ian M. Sheldon, and Stanley Thompson, "Estimating the Welfare Effects of U.S. Distortions in the Ethanol Market Using a Partial Equilibrium Trade Model," *Journal of Agricultural & Food Industrial Organization* 5, no. 2 (2007): art. 5, http://www.bepress.com/jafio/vol5/iss2/art5; Andrew Schmitz, Charles B. Moss, and Troy G. Schmitz, "Ethanol: No Free Lunch," *Journal of Agricultural & Food Industrial Organization* 5, no. 2 (2007): art. 3, http://www.bepress.com/jafio/vol5/iss2/art3; Samantha Slater, Renewable Fuels Association, *Today's U.S. Ethanol Industry* (Washington, DC: Renewable Fuels Association, April 11, 2007) (for an estimated number of U.S. distilleries and ethanol output in 2008; Wallace E. Tyner, "Policy Alternatives for the Future Biofuels Industry," *Journal of Agricultural & Food Industrial Organization* 5, no. 2 (2007): art. 2, http://www.bepress.com/jafio/vol5/iss2/art2; Paul C. Westcott, *Ethanol Expansion in the United States: How Will the Agricultural Sector Adjust?* no. FDS-07D-01 (Washington, DC: Economic Research Service, U.S. Department of Agriculture, May 2007); Bryce, "The Corn Ethanol Juggernaut"; Harry De Gorter and David R. Just, *The Law of Unintended Consequences: How the U.S. Biofuel Tax Credit with a Mandate Subsidizes Oil Consumption and Has No Impact on Ethanol Consumption*, Working Paper no. 2007-20 (Ithaca, NY: Cornell University, February 1, 2008), http://ssrn.com/abstract=1024525; Royal Society, *Sustainable Biofuels: Prospects and Challenges* (London: Royal Society, January 2008), http://royalsociety.org/sustainable-biofuels-prospects-and-challenges; John M. Urbanchuk, *Economic Contribution of the Partial Exemption for Ethanol from the Federal Excise Tax on Motor Fuel Increased Revenues and Reduced Dependence on Foreign Oil* (Emeryville, CA: LECG, November 18, 2008); Joshua A. Blonz, Shalini P. Vajjhala, and Elena Safirova, *Growing Complexities: A Cross-Sector Review of U.S. Biofuels Policies and Their Interactions*, Discussion Paper no. 08-47 (Washington, DC: Resources for the Future, December 30, 2008), http://papers.ssrn.com/sol3/papers.cfm?abstract_id=1368419; Ryan Blitstein, "ADM Does the Ethanol Shuffle," *The Big Money*, June 2, 2009, http://www.thebigmoney.com/print/2341; "Ethanol's Grocery Bill," *Wall Street Journal*, June 2, 2009, A20; and Robert Hahn and Caroline Cecot, "The Benefits and Costs of Ethanol: An Evaluation of the Government's Analysis," *Journal of Regulatory Economics* 35, no. 3 (June 2009): 275–295. Data on renewable energy sources in 2008 and 1977 are from U.S. Energy Information Administration, *Annual Energy Review 2008*, the diagram available at http://people.virginia.edu/~gdc4k/phys111/fall09/important_documents/aer_2008.pdf. The CBO study describing the costs of ethanol subsidies per gallon of gasoline and ton of carbon dioxide emissions is Congressional Budget Office, "Using Biofuel Tax Credits to Achieve Energy and Environmental Policy Goals," July, 2010, www.cbo.gov/doc.cfm?index=11477.

For details on the efforts to conserve energy in the United States during the 1970s and 1980s, see Demand and Conservation Panel of the Committee on Nuclear and Alternative Energy Systems, "U.S. Energy Demand: Some Low Energy Futures," *Science* 200, no. 4338 (April 14, 1978): 142–151; Daniel Yergin, "Conservation: The Key Energy Source," in *Energy Future*, ed. Stobaugh and Yergin, 136–182 ("using 40 percent less energy" quote on page 177); Reginald Stuart, "Auto Industry Prepares for Clash with U.S. over Fuel Standards," *New York Times*, March 12, 1979, B9; James Reston, "O Jimmy! It's Easy!" *New York Times*, June 1, 1979, A25; Lee Dembart,

"Candidates Agree Only That Energy Is a Problem," *Los Angeles Times*, May 15, 1980, B1; Hugh O'Haire, "Con Ed Battles Energy Officials on Cogeneration," *New York Times*, November 2, 1980, R1; Douglas Martin, "Gains in Saving Oil Change U.S. Outlook," *New York Times*, January 4, 1981, 1 (including "we'll be hot in the summer" quote); Douglas Martin, "Talking Business: The Bonanza of Fuel Savings," *New York Times*, January 6, 1981, D2; John Tirman, "When Markets Have Little Reason to Save Energy," *New York Times*, January 6, 1981, 30; Robert D. Hershey Jr., "U.S. Study Says Conservation Could Slash Energy Use," *New York Times*, March 15, 1981, 23; "High Court Upholds Utility Rules of U.S.," *New York Times*, May 17, 1983, D5; Don E. Kash and Robert W. Rycroft, *U.S. Energy Policy: Crisis and Complacency* (Norman: University of Oklahoma Press, 1984), 185–186, 238; Steven J. Marcus, "Small Sources Show Appeal," *New York Times*, February 13, 1984, D1; Stuart Diamond, "Cogeneration Jars the Power Industry," *New York Times*, June 10, 1984, F1; Reginald Stuart, "U.S. Alters Mileage Rules, Aiding Ford and G.M.," *New York Times*, June 27, 1985, A1; International Energy Agency, Organization for Economic Cooperation and Development (OECD), *Energy Conservation in IEA Countries* (Paris: OECD, 1987), http://www.iea.org/textbase/nppdf/free/1980/Ene_Cons_1987.pdf; Andrew Pollack, "Non-utility Electricity Rising," *New York Times*, August 12, 1987, D1; Daniel Yergin, *The Prize: The Epic Quest for Oil, Money, and Power* (New York: Simon and Schuster, 1991), 718; Matthew L. Wald, "Utilities Say '81 New York Law Is Proving Costly," *New York Times*, June 16, 1992, D15; and Agis Salpukas, "Utilities See Costly Time Warp in '78 Law," *New York Times*, October 5, 1994, D1. For a review of conservation practices across countries, see Howard Geller and Sophie Attali, *The Experience with Energy Efficiency Policies and Programmes in IEA Countries*, information paper (Paris: International Energy Agency, August 2005).

For more analysis of the future of alternative and renewable energy in the United States, see Michael E. Kraft and Regina S. Axelrod, "Political Constraints on Development of Alternative Energy Sources: Lessons from the Reagan Administration," *Policy Studies Journal* 13, no. 2 (1984): 319–330; Richard K. Worthington, Rennselaer Polytechnic Institute, "Renewable Energy Policy and Politics: The Case of the Windfall Profits Tax," *Policy Studies Journal* 13 (1984), 365; Thomas A. Starrs, "Legislative Incentives and Energy Technologies: Government's Role in the Development of the California Wind Energy Industry," *Ecology Law Quarterly* 15 (1988), 103; Dana Priest, "Putting New Energy into Alternatives," *Washington Post*, August 13, 1990, A9; Paul Gipe, "Wind Energy Comes of Age: California and Denmark," *Energy Policy* 19, no. 8 (1991), 756; Gary Lee, "Auto Fuel-Efficiency Plan Sparks a New Round of Heated Lobbying," *Washington Post*, August 5, 1991, A7; Agis Salpukas, "70's Dreams, 90's Realities," *New York Times*, April 11, 1995, D1; Office of Technology Assessment, U.S. Congress, *Renewing Our Energy Future*, OTA-ETI-614 (Washington, DC: U.S. Congress, September 1995); David L. Greene, "Why CAFE Worked," *Energy Policy* 26, no. 8 (1998): 595–613; Louise Guey-Lee, "Renewable Electricity Purchases: History and Recent Developments," in *Renewable Energy 1998: Issues and Trends*, DOE/EIA-0628(1998) (Washington, DC: U.S. Energy Information Administration, March 1999), 1; Louise Guey-Lee, "Wind Energy Developments: Incentives in Selected Countries," in *Renewable Energy 1998*, 79; Mark Gielecki, Fred Mayes, and Lawrence Prete, "Incentives, Mandates, and Government Programs for Promoting Renewable Energy," in *Renewable Energy 2000*, 1; Peter Holihan, "Technology, Manufacturing, and Market Trends in the U.S. and International Photovoltaics Industry," in *Renewable Energy 2000*, 19; Louise Guey-Lee, "Forces behind Wind Power," in *Renewable Energy 2000*, 73; John P. Holdren, written testimony, *Energy Efficiency and Renewable Energy in the United States Energy Future: Hearing before the Committee on Science, House of Representatives*, 107th Cong., 1st sess., February 28, 2001; Joel Makower and Ron Pernick, *Clean Tech: Profits and Potential* (San Francisco: Clean Edge: The Clean-Tech Market Authority, April 2001); José Goldemberg, Suani Teixeira Coelho, and Oswaldo Lucon, "How Adequate Policies Can Push Renewables," *Energy Policy* 32, no. 9 (2004): 1141–1146; International Energy Agency, Organization for Economic Cooperation and Development (OECD), *Renewable Energy: Market & Policy Trends in IEA Countries* (Paris: OECD, 2004); S. Pacala and R. Socolow, "Stabilization Wedges: Solving the Climate Problem for the Next 50 Years with Current Technologies," *Science* 305 (August 13, 2004): 968; Jeremy Leggett, *Half Gone* (London: Portobello, 2005), 154–197; Robert D. Perlack, Lynn L. Wright, Anthony F. Turhollow, Robin L. Graham, Bryce J. Stokes, and Donald C. Erbach, Oakridge National Laboratory, *Biomass as Feedstock for a Bioenergy and Bioproducts Industry: The Technical Feasibility of a Billion-Ton Annual Supply*, DOE/GO-102995–215 (Oak Ridge, TN: U.S. Department of Energy, April 2005); Staffan Jacobson and Volkmar Lauber, "The Politics and Policy of Energy System Transformation—Explaining the German Diffusion of Renewable Energy Technology," *Energy Policy* 34, no. 11 (2006): 256–276; International Energy Agency, Organization for Economic Cooperation and Development (OECD), *World Energy Outlook: 2006* (Paris: OECD, 2006), 65–84; W. Musial, S. Butterfield, and B. Ram, National Renewable Energy Laboratory, "Energy from Offshore Wind," paper presented at National Renewable Energy Laboratory conference, NREL/CP-500-39450, Golden, CO, February 2006, http://www.nrel.gov/docs/fy06osti/39450.pdf; *Renewable Global Status Report: 2006 Update* (Paris: REN21: Renewable Energy Policy Network for the 21st Century, 2006), http://www.ren21.net/pdf/RE_GSR_2006_Update.pdf; Gordon Edge, "A Harsh Envrionment: The Non–Fossil Fuel Obligation and the UK Renewable Industry," in *Renewable Energy Policy and Politics*, ed. Karl Mallon (London: Earthscan, 2006), 163–184; Karl Mallon, "Ten Features of Successful Renewable Markets," in *Renewable Energy Policy and Politics*, ed. Mallon, 35–84; Randall Swisher and Kevin Porter, "Renewable Policy Lessons from the U.S.: The Need for Consistent and Stable Policies," in *Renewable Energy Policy and Politics*, ed. Mallon, 185–198; Sven Teske and Volker U. Hoffmann, "A History of Support for Solar Photovoltaics," in *Renewable Energy Policy and Politics*, ed. Mallon, 229–240; Chris Greenwood, Alice Hohler, George Hunt, Michael Liebreich, Virginia Sonntag-O'Brien, and Eric Usher, United Nations Environment Program (UNEP), *Global Trends in Sustainable Energy Investment 2007: Analysis of Trends and Issues in the Financing of Renewable Energy and Energy Efficiency in OECD and Developing Countries* (Paris: UNEP, 2007); Anthony Perl and James A. Dunn Jr., "Reframing Automobile Fuel Economy Policy in North America: The Politics of Punctuating a Policy Equilibrium," *Transport Reviews* 27, no. 1 (January 2007): 1–35; Bill Prindle, Maggie Eldridge, Mike Eckhardt, and Alyssa Frederick, American Council for an Energy-Efficient Economy (ACEEE), *The Twin Pillars of Sustainable Energy: Synergies between Energy Efficiency and Renewable Energy Technology and Policy* (Washington, D.C.: ACEEE, May 2007); Steven Plotkin, "Examining New U.S. Fuel Economy Standards," *Environment* 49, no. 6 (July–August 2007), 8; Christopher Simon, *Alternative Energy: Political, Economic, and Social Feasibility* (Lanham, MD: Rowman and Littlefield, 2007); Bo Nelson Vattenfall, *Vattenfall's Global Climate Impact Abatement Map* (Paris: International Energy Agency, February 15, 2007); Bo Nelson Vattenfall, *Vattenfall's Climate Map 2030* (Paris: International Energy Agency, 2007); Ronald Bailey, "Alternative Energy Subsidies Make Their Biggest Comeback since Jimmy Carter," *Reason* magazine, June 2009, http://www.reason.com/news/show/133227.html; International Energy Agency (IEA), "Renewable Energy—Markets and Policy Trends in IEA Countries," press release, http://www.iea.org/

Textbase/press/pressdetail.asp?PRESS_REL_ID=128 (last visited June 11, 2009); Third Way and Greenberg Quinlan Rosner Research, "Get America Running on Clean Energy: Findings from National Focus Groups," memo, June 16, 2009; Jonathan Frahey, "Wind Power's Weird Effect," Forbes.com: OutFront, August 19, 2009, http://www.forbes.com/forbes/2009/0907/outfront-energy-exelon-wind-powers-weird-effect.html; and Todd Woody, "Alternative Energy Projects Stumble on a Need for Water," *New York Times*, September 30, 2009, B1.

Chapter 8

The events leading up to Carter's "crisis of confidence" speech and the ramifications of this speech are thoroughly explored in Kevin Mattson, *"What the Heck Are You Up to, Mr. President?": Jimmy Carter, America's "Malaise," and the Speech That Should Have Changed the Country* (New York: Bloomsbury, 2009). Mattson's account unsurprisingly focuses heavily on the energy challenges that bedeviled Carter's administration. Carter offers his own account of these challenges in Jimmy Carter, *Keeping Faith: Memoirs of a President* (New York: Bantam, 1982). Two other important works for understanding energy policy during the Carter and early Reagan administrations are Daniel Yergin, *The Prize: The Epic Quest for Oil, Money, and Power* (New York: Free Press, 1991), and James T. Patterson, *Restless Giant: The United States from Watergate to* Bush v. Gore (New York: Oxford University Press, 2005).

Mattson discusses America's experiences with gas station lines in 1979 as well as truckers and the Levittown riots in *"What the Heck?"* 109–113, 113–119. Yergin, *The Prize*, examines the effects of the Iranian Revolution on oil prices, including increased demand and stockpiling of reserve supplies (684–687), the oil wasted daily while cars idled in line for gas (692), the rise in price from $13 to $34 a barrel (685), and the figures indicating total shortfall of 10 percent of consumption (687). The skyrocketing of prices to $39.50 per barrel is recounted in Jad Mouawad, "Oil Prices Pass Record Set in '80s, but Then Recede," *New York Times*, March 3, 2008, http://www.nytimes.com/2008/03/03/business/worldbusiness/03cnd-oil.html.

On the Tokyo G8 summit and Carter's return to Washington to deal with the energy crisis, see Carter, *Keeping Faith*, 111–114. On preparations for a July 5 televised energy speech, see Mattson, *"What the Heck?"* 129–131. On Caddell's advice, the decision to broaden the speech's scope, and cancellation of the original energy speech, see Carter, *Keeping Faith*, 114–120. For more on the cancellation of the energy speech and development of the "crisis of confidence" speech, see Elizabeth Drew, "Phase: In Search of a Definition," *The New Yorker* (August 22, 1979): 45–73; the Mondale quote is found on page 59. For more on Mondale's reaction and on Caddell, see Mattson, *"What the Heck?"* 133, 21–22; Carter's 27 percent approval rating is cited on page 130.

On Carter's Camp David speech preparations and his discussions with advisors, see Carter, *Keeping Faith*, 115–120, which includes his quote about the public's doubts (117) and his diary entry on accepting criticism (118). For more on the advice of Camp David visitors, including Bill Clinton, see Mattson, *"What the Heck?"* 135–145, 147–148. On delivery of the July 15 speech and the viewing audience figure, see Carter, *Keeping Faith*, 120–121. Mattson, *"What the Heck?"* 21, cites the smaller audience for the April speech. The text of Carter's July 15, 1979, televised speech can be found at http://www.pbs.org/wgbh/amex/carter/filmmore/ps_crisis.html, from which the quotes here are drawn. See Mattson for the influence of Lasch and Wolfe on the speech (42–46), Carter's overnight approval ratings bump (121), and the postspeech cabinet firings (167–169); see also Christopher Lasch, *The Culture of Narcissism: American Life in an Age of Diminishing Expectations* (New York: W. W. Norton, 1979), and Tom Wolfe, "The 'Me' Decade and the Third Great Awakening," *New York Magazine* (August 23, 1976): 26–40. For Carter's account of the firings and public reaction, see Carter, *Keeping Faith,* 121. Drew, "Phase," 73, reports the drop in Carter's approval ratings back to 25 percent by late July. For the opinions held by Caddell, Hertzberg, and others that the cabinet purge, not the speech, was responsible for Carter's undoing, see "Jimmy Carter," episode of *American Experience* (PBS), film transcript, 2002, part 2, http://www.pbs.org/wgbh/amex/carter/filmmore/pt_2.html. Wolfe's disparaging quote about Carter is from "Entr'actes and Canapes," in *In Our Time* (New York: Farrar, Straus and Giroux, 1980), 22, quoted in Patterson, *Restless Giant,* 110.

The onset of the Iranian hostage crisis is recounted in Yergin, *The Prize,* 699–703. Carter, *Keeping Faith,* describes his learning of the hostage situation (457), his prior meetings with the shah (434–437), his observations of the shah (434–435), and his toast to Iran's stability (437). *Keeping Faith* also recounts the imposition of marshal law in Iran (438), Carter's diary quotes on the shah (440), commencement of the general strike in Iran (441), the shah's departure (446–447), the decision to allow the shah to enter the United States for treatment (454–456), and the timing of Kennedy's and Brown's candidacy announcements (463). On the ayatollah's designation of Iran as an Islamic republic, see Mattson, *"What the Heck?"* 15–16. Yergin, *The Prize,* 701, describes unrest in Saudi Arabia and the Soviet invasion of Afghanistan.

Carter, *Keeping Faith,* discusses the 1980 state of the union speech (483), the grain embargo (474–475, 476–478), the SALT II treaty (475–476), the withdrawal of U.S. athletes from Moscow Olympics (481–482), support for Afghan fighters (473–475), the Iowa caucuses (482), and the failed Iranian hostage rescue mission and its aftereffects (506–520). For more on the rescue mission, see Yergin, *The Prize,* 705, 714, and Patterson, *Restless Giant,* 125–126. Wilkins's quote on the Iranian demonstrations is found in the "Jimmy Carter" episode of *American Experience.* Gaddis Smith is quoted in an essay accompanying the transcript, "People & Events: The Iranian Hostage Crisis, November 1979–January 1981," available at http://www.pbs.org/wgbh/amex/carter/peopleevents/e_hostage.html. The Levy article quoted is Walter J. Levy, "Oil and the Decline of the West," *Foreign Affairs* 58, no. 5 (Summer 1980): 999–1015. The Schlesinger quotes are from James R. Schlesinger, "Energy Risks and Energy Futures: Some Farewell Observations," speech before the National Press Club, Washington, DC, August 16, 1979, in *Vital Speeches of the Day* 45, no. 23 (1979): 709–712; and his "B" grade for Carter is reported in Edward Cowan, "Aides Concede His Mistakes, Foes His Gains," *New York Times,* October 6, 1980, D1.

For the effect of the commencement of the Iran–Iraq War on oil supplies and prices, see Yergin, *The Prize,* 711. For the settling of oil prices after the initial shock of the invasion subsided, see Walter J. Levy, "Oil: An Agenda for the 1980s," *Foreign Affairs* 59, no. 5 (Summer 1981): 1079–1101.

Carter, *Keeping Faith,* describes unsuccessful attempts to negotiate the release of the American hostages in Iran (484–488, 496–501, 557–559), Congress's termination of his power to impose the oil-import fee (529–530), Billy Carter's dealings with Libya (446–449), how election day coincided with the anniversary of the hostage crisis (567–568), and Carter's diary observations on Reagan (549). Patterson, *Restless Giant,* recounts Reagan's "trees cause pollution" comment (176), his "Are you better off than you were four years ago?" question (148), and the October presidential debate (149). The lopsided 1980 election results and the differences between the two candidates are examined in Kenneth T. Walsh, "The Most Consequential Elections in History: Ronald Reagan and the Election of 1980," *U.S. News & World Report*, September 25, 2008, http://www.usnews.com/articles/news/politics/2008/09/25/the-most-consequential-elections-in-history-ronald-reagan-and-the-election-of-1980.html.

For discussions of Carter's legacy and the ongoing relevance of his unfulfilled energy goals, see Stephen Koff, "Was Jimmy Carter Right?" *Energy Bulletin*, October 1, 2005, http://www.energybulletin.net/node/9657; Thom Hartmann, "Carter Tried to Stop Bush's Energy Disasters—28 Years Ago," *CommonDreams.org*, May 3, 2005, http://www.commondreams.org/views05/0503-22.htm. For contemporary diagnoses of the Carter administration's failures and successes, see James Fallows, "The Passionless Presidency," *Atlantic Monthly* (May 1979): 33–48; Timothy D. Schellhardt, "The Jimmy Carter Legacy," *Wall Street Journal*, November 6, 1980, 32; Edward Cowan, "Aides Concede His Mistakes, Foes His Gains," *New York Times*, October 6, 1980, D1; and Steven V. Roberts, "Analysts Give Carter Higher Marks in Foreign Affairs Than in Domestic Policy," *New York Times*, January 19, 1981, A22.

Chapter 9

On the sharp contrast between Carter's and Reagan's energy polices, see Lee Dembart, "Candidates Agree Only That Energy Is a Problem: Policies Fit Philosophies," *Los Angeles Times*, May 15, 1980, B1; Robert Shogan, "Reagan Assaults Carter's Energy Policies: Says President Is Hiding Fact That U.S. Resources Are 'Abundant,'" *Los Angeles Times*, September 11, 1980, B17; and Douglas E. Kneeland, "Reagan Charges Carter Misleads U.S. on Threat to Energy Security," *New York Times*, September 11, 1980, D17.

The six pieces of energy legislation Carter signed in 1980 are noted in Paul L. Joskow, "Energy Policies and Their Consequences after 25 years," *Energy Journal* 24, no. 4 (2003): 17–49; the assessment of these acts as a "low point" appears on page 31. Carter's claim about reducing oil imports by 20 percent is critiqued in Albert R. Hunt, "The Quotable Mr. Carter," *Wall Street Journal*, October 7, 1980, 32. For Reagan's criticism of the Energy Department and skepticism about a fuel shortage, see Nick Thimmesch, "Reagan: Yearning to 'Take a Crack,'" *Washington Post*, July 9, 1979, A23.

For Reagan's attacks on Carter's energy policies and his promise to increase U.S. energy supplies, see Robert Shogan, "Reagan Assaults Carter's Energy Policies: Says President Is Hiding Fact That U.S. Resources Are Abundant," *Los Angeles Times*, September 11, 1980, B17, from which the quotes are drawn. See also Douglas F. Kneeland, "Reagan Charges Carter Misleads U.S. on Threat to Energy Security," *New York Times*, September 11, 1980, D17; Lou Cannon, "Reagan: Carter Discourages U.S. Fuel Production," *Washington Post*, September 11, 1980, A4; and Douglas E. Kneeland, "Reagan Presses Carter on Charge of Misleading Nation on Energy," *New York Times*, September 12, 1980, D14; and Lou Cannon, "Reagan Says Carter Using 'Half Truths,'" *Washington Post*, September 12, 1980, A4. See also Rich Jaroslovsky, "Reagan Is Easing Stand on Energy Plans He May Not Be Able to Scuttle If Elected," *Wall Street Journal*, October 16, 1980, 18, reporting that "pragmatic politics may well force him to hold his nose and continue, for at least a few years, some of the same Carter administration programs . . . he derides."

On Reagan's preferred approach, see Douglas E. Kneeland, "A Summary of Reagan's Positions on the Major Issues of This Year's Campaign," *New York Times*,

July 16, 1980, A14 (including the "price-fixing, regulating and controlling" quote); Lee Dembart, "Candidates Agree Only That Energy Is a Problem: Policies Fit Philosophies," *Lost Angeles Times*, May 15, 1980, B1 (outlining Reagan's specific proposals to increase domestic energy production); Rich Jarovslovsky, "A Look at Reagan's Energy Policy," *Wall Street Journal*, August 5, 1980, 26; and John M. Berry, "The New Energy Emphasis: Reagan Committed to Production, Not Conservation," *Washington Post*, November 9, 1980, G1.

For an overview of the Reagan administration's early approach to energy, see Larry B. Parker, Robert L. Bamberger, and Susan R. Abbasi, Congressional Research Service, *The Unfolding of the Reagan Energy Program*, Report no. 810266 ENR (Washington, DC: Congressional Research Service, December 17, 1981); and Paul F. Rothberg, Congressional Research Service, *Reagan Administration Energy Policies Affecting New Energy Technologies*, Report no. 82-7 SPR (Washington, DC: Congressional Research Service, January 19, 1982). See also John M. Berry, "Programs in Energy Targeted: Administration Acts to Reduce Effort in Three Dozen Areas," *Washington Post*, March 21, 1981, A1.

On Reagan's lifting of federal oil and gas price controls, see Ronald Reagan, "Oil Decontrol Statement," *New York Times*, January 28, 1981, D6, from which the "restrictive price controls" quote comes; James Worsham, "Reagan to Lift Last Oil Controls Today," *Chicago Tribune*, January 28, 1981, 1 (noting administration estimates about the expected rise in oil and gas prices); Lydia Chavez and Robert A. Rosenblatt, "Oil Controls Lifted; Treasury to Benefit as Gas Prices Rise," *Los Angeles Times*, January 29, 1981, G1; and Robert D. Hershey Jr., "President Abolishes Last Price Controls on U.S.-Produced Oil," *New York Times*, January 28, 1981, A1 (reporting congressional desire to block Reagan's decontrol of oil prices and the Tip O'Neill quote). The unfavorable observations on the price-control system are from Thomas L. Jackson, "Confessions of an Energy Regulator," *Wall Street Journal*, January 5, 1981, 10.

For Reagan's proposal to abolish the Energy Department, see Ronald Reagan, "Address to the Nation on the Program for Economic Recovery," televised speech, September 24, 1981, http://www.reagan.utexas.edu/archives/speeches/1981/92481d.htm. See also, Robert L. Jackson, "Edwards Aims to Dismantle Energy Dept.," *Los Angeles Times*, December 23, 1980, B6 (reporting that Reagan's pick to head the department hoped to dismantle it and "work myself out of a job"). On the difficulty of abolishing the department, see UPI, "DOE May Get Off Death Row," *Chicago Tribune*, January 6, 1981, 10; Joanne Omang, "Edwards Finds Abolishing Energy Dept. Has Lost Much of Its Appeal," *Washington Post*, January 13, 1981, A2; Peter Behr, "Liberal Democrats Hit Reagan Pledge to Speed DOE's End," *Washington Post*, September 26, 1981, F9; and "No Energy Department—or Policy," *New York Times*, December 21, 1981, A26. McClure's skepticism about abolishing the department is reported in Martha M. Hamilton, "McClure Expects Hill Resistance to Gas Deregulation," *Washington Post*, February 9, 1981, A3. See also Rich Jaroslovsky, "Senators to Study Quicker Gas Decontrol but Vote Unlikely Till '82, McClure Says," *Wall Street Journal*, March 16, 1981, 3. On McClure's support for the Synthetic Fuels Corporation and a nuclear breeder reactor, see Robert D. Hershey Jr., "New Senate Energy Chief Sees No 'Radical Change' on Issues," *New York Times*, November 26, 1980, D6.

On Reagan's withdrawal of support for solar energy research, see Mary McGory, "The President Is Pulling the Plug on Solar Energy Programs," *Washington Post*, May 4, 1982, A3; and Michael deCourcy Hinds, "U.S. Assistance Wanes for the Solar Industry," *New York Times*, November 4, 1982, C13. On the dismantling of federal efforts to encourage energy conservation, see Philip Shapecoff, "Oil-Saving Plans

Face U.S. Attack," *New York Times*, March 27, 1981, D1; UPI, "Reagan, Citing 'Excessive Regulatory Burden,' Cancels Temperature Rules for Public Buildings," *Los Angeles Times*, February 18, 1981, B10; and Robert D. Hershey, "U.S. to Ease Some Energy-Saving Rules," *New York Times*, February 19, 1981, D2. For Hayes's observations on declining support for solar energy and Edwards's "nuclear future" quote, see Denis Hayes, "Washington Decrees a Solar Eclipse," *New York Times*, August 12, 1981, A27. More on Reagan's views on solar power are cited in the bibliographic notes to chapter 7.

On the administration's support for nuclear energy, see Robert A. Rosenblatt, "Edwards Vows Big Increase in Nuclear Power Spending," *Los Angeles Times*, February 11, 1981, B1; Robert D. Hershey Jr., "Can Reagan Lift the Cloud over Nuclear Power?" *New York Times*, March 8, 1981, F1; and Mark Hertsgaard, "Nuclear Reaganomics," *New York Times*, October 9, 1981, A31. But see also Judith Miller, "U.S. Policy Dismays Nuclear Industry," *New York Times*, November 21, 1982, D4 (reporting that industry leaders "have begun to express disappointment with the Reagan Administration, on which they had once pinned their hopes for a reversal of the industry's precipitous decline").

For Reagan's NEPP, see U.S. Department of Energy, *Securing America's Energy Future: The National Energy Policy Plan, a Report to Congress Required by Title VIII of the Department of Energy Organization Act* (Washington, D.C.: U.S. Department of Energy, July 1981); the quoted passages are on page 2. For the accompanying message to Congress, see Ronald Reagan, "Message to the Congress Transmitting the National Energy Policy Plan," July 17, 1981, http://www.reagan.utexas.edu/archives/speeches/1981/71781b.htm. Carter's 1977 National Energy Plan is quoted in Parker, Bamberger, and Abbasi, *The Unfolding of the Reagan Energy Program*, 6, which also includes the characterization of Reagan's policies as a return to the "old-time religion" of the pre-1973 period.

On OPEC's November 1981 price agreement, see Douglas Martin, "OPEC Members Unite to Freeze Oil Price at $34," *New York Times*, October 30, 1981, 1. For country-specific analysis and for the reduction in daily OPEC output between 1979 and 1981, see Douglas Martin, "Price Rift in OPEC May Close," *New York Times*, October 16, 1981, D1. On the uncertainty in world oil markets in the wake of the Sadat assassination, see Douglas Martin, "World Oil Markets Uneasy," *New York Times*, October 7, 1981, D1. On the March 1983 OPEC price accord, see Richard Johns and Ray Dafter, "OPEC Accord on $29 Price," *Financial Times*, March 15, 1983, 1. On changes in oil production, consumption, and prices between 1979 and 1983, see Daniel Yergin, *The Prize: The Epic Quest for Oil, Money, and Power* (New York: Free Press, 1991), 718, which notes the rise of non-OPEC production and the worldwide surplus in 1982. Yergin also discusses the lack of attention devoted to energy policy at the 1985 Bonn summit (742–743), the plummeting price of oil by the end of 1985 (750), the December 1986 price stabilization (763–764), Vice President Bush's trip to Saudi Arabia (755–757), and the Iraq/Kuwait crisis and Desert Storm (774, 776–777).

On Carter's solar panels, see John Wihbey, "Jimmy Carter's Solar Panels: A Lost History That Haunts Today," *Yale Forum on Climate Change & the Media* (online publication), November 11, 2008, http://www.yaleclimatemediaforum.org/2008/11/jimmy-carters-solar-panels.

For a profile of James Hansen, see Elizabeth Kolbert, "The Catastrophist," *The New Yorker* (June 29, 2009), 39. For an example of Hansen's early work on climate change, see J. Hansen, D. Johnson, A. Lacis, S. Lebedeff, P. Lee, D. Rind, and G. Russell, "Climate Impact of Increasing Atmospheric Carbon Dioxode," *Science* 213, no. 4511 (August 28, 1981): 957–966. For samples of Hansen's testimony in 1987, see James Hansen, testimony, *Greenhouse Gas Effect and Global Warming: Hearings before the Committee on Energy and Natural Resources, U.S. Senate*, 100th Cong., 1st sess., November 9, 1987, http://www.lexisnexis.com/congcomp/getdoc?HEARING-ID=HRG-1987-NAR-0041; and James Hansen, testimony, *Greenhouse Gas Effect and Global Climate Change: Hearings before the Committee on Energy and Natural Resources, U.S. Senate*, Part 2, 100th Cong., 2d sess., June 23, 1988, http://www.lexisnexis.com/congcomp/getdoc?HEARING-ID=HRG-1988-NAR-0055.

Chapter 10

On the awarding of the Nobel Prize to Al Gore and the IPCC, see Walter Gibbs and Sarah Lyall, "Gore Shares Peace Prize for Climate Change Work," *New York Times*, October 13, 2007, http://www.nytimes.com/2007/10/13/world/13nobel.html (including "security of mankind" quote). For Gore's acceptance speech, see Al Gore, "Nobel Lecture," Oslo, Norway, December 10, 2007, http://nobelprize.org/nobel_prizes/peace/laureates/2007/gore-lecture_en.html. For background on the IPCC, see "Gore, U.N. Climate Panel Win Nobel Peace Prize," *msnbc.com*, October 12, 2007, http://www.msnbc.msn.com/id/21262661, and R. K. Pachauri, IPCC Chairman, "Nobel Lecture," Oslo, Norway, December 10, 2007, http://nobelprize.org/nobel_prizes/peace/laureates/2007/ipcc-lecture_en.html. See also Al Gore, *An Inconvenient Truth: The Planetary Emergence of Global Warming and What We Can Do about It* (Emmaus, PA: Rodale Press, 2006). Carl Pope's comments on Gore are found in Carl Pope, "Al Gore Wins the Nobel Peace Prize: Gore Is Rewarded for His Epic Efforts to Focus Attention on the Perils of Climate Change," *Salon.com*, October 11, 2007, http://www.salon.com/opinion/feature/2007/10/11/gore/print.html. Buchanan's dismissive comments can be found on the Media Matters Web site at http://mediamatters.org/blog/200908030045, which contains a transcript of the televised exchange.

The controversies and embarrassments that beset the IPCC and global warming activists' efforts in 2009 and early 2010 are discussed in Juliet Eilperin and David A. Farenthold, "Series of Missteps by Climate Scientists Threatens Climate-Change Agenda," *Washington Post*, February 15, 2010, http://www.washingtonpost.com/wp-dyn/content/article/2010/02/14/AR2010021404283.html; Jeffrey Ball and Keith Johnson, "Push to Oversimplify at Climate Panel," *Wall Street Journal*, February 26, 2010, http://online.wsj.com/article/NA_WSJ_PUB:SB10001424052748704188104575083681319834978.html; Seth Borenstein, "Climate Change Panel Seeks Outside Review on Reports," *Washington Post*, February 28, 2010, http://www.washingtonpost.com/wp-dyn/content/article/2010/02/27/AR2010022703449.html; "Did D.C.'s Blizzard Bury Climate Change Legislation?" *Washington Post*, February 14, 2010, A19; and Kim Chipman, "Climate-Change Fervor Cools amid Disputed Science," *Bloomberg.com*, February 22, 2010, http://www.businessweek.com/news/2010-02-22/climate-change-fervor-cools-amid-disputed-science-defections.html. Jim Inhofe's igloo and Jim DeMint's quote about Al Gore are reported in Dana Milbank, "Global Warming's Snowball Fight," *Washington Post*, February 14, 2010, http://www.washingtonpost.com/wp-dyn/content/article/2010/02/12/AR2010021203908.html. Al Gore's response to the controversies, from which the quotes are drawn, is found in Al Gore, "We Can't Wish Away Climate Change," *New York Times*, February 28, 2010, http://www.nytimes.com/2010/02/28/opinion/28gore.html.

For the article from *The Economist* quoted in the text, see "The Clouds of Unknowing," *The Economist* (March 20–26, 2010): 83–86. A good summary of the climate science is also provided by the director of Columbia Law School's Center for Climate Change Law, Michael B. Gerrard, in "Does Climate Science Warrant Greenhouse Gas Regulation?" unpublished Power Point presentation, Columbia University, May 6, 2010.

On the military's growing recognition of the significance of climate change, including the Dory and Zinni quotes, see John M. Broder, "Climate Change Seen as Threat to U.S. Security," *New York Times*, August 8, 2009, A1. See also U.S. Department of Defense, *Quadrennial Defense Review Report* (Washington, DC: U.S. Department of Defense, February, 2010), 85, http://www.defense.gov/qdr/images/QDR_as_of_12Feb2010_1000.pdf. For Obama's Nobel acceptance speech, highlighting similar national security concerns, see Barack H. Obama, "Nobel Lecture," Oslo, Norway, December 10, 2009, http://nobelprize.org/nobel_prizes/peace/laureates/2009/obama-lecture_en.html. The Senate's resistance to binding greenhouse gas limits is examined in John C. Dernbach, "U.S. Policy," in *Global Climate Change and U.S. Law*, ed. Michael B. Gerrard (Chicago: American Bar Association, 2007), 63–64. For Bush's acknowledgment of global warming "concerns," see George W. Bush, "Remarks by the President on Climate Change," Rose Garden speech, Washington, D.C., April 16, 2008, http://www.america.gov/st/texttrans-english/2008/April/20080416165403eaifas0.1469843.html. See also Andrew C. Revkin, "The [Annotated] Climate Speech," *New York Times* (Dot Earth blog), April 17, 2008, http://dotearth.blogs.nytimes.com/2008/04/17/the-annotated-climate-speech, featuring the speech transcript with commentary interspersed.

For candidate Obama's Lansing, Michigan, energy speech, see Mark Halperin, "Prepared Remarks of Obama's Energy Speech," *Time* (The Page blog), February 22, 2010, http://thepage.time.com/prepared-remarks-of-obamas-energy-speech, emphasis added. See also Andrew C. Revkin, "The Obama Energy Speech, Annotated," *New York Times* (Dot Earth blog), August 5, 2008, http://dotearth.blogs.nytimes.com/2008/08/05/the-obama-energy-speech-annotated, featuring interspersed commentary. Among the proposals included in Obama's speech were to employ 6 billion gallons of biofuels by 2022 (one-sixth the amount urged by George W. Bush), to produce and import from Canada more natural gas, to sell 70 million barrels of oil from the Strategic Petroleum Reserve, to raise automobile fuel-efficiency standards 4 percent a year, to find safer ways to use nuclear power and store nuclear waste, and to enact a windfall profits tax on oil companies to finance a $1,000 "energy rebate" for every working family in America. The "new chapter" quote is from Barack Obama, "Remarks to the Global Climate Summit," video-recorded speech, broadcast to Los Angeles, California, November 18, 2008, http://change.gov/newsroom/entry/president_elect_obama_promises_new_chapter_on_climate_change.

On Gore's Netroots Nation speech, see Garance Franke-Ruta and Jose Antonio Vargas, "Gore Speaks at Netroots Nation," *Washington Post* (The Trail blog), July 19, 2008, http://blog.washingtonpost.com/44/2008/07/19/gore_to_speak_at_netroots_nati.html. A transcript of Gore's remarks is available on the Netroots Nation Web site at http://www.netrootsnation.org/node/1037. The percentages of electricity generated by oil in 1970 and 1980 are provided in *The Energy Decade: 1970–1980*, ed. William L. Liscom (Pensacola, FL: Ballinger, 1981), 265. The 2008 figure derives from U.S. Energy Information Administration, *Annual Energy Review 2008* (Washington, DC: U.S. Energy Information Administration, June 2009), v, 241, http://www.eia.doe.gov/aer/pdf/aer.pdf; on petroleum consumption, production, and importation in 1970 and 2008, see pages v, xxiii. The drawbacks of replacing gasoline with a coal-based synthetic fuel are discussed in Robert H. Socolow, "Can We Bury Global Warming?" *Scientific American* (July 2005), 53.

On the Pickens Plan, see http://www.pickensplan.com. T. Boone Pickens's oil background is noted in Daniel Yergin, *The Prize: The Epic Quest for Oil, Money, and Power* (New York: Free Press, 1991), 727. The conversion cost estimate for passenger vehicles cited here is from Brent D. Yacobucci, Congressional Research Service, *Natural Gas Passenger Vehicles: Availability, Cost, and Performance* (Washington, DC: Congressional Research Service, October 20, 2008), http://www.policyarchive.org/handle/10207/bitstreams/19063.pdf. Pickens cited the truck conversion estimate in an August 2009 CNBC interview. The fuel mileage and carbon footprints of particular hybrid vehicles can be examined at http://www.fueleconomy.gov/feg/hybrid_sbs.shtml. Year-by-year tax credit amounts for particular hybrid vehicles are available at http://www.fueleconomy.gov/feg/tax_hybrid.shtml. On the rise (and recent decline) in hybrid automobile sales since 2000, see "Annual Hybrid Sales Drop for First Time," *HybridCars.com*, January 6, 2009, http://www.hybridcars.com/news/annual-hybrid-sales-drop-first-time-25388.html.

The 2006 documentary *Who Killed the Electric Car?* (Paine, Sony Pictures) explores the short-lived commercial life of electric cars in the 1990s, focusing on the General Motors EV1, sold as a Saturn. The film, which received several awards but was not nominated for an Oscar, proceeds mostly through newsreels and interviews featuring prominent movie personalities such as Mel Gibson and Tom Hanks, political figures such as Ralph Nader, and the environmentalist S. David Freeman, chairman of the Hydrogen Car Company, which produced the $100,000 Hydrogen Shelby Cobra. Jimmy Carter had earlier appointed Freeman to head the Tennessee Valley Authority, where Freeman shut down several nuclear plants (in his 2007 book *Winning Our Energy Independence* [Layton, Utah: Gibbs Smith], Freeman says we need to eliminate oil, coal, and nuclear power). At the end of this movie, the filmmakers pronounce guilty verdicts for the electric car's failure on the oil industry, the automobile companies, the federal government, the California Air Resources Board, the hydrogen fuel cell (an expensive alternative), and consumers, who failed to buy the cars in sufficient numbers. According to *Wikipedia*, the film's director Chris Paine says he is making a sequel titled *Revenge of the Electric Car*. On more recent plans for electric cars, see Bill Vlasic and Nick Bunkley, "G.M. Puts Electric Car's City Mileage in Triple Digits," *New York Times*, August 11, 2009, http://www.nytimes.com/2009/08/12/business/12auto.html (including "electric vehicle initially" quote); Peter Whoriskey, "GM Says New Car Is Capable of 230 MPG," *Washington Post*, August 12, 2009, http://www.washingtonpost.com/wp-dyn/content/article/2009/08/11/AR2009081101090.html, including quotes by Henderson; and Todd Woody and Clifford Krauss, "Cities Prepare for Life with the Electric Car," *New York Times*, February 15, 2010, http://www.nytimes.com/2010/02/15/business/15electric.html. A map of London's charging points for electric cars can be found at http://www.london.gov.uk/electricvehicles. A description of London's practices and goals for electric cars is contained in New York City Global Partners, "Best Practice: Electric Vehicle Development," available at http://www.nyc.gov/html/unccp. New York City's analysis is contained in PlanNYC, "Exploring Electric Vehicle Adoption in New York City," January 2010, http://www.nyc.gov/htmlplanyc2030/downloads/pdf/electric_vehicle_adoption_study_2010_01.pdf. See also Gerald F. Seib, "Time to Plug In Electric Cars," *Wall Street Journal*, June 18, 2010, A2. Tesla's initial public offering is described in "Tesla Motors' Underwriters Sell More Shares," *Reuters*, June 30, 2010, http://www.reuters.com/article/idUSN3020675220100630.

On Steven Chu's appointment as energy secretary, see his biography on the Department of Energy's Web site, http://www.energy.gov./organization/dr_steven_chu.htm. For Chu's Senate testimony, see Steven Chu, Secretary of Energy, statement to the Committee on Environment and Public Works, U.S. Senate, 111th Cong., 1st sess., July 7, 2009, http://epw.senate.gov/public/index.cfm?FuseAction=Files.View&FileStore_id=86536a0a-dcba-4db2-b51a-eeaad9c56b6e. Chu gave a press briefing at the Summit of the Americas just before Earth Day 2009 discussing the imperative of combating climate change, the transcript of which is available at "Earth Day 2009: Obama Energy Chief Lays Out Climate Doomsday Scenario," *Huffington Post*, April 19, 2009, http://www.huffingtonpost.com/2009/04/19/earth-day-2009-obama-ener_n_188721.html.

The quotes from Obama regarding the costs and benefits (including jobs) of energy independence are from the Lansing, Michigan, speech in Halperin, "Prepared Remarks of Obama's Energy Speech." For David Clark's 1987 green jobs claim, see "Labour's Green Policies 'to Bring Many Thousands of Green Jobs,'" *PR Newswire Europe*, June 9, 1987. The Friedman quotes are from Danny Shea, "Tom Friedman Calls for Green Revolution," *Huffington Post*, July 4, 2008, http://www.huffingtonpost.com/2008/07/04/thomas-friedman-the-5-key_n_110933.html. See also Thomas L. Friedman, *Hot, Flat, and Crowded: Why We Need a Green Revolution—and How It Can Renew America* (New York: Farrar, Straus and Giroux, 2007), and Thomas L. Friedman, "The Green-Collar Solution," *New York Times*, October 17, 2007, http://www.nytimes.com/2007/10/17/opinion/17friedman.html.

On the debate over green jobs, see Alex Kaplun, "'Green Jobs' at Heart of Obama's Earth Day Push on Energy," *New York Times*, April 22, 2009, http://www.nytimes.com/gwire/2009/04/22/22greenwire-green-jobs-at-heart-of-obamas-earth-day-push-o-10631.html, and Russ Juskalian, "The True Color of 'Green-Collar' Jobs: Press Wrestles with Definition and Economic Reality," *Columbia Journalism Review*, May 1, 2008, http://www.cjr.org/the_observatory/the_true_color_of_greencollar.php. Obama's visit to the Newton, Iowa, plant is reported in Sheryl Gay Stolberg, "Obama Urges Passage of Energy Legislation," *New York Times*, April 22, 2009, http://www.nytimes.com/2009/04/23/us/politics/23obama.html, and "Obama Calls for a 'New Era of Energy Exploration in America,'" *Guardian.co.uk*, April 22, 2009, http://www.guardian.co.uk/world/2009/apr/22/barack-obama-earth-day-wind-power. In Iowa, Obama cited Denmark as a country that generates almost 20 percent of its electricity from wind (of course, Denmark consumes only a small fraction of the electricity the United States does). On TPI Composites and Newton's transition to green jobs, see Peter S. Goodman, "A Splash of Green for the Rust Belt," *New York Times*, November 2, 2008, http://www.nytimes.com/2008/11/02/business/02wind.html. See also the Kit Bond study *Yellow Light on Green Jobs: A Report by the U.S. Senate Subcommittee on Green Jobs and the New Energy Economy Ranking Member Kit Bond* (Spring 2009), 28–29, http://bond.senate.gov/public/_files/BondGreenJobsReport.pdf. For the Chu/Solis op-ed on green jobs, see Steven Chu and Hilda Solis, "Building the American Clean Energy Economy," April 22, 2009, http://www.whitehouse.gov/the_press_office/Op-Ed-by-Secretary-of-Energy-Steven-Chu-and-Secretary-of-Labor-Hilda-Solis-Building-the-American-Clean-Energy-Economy. Van Jones's comments are reported in Alex Kaplun, "'Green Jobs' at Heart of Obama's Earth Day Push on Energy," *New York Times*, April 22, 2009, http://www.nytimes.com/gwire/2009/04/22/22greenwire-green-jobs-at-heart-of-obamas-earth-day-push-o-10631.html.

The material cited from Kit Bond's study on green jobs, *Yellow Light on Green Jobs*, is on pages 17, 18, and 27, and the quote from the conclusion is on page 31. For the Spanish study on renewable energy subsidies cited in Bond's report, see Gabriel Caldaza Álvares, Raquel Merino Jara, Juan Ramón Rallo Julián, and José Ignacio García Bielsa, *Study of the Effects on Employment of Public Aid to Renewable Energy Sources* (Madrid: King Juan Carlos University, March 2009), http://www.juandemariana.org/

pdf/090327-employment-public-aid-renewable.pdf; the figures cited are found on pages 2 and 25. George Will discusses the Gibbs colloquy regarding the Spanish study in George F. Will, "Tilting at Green Windmills," *Washington Post*, June 25, 2009, http://www.washingtonpost.com/wp-dyn/content/article/2009/06/24/AR20090624 03012.html. For Sharan's skeptical analysis about the job-creation potential of green technologies, see Sunil Sharan, "The Green Jobs Myth," *Washington Post*, February 26, 2010, A23.

On China's rising dominance in wind and solar energy manufacturing, see Keith Bradsher, "China Leading Global Race to Make Clean Energy," *New York Times*, January 31, 2010, http://www.nytimes.com/2010/01/31/business/energy -environment/31renew.html; and Christopher Martin and Jim Efstathiou Jr., "China's Labor Edge Overpowers Obama's 'Green' Jobs Initiative," *Bloomberg.com*, February 4, 2010, http://www.bloomberg.com/apps/news?pid=20601109&sid=adIdr mtTtyw8&pos=12 (including information about SunTech hiring 11,000 Chinese workers). The latter includes Kit Bond's comments and the suggestion that the United States might ultimately replace dependence on Middle East oil imports with dependence on Asian clean-energy equipment imports.

Tom Friedman's climate change exhortations are lampooned in Matt Taibbi, "Flat N All That," *New York Press*, January 14, 2009, http://www.nypress.com/article -19271-flat-n-all-that.html. Al Gore's household energy use is discussed in Jake Tapper, "Al Gore's 'Inconvenient Truth'?—A $30,000 Utility Bill," *ABC News*, February 26, 2007, http://abcnews.go.com/Politics/GlobalWarming/story?id=2906888& page=1. Tapper also relates Gregg Easterbrook's criticism of Laurie David's private jet use. Saul Griffith's comments about Al Gore's hypocrisy can be found in David Owen, "The Inventor's Dilemma," *The New Yorker* (May 17, 2010), 50. For the letter to the editor defending large vehicles, see Bernard P. Giroux, "When a Hybrid Can Pull a Boat, Then We'll Talk," *Wall Street Journal*, March 20, 2009, http://online.wsj .com/article/SB123751598127291711.html. The 1975 figure on miles driven by U.S. vehicles is from Terry S. T. Shelton, National Highway Traffic Safety Administration, *Research Note: Revised Vehicle Miles of Travel for Passenger Cars and Light Trucks, 1975 to 1993* (Washington, DC: National Highway Traffic Safety Administration, September 21, 1995), http://www-nrd.nhtsa.dot.gov/Pubs/95.830.PDF. For the 2006 figure, see National Highway Traffic Safety Administration, *Traffic Safety Facts: 2006 Data* (Washington, DC: National Highway Traffic Safety Administration, updated March 2008), http://www-nrd.nhtsa.dot.gov/Pubs/810809.PDF.

The biographical information about Bill McKibben can be found on his Web site at http://www.billmckibben.com/bio.html. The *Boston Globe* quote is fron Anis Shivani, "Facing Cold, Hard Truths about Global Warming," *Boston Globe*, May 30, 2010, http://www.boston.com/ae/books/articles/2010/05/30/facing_cold_hard_truths _about_global_warming. See also Bryan Walsh, "The Skimmer," *Time* magazine, April 26, 2010, http://www.time.com/time/magazine/article/0,9171,1982309,00 .html; and for the *Foreign Policy* quote, see "The FP Top 100 Global Thinkers," available at http://www.foreignpolicy.com/articles/2009/11/30/the_fp_top_100_global _thinkers?page=full. Chapter 1 of Bill McKibben's *Eaarth: Making a Life on a Tough New Planet* (New York: Times Books, 2010) recites many of the adverse consequences of climate change. McKibben's praise for the Club of Rome's *The Limits of Growth* (New York: Universe Books, 1972), and comments about the end of economic growth are in the references for chapter 2. His prescriptions can be found in chapters 3 and 4. For a review challenging McKibben's claims that limiting climate change is incompatible with economic growth, see Nicholas Stern, "Climate: What You Need to Know," *New York Review of Books*, June 24, 2010, 35–37.

Steven Chu cited the disproportionate U.S. energy consumption and the number of people without electricity worldwide in a Harvard commencement address. See "U.S. Energy Secretary Steven Chu's Address at Harvard's Afternoon Exercises," *Harvard Gazette*, June 4, 2009, http://news.harvard.edu/gazette/story/2009/06/u-s-energy-secretary-steven-chus-address-at-harvards-afternoon-exercises. On Hillary Clinton's climate-related diplomacy efforts in India, see Mehul Srivastava, "Hillary in Indian Climate Change Standoff," *Bloomberg BusinessWeek*, July 20, 2009, http://www.businessweek.com/globalbiz/blog/eyeonasia/archives/2009/07/hillary_in_indian_climate_change_standoff.html; and Glenn Kessler, "Clinton, India Minister Clash over Emissions Reduction Pact," *Washington Post*, July 20, 2009, http://www.washingtonpost.com/wp-dyn/content/article/2009/07/19/AR2009071900705.html. The per capita and energy intensity figures are the author's calculations derived from data provided in U.S. Energy Information Administration, *International Total Primary Energy Consumption and Energy Intensity*, http://www.eia.doe.gov/emeu/international/energyconsumption.html. On China's pledge to cut carbon intensity following a U.S. summit (including skeptical reaction to this pledge), see Edward Wong and Keith Bradsher, "China Joins U.S. in Pledge of Hard Targets on Emissions," *New York Times*, November 26, 2009, http://www.nytimes.com/2009/11/27/science/earth/27climate.html. I am grateful to PricewaterhouseCoopers for the calculation indicating that President Obama's Beijing pledge would return U.S. carbon dioxide emissions to their 1905 level. That calculation was based on information from the following source: Climate Analysis Indicators Tool (CAIT) version 3.0. (Washington, DC: World Resources Institute, 2005), http://cait.wri.org. WRI calculates carbon dioxide emissions from 3 sources: EIA 2004, International Energy Annual 2002, http://www.eia.doe.gov/iea/carbon.html; IEA 2004, CO2 Emissions from Fuel Combustion (2004 edition), http://data.iea.org/ieastore/co2_main.asp; and G. Marland, T. A. Boden, and R. J. Andres, 2005. Global, Regional, and National Fossil Fuel CO2 Emissions in Trends: A Compendium of Data on Global Change. Carbon Dioxide Information Analysis Center, Oak Ridge National Laboratory, U.S. Department of Energy, Oak Ridge, Tenn., http://cdiac.esd.ornl.gov/trends/emis/meth_reg.htm.

On the Copenhagen Accord and negotiations, see John M. Broder, "Many Goals Remain Unmet in 5 Nations' Climate Deal," *New York Times*, December 18, 2009, http://www.nytimes.com/2009/12/19/science/earth/19climate.html (including Obama's "breakthrough" quote and the Carl Pope quote), and James Kanter, "Copenhagen's One Real Accomplishment: Getting Some Money Flowing," *New York Times*, December 20, 2009, http://www.nytimes.com/2009/12/21/business/energy-environment/21iht-green21.html (including the Ban Ki-moon quote). The Doniger quote is from David Doniger, "The Copenhagen Accord: A Big Step Forward," *Switchboard* (blog of the National Resources Defense Council), December 21, 2009, http://switchboard.nrdc.org/blogs/ddoniger/the_copenhagen_accord_a_big_st.html. The Hansen quote is from Madeline Ostrander, "James Hansen: Good Riddance, Copenhagen. Time for Better Ideas," *YES!* December 22, 2009, http://www.yesmagazine.org/planet/james-hansen-why-copenhagens-failure-is-a-blessing. The Wen Jiabou quote and Obama's "disappointed" quote are from Tom Zeller Jr., "Fault Lines Remain after Climate Talks," *New York Times*, January 3, 2010, http://www.nytimes.com/2010/01/04/business/energy-environment/04green.html. Obama's "much further to go" quote is from Andrew C. Revkin and John M. Broder, "A Grudging Accord in Climate Talks," *New York Times*, December 19, 2009, http://www.nytimes.com/2009/12/20/science/earth/20accord.html. On Yvo de Boer's resignation, see Juliet Eilperin and Steven Mufson, "Copenhagen Emissions Pact Appears Increasingly Fragile," *Washington Post*, February 8, 2010, A3, and "Climate

Change," *New York Times*, February 22, 2010. For the fact that as of June 2010, more than 100 nations had associated themselves with the accord, see Nicholas Stern, "Climate: What You Need to Know," *New York Review of Books*, June 24, 2010, 37. See also Robert Stavins, "What Hath Copenhagen Wrought? A Preliminary Assessment of the Copenhagen Accord," December 20, 2009, http://belfercenter.ksg.harvard.edu/analysis/stavins/?p=464.

Daniel Bell is quoted in Kevin Mattson, *"What the Heck Are You Up to, Mr. President?": Jimmy Carter, America's "Malaise," and the Speech That Should Have Changed the Country* (New York: Bloomsbury, 2009), 165. See also Daniel Bell, *The Cultural Contradictions of Capitalism* (New York: Basic Books, 1976).

Chapter 11

The quoted portion of President Obama's *60 Minutes* interview is provided in Andrew C. Revkin, "Obama on the 'Shock to Trance' Energy Pattern," *New York Times* (Dot Earth blog), November 17, 2008, http://dotearth.blogs.nytimes.com/2008/11/17/obama-on-shock-to-trance-energy-pattern. Full transcript and video of the interview are available at http://www.cbsnews.com/stories/2008/11/16/60minutes/main4607893.shtml. On worldwide daily trades in oil and on the volatility of the oil market, see Daniel Yergin, "It's Still the One," *Foreign Policy* (September–October 2009): 88–95, http://www.foreignpolicy.com/articles/2009/08/17/its_still_the_one. On the role of speculative oil trading in contributing to price volatility, see Edmund L. Andrews, "U.S. Considers Curbs on Speculative Trading of Oil," *New York Times*, July 8, 2009, http://www.nytimes.com/2009/07/08/business/08cftc.html.

For a primer on the challenges that confront efforts to address climate change, see Kirsten R. W. Matthews, Lauren A. Smulcer, Amy Myers Jaffe, and Neal Lane, *Beyond Science: The Economics and Politics of Responding to Climate Change*, conference report (Houston: James A. Baker III Institute for Public Policy, February 2008), http://www.bakerinstitute.org/publications/EF-ST-pub-BeyondScienceConfReport-121008.pdf/view. An introduction to the key issues in devising a policy to combat climate change can be found in Joseph E. Aldy, Alan J. Krupnick, Richard G. Newell, Ian W. H. Perry, and William A. Pizer, *Designing Climate Change Mitigation Policy*, Discussion Paper 08-16 (Washington, DC: Resources for the Future, May 2009), http://www.rff.org/documents/RFF-DP-08-16.pdf. For an analysis of the international law that is developing in regard to climate change, see Elizabeth Burleson, "Climate Change Consensus: Emerging International Law," *William and Mary Environmental Law and Policy Review* 34 (2009): 543–588.

For more about a tax on gasoline or other fossil-fuel products, see Gilbert Metcalf and David Weisbach, *The Design of a Carbon Tax*, Working Paper no. 09-05 (Washington, DC: Reg-Markets Center, January 8, 2009), http://papers.ssrn.com/sol3/papers.cfm?abstract_id=1327260; Thomas Merrill and David Schizer, *A Proposed Petroleum Fuel Price Stabilization Plan*, Columbia Law and Economics Working Paper no. 349 (New York: Columbia University, May 5, 2009), http://papers.ssrn.com/sol3/Delivery.cfm/SSRN_ID1399706_code254274.pdf?abstractid=1399706&mirid=5; Joseph J. Thorndike, "Plans for a Refundable Gas Tax Gain Popularity," *Tax Analysts*, June 8, 2006, http://www.taxhistory.org/thp/readings.nsf/artweb/6b51901d5294bebb852571a20068ed15?opendocument; Martin A. Sullivan, "Tech Neutrality, Tax Credits, and the Gas Tax," *Tax Notes* 122, no. 5 (February 2, 2009): 619–624; Michael Waggoner, "The House Erred: A Carbon Tax Is Better Than Cap and Trade," *Tax Notes* 180, no. 11 (September 21, 2009): 1257–1262; "Putting a Price on Carbon: An

Emissions Cap or a Tax?" *Yale Environment 360*, May 7, 2009, http://e360.yale.edu/content/feature.msp?id=2148. In *Instrument Choice Is Instrument Design*, University of Chicago Law and Economics, Olin Working Paper no. 490 (Chicago: University of Chicago, October 23, 2009), http://papers.ssrn.com/sol3/papers.cfm?abstract_id=1493312, David A. Weisbach argues that in the domestic context, taxes and cap and trade are essentially equivalent instruments, but taxes operate more effectively in the international arena. Weisbach thus recommends taxes as the best option to control worldwide greenhouse gas emissions.

For the estimate on the costs of oil dependence to the U.S. economy, see David L. Greene, testimony, *Technology-Neutral Incentives for Energy-Efficient Low Greenhouse Gas Emitting Vehicles: Hearings before the Finance Committee, U.S. Senate*, 111th Cong., 1st sess., April 23, 2009, 2, http://finance.senate.gov/hearings/hearing/?id=d87515c8-c2a4-494d-c9ef-90697126dc42 (including estimate that oil dependence cost our economy between $700 and $800 billion in 2008). On the advantages and disadvantages of a carbon tax, see Congressional Budget Office, *Policy Options for Reducing CO_2 Emissions* (Washington, DC: Congressional Budget Office, February 2008), http://www.cbo.gov/ftpdocs/89xx/doc8934/02-12-Carbon.pdf. On the potential ability of a carbon tax to capture 80 to 90 percent of emissions, see Metcalf and Weisbach, *The Design of a Carbon Tax*, 1–2. For Al Gore's statement on the value of a carbon tax, see Al Gore, *An Inconvenient Truth: The Crisis of Global Warming* (Emmaus, PA: Rodale Press, 2006), 320. See also James Surowiecki, "Oil Check," *The New Yorker* (June 22, 2009), http://www.newyorker.com/talk/financial/2009/06/22/090622ta_talk_surowiecki, which discusses the use of a gas tax to stabilize prices and reduce importation of oil.

The history of gas tax proposals from the Nixon administration to the early Reagan administration is recounted in Martin A. Sullivan, "Gas Tax Politics, Part I," *Tax Analysts*, September 22, 2008, http://www.taxhistory.org/thp/readings.nsf/ArtWeb/5DDB79194769C2BF852574D5003C28D5?OpenDocument. See also "Carter Considers a Gas Tax," *Time*, December 17, 1979; and Thorndike, "Plans for a Refundable Gas Tax Gain Popularity." On Congress's aborted efforts in the early 1990s to impose a substantial energy tax, see Michael J. Graetz, *The Decline (and Fall?) of the Income Tax* (New York: W. W. Norton, 1997), 154–159.

For Obama's pledge not to raise taxes on Americans earning less than $250,000 a year, see "Transcript: Democratic Debate in Philadelphia," *New York Times*, April 16, 2008, http://www.nytimes.com/2008/04/16/us/politics/16text-debate.html?pagewanted=all. For two opinion pieces criticizing the pledge as conventional cynical politics, see Howard Gleckman, "Read Their Lips: Clinton and Obama Take the Pledge," *Tax Policy Center* (TaxVox blog), April 17, 2008, http://taxvox.taxpolicycenter.org/blog/_archives/2008/4/17/3644935.html; and David Brooks, "How Obama Fell to Earth," *New York Times*, April 18, 2008, http://www.nytimes.com/2008/04/18/opinion/18brooks.html. On the Budget Reconciliation Act of 1990, see Robert Pear, "The Struggle in Congress; Most in U.S. Will Feel Effect of Shift in Spending Priorities," *New York Times*, October 28, 1990, http://www.nytimes.com/1990/10/28/us/struggle-congress-most-us-will-feel-effect-shift-spending-priorities.html?pagewanted=1?pagewanted=1. Grover Norquist is quoted in Michael J. Graetz and Ian Shapiro, *Death by a Thousand Cuts: The Fight over Taxing Inherited Wealth* (Princeton, NJ: Princeton University Press, 2005), 9.

For a discussion of policy options to encourage energy conservation other than tax increases, see Gilbert E. Metcalf, "Federal Tax Policy towards Energy," *Tax Policy and the Economy* 21 (2007): 145–184; and Gilbert E. Metcalf, testimony before the Committee on Finance, U.S. Senate, 111th Cong., 1st sess., April 23, 2009, http://finance.senate.gov/hearings/hearing/?id=d87515c8-c2a4-494d-c9ef-90697126dc42.

Some of the inherent reasons that taxes and subsidies operate differently are explained in William J. Baumol and Wallace E. Oates, *The Theory of Environmental Policy*, 2d ed. (Cambridge, UK: Cambridge University Press, 1988), 211–234. On the practical advantages of taxes over subsidies (including efficiency and avoiding choosing "favorites" among technologies), see Staff of the Joint Committee on Taxation, U.S. Congress, *Tax Expenditures for Energy Production and Conservation: Scheduled for a Public Hearing before the Senate Committee on Finance on April 23, 2009*, JCX-25-09R (Washington, DC: U.S. Government Printing Office, April 21, 2009), 115, http://www.jct.gov/publications.html?func=select&id=17. In "The Losers in the Energy Subsidy Game," *Tax Notes* (November 3, 2008): 510–515, economist Martin A. Sullivan critiques the use of energy subsidies as "inherently flawed," characterizing the approach as a "game" that "is essentially one of picking winners and losers" (510). Sullivan's article lists 50 items that can save motor fuel that do not enjoy subsidies through the tax law. For a discussion of subsidies for wind power (one instance of subsidizing favored fuels instead of taxing disfavored fuels), see Jeffry S. Hinman, "The Green Economic Recovery: Wind Energy Tax Policy after Financial Crisis and the American Recovery and Reinvestment Tax Act of 2009," *Journal of Environmental Law and Litigation* 24, no. 1 (2009): 35–74.

On the subordination of energy research to defense and health research over the past three decades of federal R&D funding, see Genevieve J. Knezo, Congressional Research Service, *Federal Research and Development: Budget and Priority-Setting Issues: 109th Congress*, RL3511 (Washington, DC: Congressional Research Service, June 30, 2006). See also Laura Diaz Anadon, Kelly Sims Gallagher, Matthew Bunn, and Charles Jones, Energy Technology Innovation Policy Group, *Tackling U.S. Energy Challenges and Opportunities: Preliminary Policy Recommendations for Enhancing Energy Innovation in the United States* (Cambridge, MA: Kennedy School, Harvard University, February 2009), 3, fig. 1, http://belfercenter.ksg.harvard.edu/files/ERD3_Energy_Report_Final.pdf, which provides a year-by-year breakdown of energy R&D spending on individual research areas. The comparison between energy research funding in 1979 and funding in 2008 as a percentage of total federal R&D is found in Alix R. Broadfoot and Michael E. Webber, "Oil Barrel Politics," *Earth*, April 8, 2009, http://www.earthmagazine.org/earth/article/202-7d9-4-8. For the real-dollar comparison between federal energy R&D funding in 1980 and funding in 2006 as well as the decline in energy research as a percentage of total federal R&D, see Gregory F. Nemet and Daniel Kammen, "U.S. Energy Research and Development: Declining Investment, Increasing Need, and the Feasibility of Expansion," *Energy Policy* 35, no. 1 (January 2007), 747.

The varied trajectory of federal R&D spending on renewable energy from 1970 through the 1990s is discussed in Fred Sissine, Congressional Research Service, *Renewable Energy: Tax Credit, Budget, and Electricity Production Issues*, IB10041 (Washington, DC: Congressional Research Service, January 20, 2006), 3. The dollar amounts reported by Sissine are in constant 2003 dollars. The 9 percent annual increase in federal energy subsidies between 1999 and 2007 is noted in Office of Coal, Nuclear, Electric, and Alternative Fuels, U.S. Energy Information Administration, *Federal Financial Interventions and Subsidies in Energy Markets 2007*, SR/CNEAF/2008-01 (Washington, DC: U.S. Energy Information Administration, April 2008), xi–xiv, http://www.eia.doe.gov/oiaf/servicerpt/subsidy2/index.html. The Energy Information Administration report explains the changes in subsidies for particular energy sources (xii), the tripling of subsidies that take the form of tax breaks (xii), and the figures on total U.S. energy consumption and production (xiii) during this period. A good overview of federal climate change programs is provided in Congressional Budget Office, "Federal Climate Change Programs: Funding History and Policy Issues," March 2010, www.cbo.gov/doc.cfm?index=11224.

On the proliferation of congressional earmarks between 2003 and 2006 within Department of Energy programs on energy efficiency, renewable fuels, and electricity production, see Fred Sissine, Congressional Research Service, *DOE Budget Earmarks: A Selective Look at Energy Efficiency and Renewable Energy R&D Programs*, RL33294 (Washington, DC: Congressional Research Service, March 3, 2006), 2–3. Broadfoot and Webber's article "Oil Barrel Politics," *Earth*, April 8, 2009, provides the figures on 2008 congressional earmark totals. For the percentages of biomass, wind, and hydrogen projects taken up by earmarks in 2008, see American Academy for the Advancement of Science, "R&D Earmarks Hit New Record of $2.4 billion, up 13 Percent," news release, January 4, 2006, 3, http://www.aaas.org/spp/rd/earm06c.pdf; the academy quote on earmarks is on page 3.

The political appeal of using tax breaks to encourage desired behavior is discussed in Michael J. Graetz, *100 Million Unnecessary Returns: A Simple, Fair, and Competitive Tax Plan for the United States* (New Haven, CT: Yale University Press, 2008), 10–13. For the lengthy JCT report on tax benefits that subsidize energy production and conservation, see Staff of the Joint Committee on Taxation, U.S. Congress, *Tax Expenditures for Energy Production and Conservation: Scheduled for a Public Hearing before the Senate Committee on Finance on April 23, 2009*, JCX-25–09R (Washington, DC: U.S. Government Printing Office, April 21, 2009), http://www.jct.gov/publications.html?func=select&id=17. This document gives the table detailing various tax benefits (2–20); JCT's estimates of these benefits' cost between 2008 and 2012(109–112); estimates of the value of various energy production tax credits per million BTUs of fossil fuels saved(118, table 10); a comparison of two solar energy plants and the costs of each, per million BTUs(119–120); and the estimates of tax credit dollar amounts for specific hybrid motor vehicles, per million BTUs of fossil fuel saved (118–119 and table 11). For Martin Sullivan's comparison of tax benefits for the Prius, Tahoe, and other hybrid vehicles, see Sullivan, "Tech Neutrality," 621, table 1.

For a comparison of subsidies for coal, renewable fuels, and other energy sources per BTU of energy, see Metcalf, testimony before the Committee on Finance, April 23, 2009, 6, table 4. Metcalf's findings on the effective tax rates in 2007 of various energy sources are on page 5, table 3. His comparison of the subsidies for wind and geothermal energy (per ton of carbon dioxide saved) is on page 9, table 6. See also Gilbert E. Metcalf, *Taxing Energy in the United States: Which Fuels Does the Tax Code Favor?* (New York: Center for Energy Policy and the Environment at the Manhattan Institute, January 2009), http://www.manhattan-institute.org/pdf/eper_04.pdf. For the wide variation in federal subsidy amounts per megawatt hour for different sources of electricity production, see U.S. Energy Information Administration, *Federal Financial Interventions and Subsidies in Energy Markets 2007*, xvi, table ES5.

Jonathan Fahey examines the inefficiencies and unintended consequences that occur when states predetermine the portions of their electricity supplied by renewable resources. His discussion, focusing on wind power in Texas, is found in Jonathan Fahey, "Wind Power's Weird Effect," *Forbes.com*, August 19, 2009, http://www.forbes.com/forbes/2009/0907/outfront-energy-exelon-wind-powers-weird-effect.html. The "black liquor" scandal is reported in Jad Mouawad and Clifford Krauss, "Lawmakers May Limit Paper Mills' Windfall," *New York Times*, April 17, 2009, http://www.nytimes.com/2009/04/18/business/energy-environment/18sludge.html, which characterizes the use of the subsidy provision (originally meant to spur development of alternative fuels) as "an example of the law of unintended consequences." The black liquor controversy is also discussed in Martin A. Sullivan, "IRS and Congress Let Papermakers Raid the Treasury," *Tax Notes* 131, no. 3 (July 13, 2009): 105–109. In a congressional hearing held during the media publicity about "black liquor," Senator Max Baucus (D–Mont.) advocated a technology-neutral tax incentive approach while cautioning about "pitfalls" such as the black liquor loophole. See Max Baucus, Chairman, Committee on Finance, U.S. Senate, "Hearing Statement of Senator Max Baucus (D–Mont.) Regarding Technology Neutrality in Energy Tax," news release, April 23, 2009, http://finance.senate.gov/hearings/hearing/?id=d87515c8-c2a4-494d-c9ef-90697126dc42. See also Bob Ivry and Christopher Donville, "Black Liquor Tax Boondoggle May Net Billions for Papermakers," *Bloomberg.com*, April 17, 2009, http://www.bloomberg.com/apps/news?pid=20601109&sid=abDjfGgdumh4 (which describes Baucus's concerns and provides the quote from an American Forestry and Paper Association official); and Daniel Whitten and Christopher Donville, "Senators Offer Plan to End Tax Break 'Loophole' for Papermakers," *Bloomberg.com*, June 12, 2009, http://www.bloomberg.com/apps/news?pid=20601082&sid=a07xn11p5ARY.

On the many varieties of clean energy sources whose development might be enabled by government-sponsored research and on disputes over the necessary amounts of funding for such research, see Charles Weiss and William B. Bonvillian, *Structuring an Energy Technology Revolution* (Cambridge, MA: MIT Press, 2009). See also S. Pacala and R. Socolow, "Stabilization Wedges: Solving the Climate Problem for the Next 50 Years with Current Technologies," *Science* 305, no. 5686 (August 13, 2004): 968–972; Robert Socolow, Roberta Hotinski, Jeffery B. Greenblatt, and Steven Pacala, "Solving the Climate Problem: Technologies Available to Curb CO_2 Emissions," *Environment* 46, no. 10 (December 2004): 8–19; and Jeffrey D. Sachs, "Technological Keys to Climate Protection: Dramatic, Immediate Commitment to Nurturing New Technologies Is Essential to Averting Disastrous Global Warming," *Scientific American* (March 18, 2008), http://www.scientificamerican.com/article.cfm?id=technological-keys-to-climate-protection-extended. David E. Adelman and Kirsten H. Engel's article "Reorienting State Climate Change Policies to Induce Technological Change," *Arizona Law Review* 50 (2008): 835–879, argues for a federal/state partnership in which state policies focus on promoting technological change and the federal government takes primary responsibility for greenhouse gas emissions.

For a good summary of current technologies for carbon dioxide capture and storage and a discussion of existing carbon-sequestration projects, see David J. Hayes and Joel C. Beavais, "Carbon Sequestration," in *Global Climate Change and U.S. Law*, ed. Michael B. Gerrard, 691–739 (Chicago: American Bar Association, 2007). Regarding the absence, in 2009, of any commercial-scale power plant in the world that captures and sequesters its carbon emissions, see Anadon et al., *Tackling U.S. Energy Challenges and Opportunities*, 6; and Danny Cullenward, *Carbon Capture and Storage: An Assessment of Required Technological Growth Rates and Capital Investments*, Working Paper Series no. 84 (Stanford, CA: Program on Energy and Sustainable Development, Stanford University, May 2009), http://iis-db.stanford.edu/pubs/22524/PESD_cullenward_WP_84.pdf, and the sources cited therein. In "Can We Bury Global Warming?" *Scientific American* (July 2005): 49–55, Robert H. Socolow argues that using carbon sequestration to combat global warming is an attainable goal, but "only if several key challenges can be met" (49). See also Andrew C. Revkin, "Is Capturing CO_2 a Pipe Dream?" *New York Times* (Dot Earth blog), February 3, 2008, http://dotearth.blogs.nytimes.com/2008/02/03/is-capturing-co2-a-pipe-dream; Matthew L. Wald, "Mounting Costs Slow the Push for Clean Coal," *New York Times*, May 30, 2008, http://www.nytimes.com/2008/05/30/business/30coal.html; Richard Conniff, "The Myth of Clean Coal," *Yale Environment 360*, June 3, 2008, http://e360.yale.edu/content/feature.msp?id=2014; Gregg Easterbrook, "The Dirty War against

Clean Coal," *New York Times*, June 29, 2009, http://www.nytimes.com/2009/06/29/opinion/29easterbrook.html.

On the Department of Energy's initiation of the FutureGen project and its later restructuring of the project in response to rising cost estimates, see U.S. Government Accountability Office, *Clean Coal: DOE's Decision to Restructure FutureGen Should Be Based on a Comprehensive Analysis of Costs, Benefits, and Risks*, GAO-09–248 (Washington, DC: U.S. Government Accountability Office, February 2009), http://www.gao.gov/products/GAO-09-248; FutureGen Alliance, *Testimony of Paul Thompson, Chairman of the Board, FutureGen Alliance, before the U.S. Senate Committee on Appropriations, Subcommittee on Energy and Water Development* (Mattoon Township, IL: FutureGen Alliance, May 8, 2008), http://www.futuregenalliance.org/publications/FG_Thompson_Senate_EW_Testimony_050808.pdf; and FutureGen Alliance, "FutureGen Alliance Announces New Cooperative Agreement with U.S. Department of Energy," news release, September 1, 2009, http://www.futuregenalliance.org/news/releases/pr_09_01_09.pdf. For Energy Secretary Chu's statement on the revamped program, see Department of Energy, "Secretary Chu Announces Agreement on FutureGen Project in Mattoon, IL," news release, June 12, 2009, http://www.energy.gov/news2009/7454.htm.

For a gloomy appraisal of energy R&D initiatives in the 1970s and a diagnosis of those initiatives' failures, see Linda R. Cohen and Roger G. Noll, "An Assessment of R&D Commercialization Programs," in Linda R. Cohen and Roger G. Noll, with Jeffrey S. Banks, Susan A. Edelman, and William M. Pegram, *The Technology Pork Barrel* (Washington, DC: Brookings Institution Press, 1991), 365–392. See page 78 of the Cohen and Noll volume for the "unambiguous failures" quote (in the chapter "The Structure of the Case Studies," by Cohen, Noll, and Pegram), page 365 for the "hardly a success" quote, and page 384 for the "dooming the enterprise to failure" quote. More information on the 1970s synfuels program is found in chapter 5 of this book.

On the need for major institutional changes in energy R&D efforts, see Kenneth J. Arrow, Linda R. Cohen, Paul A. David, Robert W. Hahn, Charles D. Kolstad, Lee L. Lane, W. David Montgomery, Richard R. Nelson, Roger G. Noll, and Anne E. Smith, *A Statement on the Appropriate Role for Research and Development in Climate Policy*, Working Paper no. 8-12 (Washington, DC: Reg-Markets Center, December 9, 2008), http://SSRN.com/abstract=1313827. This joint statement expresses the views of ten economists and scientists on the subject. See also Weiss and Bonvillian, *Structuring an Energy Technology Revolution*, on the implications for energy R&D of Congress's committee structure (36, 186) and on needed improvements within the executive branch (185–186). For the Energy Technology Innovation Policy Group quote on raising the price of carbon emissions, see Anadon et al., *Tackling U.S. Energy Challenges and Opportunities*, 16.

Chapter 12

On government-imposed energy-efficiency standards for household appliances in the United States and abroad, see Steven Nadel, "Appliance and Equipment Efficiency Standards," *Annual Review of Energy and the Environment* 27 (2002): 159–192; Lloyd Harrington and Melissa Damnics, *Energy Labeling and Standards throughout the World*, 2004/04 (Sydney: National Appliance and Equipment Energy Efficiency Committee, Australia, December 2001), http://www.energyrating.gov.au/library/pubs/200404-internatlabelreview.pdf; Isaac Turiel, "Present Status of Residential

Appliance Energy Efficiency Standards—an International Review," *Energy and Buildings* 26 (1997): 5–15; and Hannah Choi Granade, Jon Creyts, Anton Derkach, Philip Farese, Scott Nyquist, and Ken Ostrowski, *Unlocking Energy Efficiency in the U.S. Economy*, McKinsey & Company report (Milton, VT: Villanti, July 2009), http://www.mckinsey.com/clientservice/electricpowernaturalgas/downloads/US_energy_efficiency_full_report.pdf. See also work of the energy-efficiency standards group at http://ees.ead.lbl.gov.

For a representative work in favor of mandatory energy-efficiency standards, see Jonathan G. Koomey, Susan A. Mahler, Carrie A. Webber, and James E. McMahon, *Projected Regional Impacts of Appliance Efficiency Standards of the U.S. Residential Sector*, LBNL-39511 (Berkeley, CA: Ernest Orlando Lawrence Berkeley National Laboratory, February 1998), which declares that "efficiency standards in the residential sector have been a highly cost-effective policy instrument for promoting energy efficiency" (1). For a detractor, see Ronald J. Sutherland, "The High Costs of Federal Energy Efficiency Standards for Residential Appliances," *Policy Analysis* (December 23, 2003), http://www.cato.org/pubs/pas/pa504.pdf, which argues that errors in Koomey's method lead to an underestimation of the cost to consumers of government energy standards.

On the increase in energy efficiency of refrigerators, see the energy-efficiency standards group at http://ees.ead.lbl.gov/projects/past_projects/refrigerators, and John C. Dernbach, "U.S. Policy," in *Global Climate Change and U.S. Law*, ed. Michael B. Gerrard (Chicago: American Bar Association, 2007), 70. For a summary of the debate over what has caused California's per capita electricity use to grow so much slower than the rest of the country's, see, for example, Kate Galbraith, "Deciphering California's Energy Successes," available at http://green.blogs.nytimes.com/2009/04/14/deciphering-californias-efficiency-successes, and Breakthrough Blog at http://thebreakthrough.org/blog/2009/04/is_california_a_model_for_an_e.shtml. Criticism of efficiency standards for focusing on particular appliances rather than on functions is discussed in Steven Nadel, "The Future of Standards," *Energy and Buildings* 26 (1996): 119–128. On the Energy Star labeling system and other "strong" voluntary programs that "approach mandatory standards in the degree to which they are followed," see Nadel, "Appliance and Equipment Efficiency Standards," 179–180. Nadel also notes that voluntary agreements are frequently used in the EU, where adopting regulations is difficult.

For more on energy efficiency as it relates to climate change, see Adam B. Jaffe, Richard G. Newell, and Robert N. Stavins, *Energy-Efficient Technologies and Climate Change Policies: Issues and Evidence*, Climate Issue Brief no. 19 (Washington, DC: Resources for the Future, December 1999), http://www.rff.org/rff/Documents/RFF-CCIB-19.pdf; and Trevor Houser, *The Economics of Energy Efficiency in Buildings*, Policy Brief no. PB09-17 (Washington, DC: Peterson Institute for International Economics, August 2009), http://www.iie.com/publications/pb/pb09-17.pdf.

For a good history of clean-air legislation, see Arnold W. Reitze Jr., "The Legislative History of U.S. Air Pollution Control," *Houston Law Review* 36 (1999): 679–741. On the "command-and-control" model of environmental regulation that prevailed during the 1970s and on rising criticism of this model, see Stephen Breyer, *Regulation and Its Reform* (Cambridge, MA: Harvard University Press, 1982), 262–263. Breyer discusses the profusion of litigation engendered by this approach on pages 265 and 268–270. J. H. Tietenberg, "Economic Instruments for Environmental Regulation," *Oxford Review of Economic Policy* 6, no. 1 (1990): 17–33, examines emissions trading as a promising approach that harnesses economic incentives for more efficient results than elicited by command-and-control regimes.

A readable introduction to the workings of cap-and-trade systems is Alan Durning, with Anna Fahey, Eric de Place, Lisa Stiffler, and Clark Williams-Derry, *Cap and Trade 101: A Federal Climate Policy Primer* (Seattle: Sightline Institute, July 2009), http://www.sightline.org/research/energy/res_pubs/cap-and-trade-101/Cap-Trade_online.pdf. See also Tom Tietenberg, "Editor's Introduction," in *The Evolution of Emissions Trading: Theoretical Foundations and Design Considerations*, ed. Tom Tietenberg (Aldershot, UK: Ashgate, forthcoming), http://www.colby.edu/personal/t/thtieten/Edintro.pdf. In this introduction, Tietenberg traces the intellectual development of the theory behind cap and trade from its roots in the 1960s.

For an argument in favor of a carbon tax as an alternative to cap and trade, see Gilbert E. Metcalf, *A Proposal for a U.S. Carbon Tax Swap: An Equitable Tax Reform to Address Global Climate Change*, Discussion Paper 2007-12 (Washington, DC: Brookings Institution, October 2007), http://www.brookings.edu/~/media/files/rc/papers/2007/10carbontax_metcalf/10_carbontax_metcalf.pdf. On the ability of a carbon tax to function like a cap-and-trade system, see David A. Weisbach, *Instrument Choice Is Instrument Design* Olin Working Paper no. 490 (Chicago: University of Chicago Law and Economics, October 23, 2009), http://papers.ssrn.com/sol3/papers.cfm?abstract_id=1493312, and the sources cited therein; and J. H. Tietenberg, *Emissions Trading: An Exercise in Reforming Pollution Policy* (Washington, DC: Resources for the Future, 1985). For an excellent discussion of the difficulties in using direct emissions regulations to approximate the results of a carbon tax or cap-and-trade system, see Robert N. Stavins, "Experience with Market-Based Environmental Policy Instruments," in *Handbook of Environmental Economics*, vol. 1, ed. Karl-Göran Mäler and Jeffrey R. Vincent (San Diego: Elsevier, 2003), 355–435.

The Clean Air Act is discussed as an example of the inefficiency of command regulation in Weisbach, *Instrument Choice Is Instrument Design*, and Tietenberg, *Emissions Trading*. Significant early examples of economists' promotion of pollution taxes or cap and trade (as an alternative to command regulation) include T. D. Crocker, "The Structuring of Atmospheric Pollution Control Systems," in *The Economics of Air Pollution*, ed. Harold Wolozin (New York: W. W. Norton, 1966), 61–86; and J. H. Dales, *Pollution, Property, and Prices* (Toronto: University of Toronto Press, 1968). The cost-reducing advantages of these alternatives was recognized in mid-1970s publications such as Marc J. Roberts and Michael Spence, "Effluent Charges and Licenses under Uncertainty," *Journal of Public Economics* 5 (1976): 193–208; Martin L. Weitzman, "Prices vs. Quantities," *Review of Economic Studies* 41, no. 4 (1974): 477–491; and W. David Montgomery, "Markets in Licenses and Efficient Pollution Control Programs," *Journal of Economic Theory* 5 (1972): 395–418.

On the EPA's hesitance to embrace "market-based" alternatives to command regulation, see Stavins, "Experience with Market-Based Environmental Policy Instruments," 393–394. For examples of environmentalist resistance to tradable emissions permits, see Robert E. Goodin, "Selling Environmental Indulgences," *Kyklos* 47, no. 4 (1994): 573–594; and Michael J. Sandel, "It's Immoral to Buy the Right to Pollute," *New York Times*, December 15, 1997, A23.

The EPA's development of a cap-and-trade policy for the Clean Air Act is described in Tom Tietenberg, "Tradable Permit Approaches to Pollution Control: Faustian Bargain or Paradise Regained?" in *Property Rights, Economics, and the Environment*, ed. Michael D. Kaplowitz (Stamford, CT: JAI Press, 2001), http://www2.ing.puc.cl/seminario/Documentos/TradablePermitsApproachesTietenberg.pdf. See also Tom Tietenberg, "The Tradable-Permits Approach to Protecting the Commons: Lessons for Climate Change," *Oxford Review of Economic Policy* 19, no. 3 (2003): 400–419. For the estimate of $400 million in savings from emissions-credit programs, see Stavins, "Experience with Market-Based Environmental Policy Instruments," 393.

On the EPA's successful use of a tradable credits approach to phase out leaded gasoline, see Tietenberg, "Tradable Permit Approaches to Pollution Control," available at http://www2.ing.puc.cl/seminario/Documentos/TradablePermitsApproaches Tietenberg.pdf; and Stavins, "Experience with Market-Based Environmental Policy Instruments," 395–396. On the agency's allocation of tradable permits to eliminate ozone-depleting chemicals and on the excise tax that soon followed to mitigate windfall profits, see these same two articles: Tietenberg, 5, and Stavins, 400–401. Stavins concludes that the subsequent tax was more effective than the tradable permits.

For the increase in light trucks under the CAFE fuel-efficiency standards between 1979 and 1999, see David Friedman, Jason Mark, and Patricia Monahan, *Drilling in Detroit: Tapping Automaker Ingenuity to Build Safe and Efficient Automobiles* (Cambridge, MA: Union of Concerned Scientists, June 2001), 52, http://www.ucsusa.org/assets/documents/clean_vehicles/drill_detroit.pdf. The CAFE rules were revised in the *Energy Independence and Security Act of 2007*, Public Law 110–140, 121 Stat. 1492 (2007). The Obama administration's raising of the fuel mileage standard and its new rules for tailpipe emissions are reported in Steven Mufson, "Vehicle Emission Rules to Tighten: U.S. Would Also Raise Fuel Mileage Standards by 2016," *Washington Post*, May 19, 2009, http://www.washingtonpost.com/wp-dyn/content/article/2009/05/18/AR2009051801848.html. For the increase in passenger-vehicle miles driven in the United States since 1979, see Friedman, Mark, and Monahan, *Drilling in Detroit*, 53. Alan J. Auerbach, "Public Finance in Practice and Theory," *CESifo Economic Studies* 56, no. 1 (December 7, 2009), 11–12, explains how a cap-and-trade emissions regime would improve on the current CAFE system and lower costs.

The tradable permit system established by the 1990 Clean Air Act amendments to reduce sulfur dioxide emissions and acid rain is examined in Stavins, "Experience with Market-Based Environmental Policy Instruments," 401–403, which describes this system as arguably "[t]he most important application ever made of a market-based instrument for environmental protection" (401). See also Richard E. Cohen, *Washington at Work: Back Rooms and Clean Air*, 2d ed. (New York: Longman, 2009), which offers a case study of the legislative history behind the 1990 Clean Air Act amendments. The story of how "an unlikely mix of environmentalists and free-market conservatives hammered out the strategy known as cap-and-trade" is told in Richard Conniff, "The Political History of Cap and Trade," *Smithsonian* (August 2009), http://www.smithsonianmag.com/science-nature/Presence-of-Mind-Blue-Sky-Thinking.html. For a discussion of lessons to be drawn from the 1990 experience for the current impasse on climate issues, see David Schoenbrod, Richard B. Stewart, and Katrina M. Wyman, *Breaking the Logjam: Environmental Reform for the New Congress and Administration*, project report (New York: New York University Environmental Law Journal, February 2009), 4. On the overallocation issue, see Lesley K. McAllister, "The Overallocation Problem in Cap and Trade: Moving toward Stringency," *Columbia Journal of Environmental Law* 34 (2009), 396. Tom Tietenberg reports his students' purchases of emissions allowances as gifts in "Tradable Permit Approaches to Pollution Control," 6, http://www2.ing.puc.cl/seminario/Documentos/TradablePermitsApproachesTietenberg.pdf.

For more on the Regional Greenhouse Gas Initiative, see its Web site at http://www.rggi.org and Alan Krueger, Assistant Secretary for Economic Policy Designate and Chief Economist, U.S. Department of the Treasury, statement to the Finance Committee, U.S. Senate, 111th Cong., 1st sess., May 7, 2009, 4–5, http://finance.senate.gov/hearings/hearing/?id=d8bf9810-9806-b79e-9989-5473d579d9fb. The results of the initiative's permit auctions can be viewed at http://www.rggi.org/market/co2_auctions/results. For the Western Climate Initiative, see http://www

.westernclimateinitiative.org, and for the Midwestern Greenhouse Gas Accord see http://www.midwesternaccord.org/index.html.

On the EU's emissions-trading regime and the Kyoto Protocol, see U.S. Government Accountability Office, *International Climate Change Programs: Lessons Learned from the European Union's Emissions Trading Scheme and the Kyoto Protocol's Development Mechanism*, GAO-09–151 (Washington, DC: U.S. Government Accountability Office, November 2008), http://www.gao.gov/new.items/d09151.pdf; pages 13–26 describe the difficulties encountered during the first phase of the trading system, including price volatility. See also Kyle W. Danish, "The International Regime," in *Global Climate Change and U.S. Law*, ed. Gerrard, 31–60; Dennis Hirsch, Andrew Bergman, and Michael Heintz, "Emissions Trading—Practical Aspects," in *Global Climate Change and U.S. Law*, ed. Gerrard, 627–690; Congressional Budget Office, *Policy Options for Reducing CO_2 Emissions*, pub. no. 2930 (Washington, DC: Congressional Budget Office, February 2008), 24–25, http://www.cbo.gov/ftpdocs/89xx/doc8934/02-12-Carbon.pdf; and Jos Delbeke, Deputy Director-General, Directorate General for Environment, European Commission, Brussels, written statement, *Auctioning under Cap and Trade: Design, Participation, and Distribution of Revenues: Hearing before the Committee on Finance, U.S. Senate*, 111th Cong., 1st sess., May 7, 2009, http://finance.senate.gov/hearings/hearing/?id=d8bf9810-9806-b79e-9989-5473d579d9fb, 5–11.

Fraud in the EU trading system and the Lohmann quote are reported in James Kanter, "Fraud Besets E.U. Carbon Trade System," *New York Times* (Green Inc. blog), February 8, 2010, http://www.nytimes.com/2010/02/08/business/energy-environment/08green.html. The finding that European emissions increased between 2005 and 2007 is found in Environmental Audit Committee, House of Commons, *The Role of Carbon Markets in Preventing Dangerous Climate Change: Fourth Report of Session 2009–10*, HC 290 (London: Her Majesty's Printing Service, January 26, 2010), 10, http://www.publications.parliament.uk/pa/cm200910/cmselect/cmenvaud/290/290.pdf. For U.K. concerns about the EU trading system, see Environmental Audit Committee, "EU Carbon Market Failing to Deliver Vital Green Investment—MP's Warn," news release, February 8, 2010, http://www.parliament.uk/parliamentary_committees/environmental_audit_committee/eacpn080210.cfm, which includes the Tim Yeo quotes and which declares that the system "is failing to deliver vital green investment, after a collapse in carbon prices magnified by the recession." Some analysts would regard a collapse in prices of permits during a recession an advantage, not a disadvantage, of cap and trade. Sarah-Jayne Clifton is quoted in Terry Macalister, "MPs Propose Carbon Tax to Boost Green Investment," *Guardian.co.uk*, February 8, 2010, http://www.guardian.co.uk/environment/2010/feb/08/carbon-emissions-trading-system. For an assessment of the impact of general financial market turmoil on carbon markets, see Takashi Kanamura, "Financial Turmoil in Carbon Markets," mimeo August 3, 2010, http://ssrn.com/abstract=1652735.

On the economic and distributional impacts of 2009 cap-and-trade proposals, see Congressional Budget Office, *The Economic Effects of Legislation to Reduce Greenhouse-Gas Emissions* (Washington, DC: Congressional Budget Office, September 2009), http://www.cbo.gov/ftpdocs/105xx/doc10573/09-17-Greenhouse-Gas.pdf. For a discussion of how a greenhouse gas cap-and-trade system would likely affect different industries, regions, and income groups, see Douglas W. Elmendorf, statement on *The Distribution of Revenues from a Cap-and-Trade Program for CO_2 Emissions: Hearing before the Committee on Finance, U.S. Senate*, 111th Cong., 1st sess., May 7, 2009, http://finance.senate.gov/hearings/hearing/?id=d8bf9810-9806-b79e-9989-5473d579d9fb. The potential interaction between a cap-and-trade system and command-and-control regulation is examined in Congressional Budget Office, *How Regulatory Standards Can Affect a Cap-and-Trade Program for Greenhouse Gases* (Washington, DC: Congressional Budget Office, September 16, 2009), http://www.cbo.gov/ftpdocs/105xx/doc10562/09-16-CapandStandards.pdf.

On the allocation of emissions permits under cap and trade, see *Written Testimony of A. Denny Ellerman, Senate Committee on Energy and Natural Resources,* 111th Cong., 1st sess., October 21, 2009, http://globalchange.mit.edu/files/document/SenateTestimony_Ellerman_Oct09.pdf, which includes a bibliography on emissions trading; Gilbert E. Metcalf, testimony, *Allocation Issues in Greenhouse Gas Cap and Trade Systems: Hearing before the Committee on Energy and Natural Resources, U.S. Senate,* 111th Cong., 1st sess., October 21, 2009, http://globalchange.mit.edu/files/document/SenateTestimony_Metcalf_Oct09.pdf, which identifies essential considerations in designing an allocation system, emphasizing "simplicity, transparency, efficiency and distributional outcomes"; and Karen Palmer, testimony, *Costs and Benefits for Consumers and Energy Price Effects Associated with the Allocation of Greenhouse Gas Emissions Allowances: Hearing before the Committee on Energy and Natural Resources, U.S. Senate,* 111th Cong., 1st sess., October 21, 2009, http://www.rff.org/RFF/Documents/RFF-CTst-09-Palmer.pdf, which discusses regional distributional disparities and economic efficiency, arguing that "an auction approach to allocation will yield the most efficient outcome" by passing along the full cost of carbon use to consumers.

For more on the "important" role that allocation auctions can play in a cap-and-trade system and "how auctions have been used in some existing efficient greenhouse gas cap-and-trade programs in the United States and abroad," see Alan Krueger, Assistant Secretary for Economic Policy Designate and Chief Economist, U.S. Department of the Treasury, statement to the Finance Committee, U.S. Senate, 111th Cong., 1st sess., May 7, 2009, http://finance.senate.gov/hearings/hearing/?id=d8bf9810-9806-b79e-9989-5473d579d9fb. See also Anne E. Smith, Ph.D., prepared statement, *Auctioning under Cap and Trade,* hearing, 111th Cong., 1st sess., May 7, 2009, http://finance.senate.gov/hearings/hearing/?id=d8bf9810-9806-b79e-9989-5473d579d9fb, which discusses practical differences between auctions and free allocations in regard to "the distribution of carbon value that can be accomplished"; Delbeke, written statement, *Auctioning under Cap and Trade,* hearing, May 7, 2009, 111th Cong., 1st sess., which describes how allocating free emissions allowances would likely lead to windfall profits for regulated companies, at the expense of the public budget; and Alan D. Viard, testimony, *Climate Change Legislation: Allowance and Revenue Distribution: Hearing before the Committee on Finance, U.S. Senate,* 111th Cong., 1st sess., August 4, 2009, http://finance.senate.gov/hearings/hearing/?id=1f3ebe07-009e-744b-93a4-67168d05fc4b, which argues that free allocations of emissions allowances harm economic efficiency and produce negative distributional concerns, both of which can be avoided through an auction system.

For an argument that the choice between auctions and free allocations is "less important than it seems" and "doesn't affect the environmental efficacy or cost-effectiveness of the program," see Nathaniel O. Keohane, Ph.D., Director of Economic Policy and Analysis, Environmental Defense Fund, testimony before the Committee on Finance, U.S. Senate, 111th Cong., 1st sess., August 4, 2009, http://www.edf.org/documents/10295_Testimony_Keohane_Senate_080409.pdf.

For IPCC estimates about the likely trajectory of climate change during this century, see Intergovernmental Panel on Climate Change, *Climate Change 2007: The Physical Science Basis* (Cambridge, UK: Cambridge University Press, 2007); Intergovernmental Panel on Climate Change, *Climate Change 2007: Impacts, Adaptation, and*

Vulnerability (Cambridge, UK: Cambridge University Press, 2008); and Intergovernmental Panel on Climate Change, *Climate Change 2007: Mitigation of Climate Change* (Cambridge, UK: Cambridge University Press, 2007). The IPCC's estimates regarding the probable rise in global temperatures are discussed in Martin L. Weitzman, "Some Dynamic Economic Consequences of the Climate Sensitivity Inference Dilemma," unpublished paper, February 2008; the controversy surrounding the IPCC's work is discussed in chapter 10. There have been many ongoing reviews and analyses. For one that regards the IPCC temperature estimates as overly optimistic, see United Nations Environment Program (UNEP), *Climate Change Science Compendium* (Paris: UNEP, 2009), http://unep.org/COMPENDIUM2009.

For Hansen's opinions about cap and trade, see James E. Hansen, "Worshipping the Temple of Doom," letter to the Australian government, preceded by a two-page commentary, May 5, 2009, http://www.columbia.edu/~jeh1/mailings/2009/20090505_TempleOfDoom.pdf. See also Elizabeth Kolbert, "The Catastrophist," *The New Yorker* (June 29, 2009), 39; Bill McKibben, *Eaarth: Making Life on a Tough New Planet* (New York: Times Books, 2010), chaps. 1, 3, and 4.

For President Obama's United Nations speech on climate change, see "Remarks by the President at United Nations Secretary General Ban Ki-Moon's Climate Change Summit," speech at United Nations Headquarters, New York, September 22, 2009, http://www.whitehouse.gov/the_press_office/Remarks-by-the-President-at-UN-Secretary-General-Ban-Ki-moons-Climate-Change-Summit. The quote from Nicholas Stern is from "Climate: What You Need to Know," *New York Review of Books*, June 24, 2010, 37. CBO director Elmendorf's observations on climate change and its likely effects are from Douglas W. Elmendorf, Director, CBO, statement to the Committee on Energy and Natural Resources, U.S. Senate, 111th Cong., 1st sess., October 14, 2009, http://www.cbo.gov/ftpdocs/105xx/doc10561/10-14-Greenhouse-GasEmissions.pdf. The estimate of a 3 percent decline in GDP is on page 3; the "price path" quote is on page 6.

There is enormous debate in the climate change literature about how much of a change in emissions will be required and over what time period. Compare, for example, Nicholas Stern, *The Economics of Climate Change: The Stern Review* (Cambridge, UK: Cambridge University Press, 2007), with William D. Nordhaus, *A Question of Balance: Weighing the Options on Global Warming Policies* (New Haven, CT: Yale University Press, 2008). See also Robert S. Pindyck, *Uncertain Outcomes and Climate Change Policy*, Working Paper 4742-09 (Cambridge, MA: MIT Sloan School, August 6, 2009), http://papers.ssrn.com/sol3/papers.cfm?abstract_id=1448683.

The basic agreement between President Obama and Democratic congressional leaders on the appropriate rate of progress in combating climate change is reflected in H.R. 2454, known as the Waxman–Markey bill and the American Clean Energy and Security Act of 2009. The Senate failed to enact such legislation before the 111th Congress expired. For a detailed guide to the bill's key provisions and the current regulatory context, see Davis Polk and Wardwell, LLP, "U.S. House of Representatives Approves Landmark Climate Change Legislation," client memorandum, July 16, 2009, http://www.davispolk.com/files/Publication/f78abb11-af6f-4ade-a4bb-0462a1a49cb5/Presentation/PublicationAttachment/0d25f563-1698-4e74-b731-0d6a9ff52652/071609_Climate_Bill.pdf. Analysis of the bill's features, including discussion of the president's goals, can also be found in Larry Parker and Brent D. Yacobucci, Congressional Research Service, *Climate Change: Costs and Benefits of the Cap-and-Trade Provisions of H.R. 2454*, R40809 (Washington, DC: Congressional Research Service, September 14, 2009); Jonathan L. Ramseur, Larry Parker, and Brent D. Yacobucci, Congressional Research Service, *Market-Based*

Greenhouse Gas Control: Selected Proposals in the 111th Congress, R40556 (Washington, DC: Congressional Research Service, May 20, 2009), 1–2; and Congressional Budget Office, *Cost Estimate: H.R. 2454, American Clean Energy and Security Act of 2009* (Washington, DC: Congressional Budget Office, June 5, 2009), 5.

Writing in *The New Yorker* in late 2009, Elizabeth Kolbert complained that U.S. commitments to reducing greenhouse gas emissions were lower than those of either Japan or Europe, adding that Japan and Europe are using 1990, not 2005, as a baseline. See Elizabeth Kolbert, "Leading Causes," *The New Yorker* (October 5, 2009), http://www.newyorker.com/talk/comment/2009/10/05/091005taco_talk_kolbert. For an example of an economist who argues that emissions reductions should be implemented slowly, see Pindyck, *Uncertain Outcomes and Climate Change Policy*.

On the role of offsets in a cap-and-trade system, see Congressional Budget Office, *The Use of Offsets to Reduce Greenhouse Gases* (Washington, DC: Congressional Budget Office, August 3, 2009), http://www.cbo.gov/ftpdocs/104xx/doc10497/08-03-Offsets.pdf; and U.S. Government Accountability Office, *Carbon Offsets: The U.S. Voluntary Market Is Growing, but Quality Assurance Poses Challenges for Market Participants*, GAO-08–1048 (Washington, DC: U.S. Government Accountability Office, August 2008), http://www.gao.gov/new.items/d081048.pdf, which discusses the difficulties of predicting what emissions reductions would have occurred in the absence of an offset payment. Jason Scott Johnston, "Problems of Equity and Efficiency in the Design of International Greenhouse Gas Cap-and-Trade Schemes," *Harvard Environmental Law Review* 33 (2009): 405–430, argues that emissions reductions in an international system utilizing offsets may be undermined by problems of enforceability and verifiability. For an argument that the United States should rely on more explicit cost-control mechanisms instead of on offsets, see Michael W. Wara and David G. Victor, *A Realistic Policy on International Carbon Offsets*, Working Paper no. 74 (Stanford, CA: Program on Energy and Sustainable Development, April 18, 2008), http://iis-db.stanford.edu/pubs/22157/WP74_final_final.pdf.

For the Congressional Research Service quote on "infinite" scenarios for offsets, see Parker and Yacobucci, *Climate Change*, 18. The CBO's estimate on the degree to which the use of offsets would lower emissions allowance prices is found in Congressional Budget Office, *Cost Estimate*, 18. A consensus is that offsets will reduce costs by more than half (see Parker and Yacobucci, *Climate Change*, 47).

For a description of the Clean Development Mechanism under the Kyoto Protocol, see Kyle W. Danish, "The International Regime," in *Global Climate Change and U.S. Law*, ed. Gerrard, 46–51. For criticisms of the offset credit program under Kyoto and similar European initiatives, see, for example, Madhusree Mukerjee, "Is a Popular Carbon-Offset Method Just a Lot of Hot Air?" *Scientific American* (June 2009), http://www.scientificamerican.com/article.cfm?id=a-mechanism-of-hot-air; and Lambert Schneider, *Is the CDM Fulfilling Its Environmental and Sustainable Development Objectives? An Evaluation of the CDM and Options for Improvement* (Freiburg, Germany: Öko-Institut, prepared for the World Wildlife Fund, November 5, 2007), http://www.oeko.de/oekodoc/622/2007-162-en.pdf. Current proposals for U.S. cap-and-trade legislation contemplate a robust ability of firms subject to emissions caps to purchase both domestic and international offset credits. International projects in developing countries would qualify only if the country has an agreement with the United States. On "banking" and "borrowing" as mechanisms to mitigate price volatility, see Parker and Yacobucci, *Climate Change*, 7, 23; and Durning et al., *Cap and Trade 101*, 17. The use of a price ceiling or collar to stability volatility is discussed in Warwick J. McKibbin, Adele Morris, Peter J. Wilcoxen, and Yiyong Cai, *Consequences of Alternative U.S. Cap-and-Trade Policies: Controlling Both Emissions and Costs* (Washington, DC: Brookings Institution, July 24, 2009), 11–14.

For a discussion of how low-income households may be affected by the allocation of greenhouse gas emissions allowances and the need for well-designed policies that address this concern, see Chad Stone, Chief Economist, Center on Budget and Policy Priorities, testimony before the Committee on Energy and Natural Resources, U.S. Senate, 111th Cong., 1st sess., October 21, 2009, http://www.cbpp.org/files/10-21-09climate-testimony.pdf. See also Andrew Chamberlain, *Who Pays for Climate Policy? New Estimates of the Household Burden and Economic Impact of a U.S. Cap-and-Trade System,* Working Paper no. 6 (Washington, DC: Tax Foundation, March 16, 2009), http://papers.ssrn.com/sol3/papers.cfm?abstract_id=1361257.

Federalism and the role of the states in climate change policy is discussed in Daniel A. Farber, *Climate Adaptation and Federalism: Mapping the Issues,* Public Law Research Paper no. 1468621 (Berkeley: University of California, September 9, 2009), http://papers.ssrn.com/sol3/papers.cfm?abstract_id=1468621. Lesley K. McAllister, "Regional Climate Regulation: From State Competition to State Collaboration," *San Diego Journal of Climate and Energy Law* 1 (2009): 81–102, examines collaborative initiatives among states and how they have been prompted in part by federal inaction. Ann E. Carlson, "Iterative Federalism and Climate Change," *Northwestern University Law Review* 103 (2009): 1097–1162, examines significant state efforts on global warming and how they spring from and interact with federal law.

The MIT quote on designing U.S. policy targets with the goal of influencing other nations is found in Sergey Paltsev, John M. Reilly, Henry D. Jacoby, Angelo C. Gurgel, Gilbert E. Metcalf, Andrei P. Sokolov, and Jennifer F. Holak, *Assessment of U.S. Cap-and-Trade Proposals,* Joint Program on the Science and Policy of Global Change Report no. 146 (Cambridge, MA: MIT, April 2007), 55. Lincoln Moses is quoted in Parker and Yacobucci, *Climate Change,* i. Ireland's new carbon tax was introduced in Finance Bill 2010 (Act no. 9/2010), http://www.finance.gov.ie/documents/publications/Finance%20Bill%202010/Bill2010.pdf.

Chapter 13

On the vast number of interest groups lobbying Congress on climate change, see Marianne Lavelle, "Tally of Interests on Climate Bill Tops a Thousand," Center for Public Integrity, August 10, 2009, http://www.publicintegrity.org/articles/entry/1609. Similar lobbying efforts over energy legislation during the late 1970s are reported in Charles Mohr, "Business Using Grass-Roots Lobby," *New York Times,* April 17, 1978, A1; Charles Mohr, "Opposing Sides Agree on Ways to Shift to Coal," *New York Times,* February 10, 1978, A1; Steven V. Roberts, "Activists Prod Senators on Energy, *New York Times,* October 1, 1977, 25; and John R. Emshwiller, "Shareholders United to Help Utility Firms Battle Regulators and Consumer Groups," *Wall Street Journal,* April 13, 1978, 46. The House climate bill, known as Waxman–Markey, is the American Clean Energy and Security Act of 2009, H.R. 2454, 111th Cong., 1st sess., 2009. It was passed on June 26, 2009, and was received in the Senate on July 6.

For a comparative analysis of how different political systems vary in the extent to which they grant veto power to political actors and the ramifications of these differences, see George Tsebelis, *Veto Players: How Political Institutions Work* (Princeton, NJ: Princeton University Press, 2002). For David Wessel's comments, see David Wessel, "The Californization of Washington," *Wall Street Journal,* March 4, 2010, http://online.wsj.com/article/NA_WSJ_PUB:SB100014240527487045413045750993

71249822654.html. Richard Lazarus comprehensively reviews the history of environmental law in *The Making of Environmental Law* (Chicago: University of Chicago Press, 2004). The statements quoted are from his article "Congressional Descent: The Demise of Deliberative Democracy in Environmental Law," *Georgetown Law Journal* 94 (2006), 620–622. For Mann and Ornstein's observations on Congress, see Thomas E. Mann and Norman J. Ornstein, *The Broken Branch: How Congress Is Failing America and How to Get It Back on Track* (New York: Oxford University Press, 2006), 242. For a discussion of how economic conditions in the 1960s and early 1970s encouraged Americans to take economic abundance for granted, contributing to "an upsurge of citizen activism" and faith in the government's ability to manage economic problems, see David Vogel, *Fluctuating Fortunes: The Political Power of Business in America* (New York: Basic Books, 1989), 96–97.

On Congress's assertion of its power vis-à-vis the president during the early 1970s, see Mann and Ornstein, *The Broken Branch*, 59–60. The two statutes discussed are the *War Powers Resolution*, Public Law 93-148, 87 Stat. 555 (1973), and *Impoundment Control Act of 1974*, Public Law 93-344, title X, 88 Stat. 297 (1974). Richard E. Cohen, *Washington at Work: Back Rooms and Clean Air*, 2d ed. (New York: Longman, 2009), 12–23, examines the disproportionate influence exerted by a few senators on the Clean Air Act and other contemporary legislation. Lazarus, "Congressional Descent," discusses changes in the House (665–666), the "demise of bipartisanship" (668–677), and the pattern of partisan voting in the early 1970s and mid-1990s (670). The great influence of a small number of committee chairs during this period and the breakdown of this power structure following the elections of 1974 is explored n Mann and Ornstein, *The Broken Branch*, 51, 52–53, 61–63. Tommy Boggs is quoted in Vogel, *Fluctuating Fortunes*, 206 (which cites Philip Shabecoff, "Big Business on the Offensive," *New York Times Magazine*, December 9, 1979, 91).

For John Dingell's background and constituency, see Cohen, *Washington at Work*, 29–33. In 1981, Congressman Dingell married Deborah Insley, a General Motors executive whose family roots go back to the Fisher Body Corporation, which was acquired by General Motors in 1919. Washington insiders, however, insist that John Dingell, whose district was once the home to 80,000 auto industry jobs, was "married to General Motors long before he was married" to Deborah Insley (Cohen, *Washington at Work*, 29). Cohen's book also covers Henry Waxman's background (31) as well as Dingell's record defending the auto industry and its workers and Waxman's victory over him in the 1977 Clean Air Act amendments (31–44). The story of how Waxman later wrested Dingell's committee chairmanship from him is told in John M. Broder and Carl Hulse, "Behind House Struggle, Long and Tangled Roots," *New York Times*, November 23, 2008, A26 (including President Obama's and House Speaker Pelosi's tacit support for Waxman); Naftali Bendavid and Stephen Power, "Waxman Takes Over Energy Panel," *Wall Street Journal*, November 21, 2008, A6; and Janet Hook and Richard Simon, "Waxman Gets Key Energy Position," *Los Angeles Times*, November 21, 2008, A1. In February 2009, Dingell became the longest-serving member in the history of the House of Representatives and with the death of Senator Robert Byrd in June 2010 the senior member of Congress (see Bendavid and Power, "Waxman Takes Over Energy Panel").

For a sample of empirical research that attempts to elucidate the roles of ideology, political party, regionalism, and constituent interests in decision making by representatives, see Charles R. Shipen and William R. Lowry, "Environmental Policy Party Divergences in Congress," *Political Research Quarterly* 54 (2001): 245–263; Steven D. Levitt, "How Do Senators Vote? Disentangling the Role of Voter Preferences, Party Affiliation, and Senator Ideology," *American Economic Review* 86 (1996): 425–441;

Joseph P. Kalt and Mark A. Zupan, "The Apparent Ideological Behavior of Legislators: Testing for Principal-Agent Slack in Political Institutions," *Journal of Law and Economics* 33 (1990): 103–131; Sam Peltzman, "Constituent Interest and Congressional Voting," *Journal of Law and Economics* 27 (1984): 181–210; James W. Riddlesperger Jr. and James D. King, "Energy Votes in the U.S. Senate, 1973–1980," *Journal of Politics* 44 (1982): 838–847; Robert Bernstein and Seven R. Horn, "Explaining House Voting on Energy Policy: Ideology and the Conditional Effects of Party and District Economic Interests," *Western Political Quarterly* 34 (1981): 235–245; Linda R. Cohen, Roger G. Noll, and William M. Pegram, "The Structure of the Case Studies," in Linda R. Cohen and Roger G. Noll, with Jeffrey S. Banks, Susan A. Edelman, and William M. Pegram, *The Technology Pork Barrel* (Washington, D.C.: Brookings Institution Press, 1991), 83–96; and Edward J. Mitchell, "Energy, Ideology, Not Interest," *Wall Street Journal*, July 7, 1977, 12. One study estimates that party affiliation and regional loyalty explain about 74 percent of measured ideology. See John P. Nelson, "Green Voting and Ideology: LCV Scores and Roll Call Voting in the U.S. Senate, 1988–1998," *Review of Economics and Statistics* 84 (2002): 518–529. For research suggesting that politicians tend to support legislation that imposes costs on districts or states other than their own, see Anwar Hussein and David N. Laband, "The Tragedy of the Political Commons: Evidence from U.S. Senate Roll Call Votes on Environmental Legislation," *Public Choice* 124 (2005): 353–364.

The information on regional variation in energy consumption is taken from Paul L. Joskow, "U.S. Energy Policy during the 1990s," unpublished manuscript, July 11, 2001, http://www.hks.harvard.edu/hepg/Papers/EnergyPolicy-CEEPR-7-11-01withTables+Figure.pdf. Joseph P. Kelt, *The Economics and Politics of Oil Price Regulation: Federal Policy in the Post Embargo Era* (Cambridge, MA: MIT Press, 1981), examines how conflict between legislators from oil-producing and oil-consuming states made it difficult to remove price controls on oil and natural gas. On the use of televised congressional hearings to stoke anger at oil companies and the effect on public opinion, see Vogel, *Fluctuating Fortunes*, 126–128.

On the channeling of synfuels subsidies toward programs utilizing eastern coal, see Linda R. Cohen and Roger G. Noll, "Synthetic Fuels from Coal," in *The Technology Pork Barrel*, 260.

Except where otherwise noted, the public-opinion polling data discussed in this chapter is drawn from the Web site PollingReport.com at http://pollingreport.com: http://pollingreport.com/enviro.htm and http://pollingreport.com/energy.htm. The poll showing reduced support for a cap-and-trade program with increased utility bills was reported in "Who Cares? Don't Count on Public Opinion to Support Mitigation," *The Economist* (December 5, 2009), 15. For the 2009 poll revealing widespread public confusion about the meaning of cap and trade, see "Congress Pushes Cap and Trade, but Just 24% Know What It Is," *Rasmussen Reports*, May 11, 2009, http://www.rasmussenreports.com/public_content/politics/current_events/environment_energy/congress_pushes_cap_and_trade_but_just_24_know_what_it_is. For the poll indicating lowered support for global warming legislation that requires higher electric bills, see "Washington Post–ABC News Poll," conducted June 18–21, 2009, *Washington Post*, http://www.washingtonpost.com/wp-srv/politics/polls/postpoll_062209.html?sid=ST20090624040004. The 2009 Gallup poll showing that only one percent of Americans see the environment as the nation's biggest problem is reported in "Cap and Tirade: America Struggles with Climate-Change Legislation," *The Economist* (December 5, 2009), 14–15. On the concept of public opinion providing "running room" for Congress, see Michael J. Graetz and Ian Shapiro, *Death by a Thousand Cuts: The Fight over Taxing Inherited Wealth* (Princeton, NJ: Princeton University Press, 2005), chap. 12.

For the Yale/George Mason study, see Yale Project on Climate Change and George Mason Center for Climate Change Communication, *Global Warming's Six Americas 2009: An Audience Segmentation Analysis* (Ithaca, NY, and Fairfax, VA: Yale University and George Mason University, May 20, 2009), http://environment.yale.edu/uploads/6Americas2009.pdf; the quotes are from page 2. This study divided the public into "six Americas," ranging from the 18 percent of the population who are "Alarmed," the group most convinced that global warming is happening and is a serious and immediate threat, to the 7 percent who are "Dismissive," the group convinced that global warming is not happening and actively opposing any national policy response. In between, the largest single group is "Concerned," the 33 percent who believe global warming is a problem and support some response but feel much less urgency about it than the "Alarmed" group. The three other groups are "Cautious," who are somewhat convinced that global warming is happening but do not consider it personally important; "Disengaged," who are not at all sure global warming is happening and do not consider it personally important; and "Doubtful," who believe that if global warming is happening, it is due to natural, not human, causes and that it will not start harming people in the United States for at least 100 years. The "Alarmed" and the "Concerned" together constitute 51 percent of the population, whereas the "Disengaged, " "Doubtful," and "Dismissive" add up to about 30 percent. When the "Cautious" are added in, 49 percent of the population feels that there is no urgency to address climate change. As Anthony Leiscrowitz, director of the Yale Project on Climate Change and one of the report's coauthors, remarked, it is a "misnomer" to "talk about the 'American Public' and its views on global warming." In fact, "there is no single American voice on this issue." See Suzanne Taylor Muzzin, "Report: On Climate Change, There Are 'Six Americas,'" *Yale Bulletin & Calendar*, June 12, 2009, 10.

For the Supreme Court's 2007 decision on EPA regulation of greenhouse gases, see *Massachusetts v. EPA*, 549 U.S. 497 (2007); the quotes are from pages 505 and 533. For EPA administrator Jackson's proposed findings on greenhouse gases, see "Proposed Endangerment and Cause or Contribute Findings for Greenhouse Gases under Section 202(a) of the Clean Air Act," *Federal Register* 74, no. 78 (April 24, 2009): 18886–18910. Jackson's announcement that the EPA would begin regulating some greenhouse gases is reported in John M. Broder, "EPA Moves to Curtail Greenhouse Gas Emissions," *New York Times*, October 1, 2009, A1; and Siobhan Hughes and Ian Talley, "EPA Proposes Tough Greenhouse-Gas Rules for Big Industries," *Wall Street Journal*, October 1, 2009, A5 (including Jackson's "business as usual" quote). For Jackson's claim about the announcement's timing, see Lisa Lerer, "EPA Chief: Timing Is a Coincidence," *Politico*, December 11, 2009, http://www.politico.com/news/stories/1209/30390.html.

For the Second Circuit ruling on greenhouse gases as a "public nuisance," see *Connecticut v. American Electric Power Co.*, 582 F.3d 309 (2d Cir. 2009). The quote assessing the ruling's significance is from Hannah McCrea, "Why the Second Circuit 'Nuisance' Case Brings Good News, and Bad (Part II)," *Grist*, September 24, 2009, http://www.grist.org/article/2009-09-24-why-second-circuit-nuisance-case-brings-good-news-and-bad-part-2. This kind of litigation has many procedural and substantive obstacles to overcome, and if appellate courts disagree, the Supreme Court may put a halt to such lawsuits.

The emergence of post-Kyoto regional carbon-trading schemes is examined in Jennifer Weeks, "Carbon Trading: Will It Reduce Global Warming?" *CQ Global Researcher* 2, no. 11 (November 2008): 295–320. See also *Regional Greenhouse Gas Initiative, an Initiative of the Northeast and Mid-Atlantic States of the U.S.*, http://www.rggi.org/home. The program summary can be found at http://rggi.org/docs/program

_summary_10_07.pdf. On the Western Climate Initiative, see Western Climate Initiative, "Designing the Program," available at http://www.westernclimateinitiative.org/designing-the-program. The original partner members of the Western Climate Initiative were Arizona, British Columbia, California, Manitoba, Montana, New Mexico, Ontario, Oregon, Quebec, Utah, and Washington. See Western Climate Initiative, "WCI Provincial and State Partner Contracts," available at http://www.westernclimateinitiative.org/wci-partners. In February 2010, Arizona announced that it was dropping out. See Sindya N. Bhanoo, "Arizona Quits Western Cap-and-Trade Program," *New York Times*, February 12, 2010, http://www.nytimes.com/2010/02/12/science/earth/12climate.html. See also the references listed on this topic in chapter 12.

On Governor Schwarzenegger's renewable sources executive order, see Bryan Walsh, "Are the Governors Our Best Hope for the Climate?" *Time*, October 2, 2009, http://www.time.com/time/health/article/0,8599,1927621,00.html. The "statement of principles" on global warming released by 30 governors is reported in Avery Palmer, "Global Warming Bill Gets Energy Panel OK," *CQ Weekly*, May 25, 2009, 1217–1218. The Kyoto pledge signed by 1,000 mayors is reported in Kim Murphy, "1000 Mayors Agree to Reduce Greenhouse Gas Emissions," *Los Angeles Times*, October 3, 2009, http://articles.latimes.com/2009/oct/03/nation/na-mayors-climate3.

The alliance between environmentalists and business interests in support of cap and trade is examined in Julie Kosterlitz, "Corporate–Environmental Alliance Breaks the Mold," *National Journal*, May 30, 2009, http://www.nationaljournal.com/njmagazine/nj_20090530_9442.php, and sharply criticized in Johann Hari, "The Wrong Kind of Green," *The Nation*, March 4, 2010, http://www.thenation.com/article/wrong-kind-green, and Christina McDonald, *Green, Inc: An Environmental Insider Reveals How a Good Cause Has Gone Bad* (Guilford, CT: Lyons Press, 2000). Both of these criticisms emphasize the financial relevance to certain environmental organizations of corporate financial support. Daniel Gross, "The New Progressive CEOs," *Slate*, October 10, 2009, http://www.slate.com/id/2231925, examines how the pragmatism of business leaders (particularly during a recession and a Democratic presidency) has led them to support efforts to deal with global warming; the Gross quote comes from this article. For more on the divide-a-dollar game, see Dennis C. Mueller, *Public Choice II: A Revised Edition of* Public Choice (Cambridge, UK: Cambridge University Press, 1989), 63–65.

For a comprehensive examination of the ebbs and flows in the political power of business interests during the 1970s, see Vogel, *Fluctuating Fortunes*; the influence of labor unions, consumer advocacy groups, and environmentalists on business power are discussed on pages 58–60. The role of Energy Action is reported in Steven V. Roberts, "Activists Prod Senators on Energy," *New York Times*, October 1, 1977, 25. For a discussion of the growth in congressional regulation of business during this period, see Vogel, *Fluctuating Fortunes*, 98–104; the shifting winds that led to a resurgence of business power, abetted by the onset of stagflation, are described on pages 115–118.

For the Supreme Court decision striking down individual contribution and candidate spending limits, see *Buckley v. Valeo*, 424 U.S. 1 (1976); see also pages 117–120 of Vogel, *Fluctuating Fortunes*. For the growth in PACs between the early 1970s and 1980, see page 207 of *Fluctuating Fortunes*. For the recent case striking down more recent restrictions on "electioneering communications" by corporations and unions, see the U.S. Supreme Court decision in *Citizens United v. Federal Election Commission*, 130 S.Ct. 876 (2010), http://www.supremecourt.gov/opinions/09pdf/08-205.pdf. Many corporate PACs had benign-sounding names such as the "Good Government

Group" (Delta Airlines) and the "Voluntary Non-Partisan Political Fund" (Dow Chemical). Labor groups were not to be outdone; the United Auto Workers, for example, funded the Committee for Good Government. See Warren Weaver Jr., "What Is a Campaign Contributor Buying?" *New York Times*, March 13, 1977, 154. Douglas Ross and Harold Wolman, "Congress and Pollution: The Gentleman's Agreement," *Washington Monthly* (September 1970): 13–20, examines some of the institutional and financial factors that prevented environmental legislation from having greater success throughout the 1970s; the Nader "virtually irresistible political force" quote is from page 18.

On the mobilization of "grasstops" constituents in another context, see Graetz and Shapiro, *Death by a Thousand Cuts*, chap. 6. On the business lobby's use of "sit-downs" with legislators, see Charles Mohr, "Business Using Grass-Roots Lobby," *New York Times*, April 17, 1978, A1 (including Mark Green quote). The U.S. Chamber of Commerce's use of computerized databases and the mobilization of shareholders and employees by Atlantic Richfield and other companies are described in Vogel, *Fluctuating Fortunes*, 205–206. On the formation of shareholder groups to lobby on behalf of utilities, see John R. Emshwiller, "Shareholders Unite to Help Utilities Battle Regulators and Consumer Groups," *Wall Street Journal*, April 13, 1978, 46. On the comparative ineffectiveness of environmental organizations in mobilizing local constituencies at this time, see Robert Gottlieb, *Forcing the Spring: The Transformation of the American Environmental Movement* (Washington, DC: Island Press, 1993), 132. For background on the rise and influence of conservative think tanks, see Graetz and Shapiro, *Death by a Thousand Cuts*, chap. 9.

Duke Energy CEO Jim Rogers and his advocacy for climate change legislation are examined in Sharon Begley, "The Global Elite, 50: Jim Rogers, the CEO of Duke Energy Could Make Dreams of Renewable Power a Reality," *Newsweek*, January 5, 2009, http://www.newsweek.com/2008/12/19/50-jim-rogers.html; and Clive Thompson, "A Green Coal Baron?" *New York Times Magazine*, June 22, 2008, http://www.nytimes.com/2008/06/22/magazine/22Rogers-t.html. The latter article contains Eileen Claussen's praise of Rogers and environmentalist accusations that he is "greenwashing" his company's polluting practices. As noted in the same article, Rogers's history of environmental activism is rather lengthy, dating back to his support for America's first cap-and-trade regime: the sulfur dioxide limits imposed by the Clean Air Act Amendments of 1990. Bruce E. Nilles's criticism of Rogers is reported in Melanie Warner, "Is America Ready to Quit Coal?" *New York Times*, February 15, 2009, BU1. Frank O'Donnell's criticism is reported in John Carey, "The Clash over Clean Power," *Business Week*, October 12, 2009, http://www.businessweek.com/magazine/content/09_41/b4150055757494.htm. See also Dina Cappiello, "A Green Business Divide over Warming Plan," *CQ Weekly*, January 28, 2008, 238.

Rogers attributes Duke Energy's stance on climate change legislation to its recognition that the firm was going to require significant assistance and a long period of time to reduce its emissions: "We had one of the largest carbon footprints in the country—in the world—so we had a special responsibility to address this issue. We recognized that we needed to start making the transition, because it would take longer and be harder and more costly for our customers. So the sooner we went to work on it, the better"; in Michelle Nijhuis, "Energy in Our Lives: Interview with Jim Rogers, Industry Executive," *National Geographic* (March 2009), http://ngm.nationalgeographic.com/2009/03/energy-issue/rogers-field-notes.

For Rogers's congressional testimony, including background information on Duke Energy, see James E. Rogers, Chairman, CEO, and President, Duke Energy Corporation, testimony before the Committee on Energy and Commerce, U.S. House of

Representatives, 111th Cong., 1st sess., April 22, 2009, http://energycommerce.house.gov/Press_111/20090422/testimony_rogers.pdf; and James E. Rogers, Chairman, President, and CEO, Duke Energy Corporation, testimony before the Committee on Foreign Relations, U.S. Senate, 111th Cong., 1st sess., May 19, 2009, http://foreign.senate.gov/hearings/hearing/?id=7159ba96-e76a-924d-33db-b767345e779d.

For Rogers's advocacy of free emissions allowances, see Jim Rogers, "Climate Change Legislation Should Not Be Punitive," *The Energy Daily*, February 29, 2008, http://www.duke-energy.com/pdfs/Rogers-Energy-Daily-2-29-08.pdf. Rogers writes: "Building coal plants was a key part of our energy policy in the 1960s and 1970s. The 1974 Arab oil embargo pushed our country toward energy independence, and discouraged the use of oil in power plants. The federal Fuel Use Act of 1978 prohibited using natural gas for new power plants until its repeal in 1985. Three Mile Island stopped the nuclear industry in its tracks in 1979." See also John Carey, "The Clash over Clean Power," *Business Week*, October 12, 2009, http://www.businessweek.com/magazine/content/09_41/b4150055757494.htm, in which Rogers presents an argument for free allowances on behalf of southern and midwestern residents. A Duke Energy fact sheet includes a subsection entitled "Punishing America's Heartland," which notes: "Half of the states in the U.S. rely on coal generation for most of their electric power. Most of those states also fall below the national median for household income." See Duke Energy, "Carbon Cap and Trade: We Need to Get It Right," available at http://www.duke-energy.com/pdfs/CapNTrade_Fact_Sheet_FINAL.PDF.

For Rogers's advocacy of nuclear technologies and his doubt about the role of renewable energy sources, see Jim Rogers, "Why Nuclear Power Is Part of Our Future," *Wall Street Journal*, August 4, 2009, http://online.wsj.com/article/NA_WSJ_PUB:SB10001424052970204619004574324313479837876.html. On Duke Energy's use of solar energy, see Sharon Begley, "The Global Elite, 50: Jim Rogers, the CEO of Duke Energy Could Make Dreams of Renewable Power a Reality," *Newsweek*, January 5, 2009, http://www.newsweek.com/2008/12/19/50-jim-rogers.html; and on its purchase of a wind power company, see Clive Thompson, "A Green Coal Baron?" *New York Times Magazine*, June 22, 2008, http://www.nytimes.com/2008/06/22/magazine/22Rogers-t.html. On the company's plans to construct a new coal-fired power plant in North Carolina, see Melanie Warner, "Is America Ready to Quit Coal?" *New York Times*, February 15, 2009, BU1. Rogers has explained: "We're going to build an 830-megawatt plant, which will be very efficient. As a consequence, we're going to shut down 1,000 megawatts of old, very inefficient plants that have a large carbon footprint for every kilowatt-hour they produce. I don't view the plant as the solution. I view it as a transition plant—a bridge, if you will, to the low-carbon world" (in Nijhuis, "Interview with Jim Rogers," which also contains Rogers's optimism about the potential of carbon capture and sequestration). Duke Energy's experiment with this technology at its Edwardsport, Indiana, plant is reported in Matthew L. Wald, "Stimulus Money Puts Clean Coal Projects on a Faster Track," *New York Times*, March 17, 2009, B1.

The profit that companies such as Exelon Energy and Florida Power and Light stand to gain from cap-and-trade legislation is examined in Richard Schlesinger, "Fortunes in Cap and Trade," *Energy Central*, November 25, 2009, http://www.energyconferences.net/articles/energybizinsider/ebi_detail.cfm?id=776, including the quote from Hay.

USCAP is profiled in Kosterlitz, "Corporate–Environmental Alliance Breaks the Mold." The 25 businesses participating in USCAP are AES, Alcoa, Alstom, Boston Scientific Corporation, BP America, Caterpillar, Chrysler, ConocoPhillips, Deere & Company, the Dow Chemical Company, Duke Energy, DuPont, Exelon

Corporation, Ford Motor Company, FPL Group, General Electric, General Motors Corporation, Johnson & Johnson, NRG Energy, PepsiCo, PG&E Corporation, PNM Resources, Rio Tinto, Shell, and Siemens Corporation. These businesses are joined by five leading environmental groups: the Environmental Defense Fund, the Natural Resources Defense Council, the Nature Conservancy, the Pew Center on Global Climate Change, and the World Resources Institute. See the USCAP Web site at http://www.us-cap.org.

For USCAP's 2007 statement of principles on greenhouse gas emissions, see United States Climate Action Partnership (USCAP), *A Call for Action: Consensus Principles and Recommendations from the U.S. Climate Action Partnership, a Business and NGO Partnership* (Washington, DC: USCAP, January 2007), 2, 4, http://www.us-cap .org/USCAPCallForAction.pdf. For USCAP's 2009 legislative proposals, see United States Climate Action Partnership (USCAP), *A Blueprint for Legislative Action: Consensus Recommendations for U.S. Climate Protection Legislation* (Washington, DC: USCAP, January 2009), http://www.us-cap.org/pdf/USCAP_Blueprint.pdf. See pages 7, 8, and 16–24 for the specific goals and targets cited. Henry Waxman's laudatory quote about USCAP is reported in Ronald Brownstein, "Waxman: 'We Have a Mandate,'" *National Journal*, July 11, 2009, http://www.nationaljournal.com/njmagazine/ cg_20090711_1585.php.

For the CBO estimate on revenue that might potentially be generated by an auction of emissions allowances, see Douglas W. Elmendorf, statement, *The Distribution of Revenues from a Cap-and-Trade Program for CO_2 Emissions: Hearing before the Committee on Finance, U.S. Senate*, 111th Cong., 1st sess., May 7, 2009, http://finance .senate.gov/hearings/hearing/?id=d8bf9810-9806-b79e-9989-5473d579d9fb. For the CBO's cost estimate of an early cap-and-trade bill (projecting nearly $850 billion in auction revenue over a decade), see Congressional Budget Office, *H.R. 2454, American Clean Energy and Security Act of 2009, As Reported by the House Committee on Energy and Commerce on May 21, 2009*, cost estimate (Washington, DC: Congressional Budget Office, June 5, 2009). On the ultimate similarity between the results of an auction and a free-allowance system in reducing greenhouse gas emissions, see, for example, *Written Testimony of A. Denny Ellerman, Senate Committee on Energy and Natural Resources,* 111th Cong., 1st sess., October 21, 2009, 9, http://globalchange .mit.edu/files/document/SenateTestimony_Ellerman_Oct09.pdf.

For the estimate that 15 to 20 percent of the emissions allowances are sufficient to compensate industry, see Joseph E. Aldy, Alan J. Krupnick, Richard G. Newell, Ian W. H. Perry, and William A. Pizer, *Designing Climate Mitigation Policy*, Discussion Paper 08-16 (Washington, DC: Resources for the Future, May 2009), 20, http://www .rff.org/documents/RFF-DP-08-16.pdf. For agreement that the value of the permits in the early years far exceeds the costs of limiting emissions, see Gilbert E. Metcalf, statement, U.S. Senate, *Allocation Issues in Greenhouse Gas Cap and Trade Systems: Hearing before the Committee on Energy and Natural Resources*, 111th Cong., 1st sess., October 21, 2009, http://globalchange.mit.edu/files/document/SenateTestimony _Metcalf_Oct09.pdf; and Larry Parker and Brent D. Yacobucci, Congressional Research Service, *Climate Change: Costs and Benefits of the Cap-and-Trade Provisions of H.R. 2454*, R40809 (Washington, DC: Congressional Research Service, September 14, 2009). The latter work contains the CBO's prediction that $700 billion in allowances would be disbursed between 2011 and 2019. It also illustrates how electric, oil, and gas companies fare during the first round of emissions allowance distributions in the Waxman–Markey bill, as passed by the House. Figures 1 and 2 at page 6 of Parker and Yacobucci, *Climate Change*, are available at http://www.heartland .org/custom/semod_policybot/pdf/26007.pdf.

For the quoted USCAP sentiment, which is reflected in the Waxman–Markey bill, see USCAP, *A Blueprint for Legislative Action*, 11. The CBO estimated that a bill introduced by Senators Joe Lieberman (D–Conn.) and Mark Warner (R–Va.) in 2004 would raise $1.2 trillion the following decade, nearly twice the Obama budget estimate.

Peter Orszag's remarks to Congress on permit auctions and President Obama's "historic leap" comment are quoted in David Wessel, "Pollution Politics and the Climate-Bill Giveaway," *Wall Street Journal*, May 23, 2009, A2. For President Obama's 2010 budget, see Office of Management and Budget, *Budget of the U.S. Government, Fiscal Year 2011* (Washington, DC: Office of Management and Budget, February 2010). This budget contains a footnote that says, "A comprehensive market-based climate change policy will be deficit neutral because proceeds from emissions allowances will be used to compensate vulnerable families, communities, and businesses during the transition to a clean energy economy. Receipts will also be reserved for investments to reduce greenhouse gas emissions, including support of clean energy technologies and in adapting to the impacts of climate change, both domestically and in developing countries" (171, table S-8, n. 4).

The inevitability of regional variations in energy prices after disbursal of emissions allowances is discussed in Karen Palmer, testimony, U.S. Senate, *Costs and Benefits for Consumers and Energy Price Effects Associated with the Allocation of Greenhouse Gas Emissions Allowances: Hearing before the Committee on Energy and Natural Resources*, 111th Cong., 1st sess., October 21, 2009, http://www.rff.org/RFF/Documents/RFF-CTst-09-Palmer.pdf. For the "very well hidden" quote by an MIT economist regarding free allocations, see *Written Testimony of A. Denny Ellerman, Senate Committee on Energy and Natural Resources*.

The quotes from Johann Hari are from his article "The Wrong Kind of Green," *The Nation*, March 4, 2010, http://www.thenation.com/article/wrong-kind-green. The withdrawal of BP, ConocoPhillips, and Caterpillar from USCAP are reported in Stephen Power and Ben Casselman, "Defections Shake Up Climate Coalition," *Wall Street Journal*, February 17, 2010, http://online.wsj.com/article/NA_WSJ_PUB:SB10001424052748704804204575069440096420212.html; and Sheila McNulty and Ana Fifield, "BP Deals Blow to Obama Fight on Climate," *Financial Times*, February 17, 2010, http://www.businessinsider.com/bp-deals-blow-to-obama-fight-on-climate-2010-2. The latter article includes Conoco vice president Red Cavaney's comments. For background on Business for Innovative Climate and Energy Policy, see its Web site at http://ceres.org/bicep. The group's open letter to the president and the Senate calling for their leadership on climate change is quoted in Mike Allen, "Big Business Pushes for Climate Action," *Politico*, October 5, 2009, http://www.politico.com/news/stories/1009/27896.html. The group's lobbying efforts in Washington, D.C., are described in Business for Innovative Climate & Energy Policy, "More Than 150 Companies from 30 Plus States to Support Comprehensive Energy and Climate Legislation," news release, October 5, 2009, http://www.ceres.org/Page.aspx?pid=1136.

For information on the Investor Network on Climate Risk and its lobbying efforts, see Erin McNeill, "Business Divided over Approach to Climate Change," *CQ Weekly*, May 26, 2008, 1390; Coral Davenport, Benton Ives, and Phil Mattingly, "Carbon, from the Ground Up," *CQ Weekly*, August 3, 2009, 1836–1844; and the Investor Network's Web site at http://www.incr.com/Page.aspx?pid?=198.

The role of farmers and agricultural interests in climate change lobbying is reported in "Farmers v Greens," *The Economist*, November 12, 2009, 44; and Lavelle, "Tally of Interests on Climate Bill Tops a Thousand." The potentially counterproductive effects of promoting wood ethanol to combat climate change are explored in

Roger A. Sedjo and Brent Sohngen, "An Inconvenient Truth about Cellulosic Ethanol," *Miliken Institute Review* 11, no. 4 (Fourth Quarter 2009): 51–55.

For the Business Roundtable's key publication on climate change, see Business Roundtable, *The Balancing Act: Climate Change, Energy Security, and the U.S. Economy* (Washington, DC: Business Roundtable, June 2009), vii, http://www.businessroundtable.org/sites/default/files/2009.06%20The%20Balancing%20Act_FINAL.pdf. The quote recommending technological support instead of a carbon price is from page ix; the passage addressing "emissions leakage" is on page xi. As explained on the organization's Web site, "Business Roundtable is an association of chief executive officers of leading U.S. companies with a total of more than $5 trillion in annual revenues and more than 10 million employees." See Business Roundtable, "About Us: Overview," at http://businessroundtable.org/about. In generic terms, the Roundtable's policy recommendations include: "(1) more efficiently consume electricity and heating fuels in homes and businesses; (2) leverage domestic resources to produce cost-effective, low-carbon electricity; and (3) modernize the transportation fleet and diversify the transportation fuel mix" (Business Roundtable, *The Balancing Act*, vii).

For the projections by the American Council for Capital Formation and the National Association of Manufacturers on reduction in GDP and decline in industrial production, see American Council for Capital Formation (ACCF) and the National Association of Manufacturers (NAM), *Analysis of the Waxman-Markey Bill "The American Clean Energy Act of 2009" (H.R. 2454) Using the National Energy Modeling System*, NEMS/ACCF-NAM 2 (Washington, DC: ACCF and NAM, August 2009), http://www.accf.org/media/dynamic/3/media_381.pdf. For the CBO estimate of the carbon emissions allowance price in 2019, see Congressional Budget Office, *H.R. 2454, American Clean Energy and Security Act of 2009*.

For the U.S. Chamber of Commerce's opposition to greenhouse gas regulation, see Kate Galbraith, "Apple Resigns from Chamber over Climate," *New York Times* (Green Inc. blog), October 5, 2009, http://greeninc.blogs.nytimes.com/2009/10/05/apple-resigns-from-chamber-over-climate; and Erin McNeill, "Business Divided over Approach to Climate Change," *CQ Weekly*, May 26, 2008, 1390. Chamber vice president Kovacs's call for a "Scopes Monkey Trial" is reported in Jim Tankersley, "U.S. Chamber of Commerce President Shrugs Off Defections," *Los Angeles Times*, October 9, 2009, http://articles.latimes.com/2009/oct/09/nation/na-chamber-climate9; and Michael Burnham, "Chamber Threatens Lawsuit If EPA Rejects Climate Science 'Trial,'" *New York Times*, August 25, 2009, http://www.nytimes.com/cwire/2009/08/25/25climatewire-chamber-threatens-lawsuit-if-epa-rejects-cli-62828.html. The defection of companies from the chamber due to disagreements about climate change is reported in Stephen Power, "Political Alliances Shift in Fight over Climate Bill," *Wall Street Journal*, October 6, 2009, 35; and Anne C. Mulkern, "'Hot Button' Climate Issue Spotlights How U.S. Chamber Sets Policy," *New York Times*, October 6, 2009, http://www.nytimes.com/gwire/2009/10/06/06greenwire-hot-button-climate-issue-spotlights-how-us-cha-24103.html. For suggestions that chamber vice president Tom Donohue's conflict of interest is at fault, see Josh Harkinson, "Inside the Chamber of Carbon," *Mother Jones*, October 7, 2009, http://motherjones.com/environment/2009/10/chamber-commerce-vs-climate-change. Donohue sits on the board of Union Pacific Railroad, a company that earns a substantial portion of its revenue from transporting coal, and he has earned millions in payments and stock from the railroad.

For Greenpeace's opposition to Waxman-Markey, see "Greenpeace: Waxman-Markey Climate Change Bill Not Strong Enough to Stop Global Warming," available at http://www.greenpeace.org/usa/en/news-and-blogs/news/greenpeace-waxman-markey-clim. For the criticism of Waxman-Markey by Friends of the Earth, see "Friends of the Earth Statement on House Passage of American Clean Energy and Security Act," available at http://www.foe.org/flawed-climate-and-energy-bill-passes-house. The Community Coalition for Environmental Justice's statement in opposition to Waxman-Markey was viewed at www.ccej.org, last accessed October 1, 2009.

The Center for Public Integrity's figures on the number of companies and organizations lobbying Congress regarding climate change legislation are provided in Lavelle, "Tally of Interests on Climate Bill Tops a Thousand." The shortcomings of the Lobbying Disclosure Act and the expenditures of the American Coalition for Clean Coal Electricity are discussed in Marianne Lavelle, "Less Clean Coal Lobbying? Not Exactly," *Paper Trail* (blog for the Center for Public Integrity), April 28, 2009, http://www.publicintegrity.org/blog/entry/1304. See also Martin Kady II, "Keeping Grass-Roots Lobbying under Wraps," *CQ Weekly*, March 26, 2007, 877–878. According to *Roll Call*, in the first three quarters of 2009, ExxonMobil alone spent $20.7 million on lobbying, compared to about $1.7 million each spent by the World Wildlife Fund and the Environmental Defense Fund during the same period. The lobbying expenditures of ExxonMobil, the World Wildlife Fund, and the Environmental Defense Fund are reported in Bennett Roth, "Climate Stakeholders Target Senators," *Roll Call*, November 9, 2009, http://www.rollcall.com/issues/55_54/lobbying/40376-1.html. The cooperation of unions and environmental organizations in the "BlueGreen Alliance" is reported in Matthew Murray, "Unions Set Their Sights on Climate Change Bill," *Roll Call*, July 8, 2009, http://www.rollcall.com/issues/55_3/lobbying/36555-1.html.

On the efforts of interest groups to mobilize members and create the appearance of support for their positions, see Ken Kollman, *Outside Lobbying: Public Opinion and Interest Group Strategies* (Princeton, NJ: Princeton University Press, 1998). The American Petroleum Institute's letter to its member companies urging opposition to climate legislation is available at http://www.desmogblog.com/sites/beta.desmogblog.com/files/GP%20API%20letter%20August%202009-1.pdf. See also "Another Astroturf Campaign," *New York Times*, September 3, 2009, http://www.nytimes.com/2009/09/04/opinion/04fri2.html. The Federation for American Coal and Energy Security's description of itself as a group of "people from all walks of life" can be found on its Web site, "About Us," at http://www.facesofcoal.org/about-us. For the discovery that the federation's Web site features stock photos and is hosted by a Washington lobbying firm, see J. W. Randolph, "FACES of Coal Are iStockphotos?!" *Grist*, August 27, 2009, http://www.grist.org/article/2009-08-27-faces-of-coal-are-istockphotos.

For more on the fraudulent letter scheme originating from Bonner & Associates on behalf of the American Coalition for Clean Coal Electricity, see Jack Bonner, testimony before the Select Committee on Energy Independence and Global Warming, U.S. House of Representatives, 111th Cong., 1st sess., October 15, 2009, http://globalwarming.house.gov/files/HRG/102909Letters/bonner.pdf; the quoted passages are from page 4 of the testimony. See also David A. Fahrenthold, "Coal Group Reveals Six More Forged Lobbying Letters," *Washington Post*, August 5, 2009, A4, which includes the quote from one of the legislators who voted against the bill after receiving the letter. On the congressional inquiry into the letters and the reactions by the Sierra Club and the American Coalition for Clean Coal Electricity, see Stephanie Strom, "Coal Group Is Linked to Fake Letters on Climate Bill," *New York Times*, August 5, 2009, A12. For Bonner & Associates' failure to notify Congress immediately and for the Representative Markey quote, see Alex Kaplun, "Lobbyist Apologizes to House Climate Panel for Forged-Letter 'Scheme,'" *New York Times*,

October 29, 2009, http://www.nytimes.com/gwire/2009/10/29/29greenwire-lobbyist-apologizes-to-house-climate-panel-for-89713.html.

The statutory text of the Kerry-Lieberman "American Power Act" is available at http://lieberman.senate.gov/assets/pdf/APA_full.pdf. A section-by-section summary is available at http://lieberman.senate.gov/assets/pdf/APA_sect.pdf. The senators' summary of the bill can be found at http://lieberman.senate.gov/assets/pdf/APA_sum.pdf. The Pew Center for Global Climate Change's side-by-side comparison of this legislation and the House bill is available at http://www.pewclimate.org/federal/analysis/congress/111/comparison-waxman-markey-and-kerry-lieberman.

The "backroom deals" quote from Senators Cantwell and Collins is from Maria Cantwell and Susan Collins, "A Cap-and-Dividend Way to a Cleaner Nation and More Jobs," *Washington Post*, June 18, 2010, A27. The statute is available at http://cantwell.senate.gov/issues/Leg_Text.pdf, and a section-by-section description is available at http://cantwell.senate.gov/issues/Section_by_section.pdf. A video explanation and other explanatory documents can be found at http://cantwell.senate.gov/issues/CLEARAct.cfm. See also Christa Marshall, "Cantwell–Collins Bill Generates Lobbying Frenzy," *New York Times Climate Wire*, February 15, 2010, http://www.nytimes.com/cwire/2010/02/15/15climatewire-cantwell-collins-bill-generates-lobbying-fre-54450.html; and Kate Sheppard, "The Other Climate Bill," *Mother Jones*, March 25, 2010, http://motherjones.com/blue-marble/2010/03/cantwell-collins-climatebill?page=2.

For the EPA projection of $1.4 billion sent abroad in international offsets, see "Cap and Tirade: America Struggles with Climate-Change Legislation," *The Economist*, December 5, 2009, 14. The G-77 walkout at the Copenhagen summit and the Sudanese chairman's comments are discussed in "The Copenhagen Shakedown," *Wall Street Journal*, December 16, 2009, http://online.wsj.com/article/NA_WSJ_PUB:SB10001424052748704398304574597900307712862.html. The World Bank's estimate that $475 billion will be needed in developing countries for emissions reduction is reported in "Closing the Gaps: How the World Divides on a Global Deal," *The Economist*, December 5, 2009, 18. For the plight of the Maldives, the statement by its president, and video of the underwater meeting, see "Maldives Cabinet Makes a Splash," *BBC News*, October 17, 2009, http://news.bbc.co.uk/2/hi/south_asia/8311838.stm.

A detailed political discussion of the climate change debate is provided in *The Climate War: True Believers, Power Brokers, and the Fight to Save the Earth* (New York: Hyperion, 2010). The book covers events from Spring 2007 to March 2010 based on in-depth interviews of political actors on all sides of the debate.

Chapter 14

A transcript and video of President Obama's June 15, 2010, Oval Office address on the BP oil spill can be found at http://www.whitehouse.gov/the-press-office/remarks-president-nation-bp-oil-spill. Jon Stewart's montage of eight presidents is available at http://www.thedailyshow.com/watch/wed-june-16-2010/an-energy-independent-future. For the $100 billion estimate of the costs for restoring the Gulf of Mexico region and a 25-year time horizon for doing so, see Albert Hunt, "BP Oil Spill Cleanup Alone Won't End Gulf's Suffering," available at http://www.bloomberg.com/news/2010-06-27/bp-spill-cleanup-alone-won-t-end-gulf-s-suffering-albert-hunt.html. For a discussion of Jimmy Carter's speech, see chapter 8. For the idea that Carter's question was the one that Obama should address and the unrequited hope that the president would not plead his case for cap-and-trade legislation, see

Walter Shapiro, "Obama's Oval Office Address: Look to Jimmy Carter (No Kidding)," on the *Politics Daily* blog at http://www.politicsdaily.com/2010/06/14/obamas-oval-office-address-look-to-jimmy-carter-no-kidding. For a summary of criticism of the president's speech, see Laura Meckler and Neil King Jr., "Obama Address Gets Low Marks," *Wall Street Journal*, June 17, 2010, A5. The claim that deepwater drilling technology rivals that of space exploration can be found in Jad Mouawad, "The Spill vs. a Need to Drill," *New York Times*, April 30, 2010, http://www.nytimes.com/2010/05/02/weekinreview/02jad.html. Media coverage of the Deepwater Horizon explosion and its aftermath was extensive and no attempt to cite it is made here. Congressman Markey's quote about BP giving itself the "smallest piece" of the blame is from "BP Disaster Report Sparks Outrage from Contractors and Lawmakers," *Financial Times*, September 9, 2010, A1. See also Ian Urbina, "In Report on Gulf Spill, BP Sheds Some Light and Casts Much Blame," *New York Times*, September 9, 2010, A 14.

For background on the Massey Coal mining disaster, see Ian Urbina and Michael Cooper, "Deaths at West Virginia Mine Raise Issues About Safety," *New York Times*, April 6, 2010, http://www.nytimes.com/2010/04/07/us/07westvirginia.html?scp=1&sq=massey%20coal%20mine&st=cse, Ian Urbina, "Toll Mounts in West Virginia Coal Mine Explosion," *New York Times*, April 5, 2010, http://www.nytimes.com/2010/04/06/us/06westvirginia.html?scp=3&sq=massey%20coal%20mine&st=cse, Tom Zellner Jr., "Coal Mine's Safety Record Under Scrutiny," *New York Times*, April 6, 2010, http://green.blogs.nytimes.com/2010/04/coal-mines-safety-record-under-scrutiny/?scp=4&sq=massey%20%mine&st=cse, Kris Maher, "Officials Put Workers at Risk, Report Says," *Wall Street Journal*, June 24, 2010, http://online.wsj.com/article/SB10001424052748703900004575325410525577950.html (including Solis quote), and David A. Farenthold, "Officials Say Cause of Upper Big Branch Mine Disaster in West Virginia Still Unclear," *Washington Post*, August 12, 2010, http://www.washingtonpost.com/wp-dyn/content/article/2010/08/11/AR2010081103332.html.

Judge Feldman's decision, *Hornbeck Offshore Services L.L.C. v Salazar* (2010 WL 3219469), is available at http://www.scribd.com/doc/33448667/Hornbeck-Offshore-Services-v-Kenneth-Lee-Ken-Salazar. See also Stephen Power, "Judge Overturns Drilling Ban," *Wall Street Journal*, June 23, 2010, A1 (containing quote from Congressman Markey.)

References for many of the facts and points in this chapter come from earlier chapters and are not repeated here. The data on oil imports are from the U.S. Energy Information Administration at http://www.eia.doe.gov/dnav/pet/pet_move_impcus_a2_nus_ep00_im0_mbbl_m.htm. See also Lester Thurow, *Zero-Sum Society: Building a World-Class American Economy* (New York: Viking, 1980), and John Updike's quote is from *Rabbit Is Rich* (New York: Alfred A. Knopf, 1981), 7. On the use of oil dollars to fund radical Islamic schools, see Tom Friedman, "It's All about Schools," *New York Times*, February 10, 2008, http://www.nytimes.com/2010/02/10/opinion/10friedman.html.

On the electricity sector, see Paul L. Joskow, "Challenges for Creating a Comprehensive National Energy Policy," mimeo, September 26, 2008, and other of his works available at http://econ-www.mit.edu/faculty/pjoskow/papers.

The estimate that a 25 cents per gallon gasoline tax would have been as effective as CAFE at one-third the cost can be found in Pietro S. Nivola and Robert W. Crandall, *The Extra Mile: Rethinking Energy Policy for Automotive Transportation* (Washington, DC: The Brookings Institution, 1995).

For the information on nuclear power, see chapter 4 and the references for it as well as Paul L. Joskow and Robert W. Crandall, "The Future of Nuclear Power,"

Daedalus (Fall 2009), 46; John W. Rowe, "Nuclear Power in a Carbon-Constrained World," *Daedalus* (Fall 2009), 86; Robert H. Socolow and Alexander Glaser, "Nuclear Energy and Climate Change," *Daedalus* (Fall 2009): 31–44; Harold A. Feivson, "A Skeptic's View of Nuclear Power," *Daedalus* (Fall 2009): 60–70; Michael D. Shear and Steven Mufson, "Obama Offers Loan to Help Fund Two New Nuclear Reactors," *Washington Post*, February 17, 2010, http://www.washingtonpost.com/wp-dyn/content/article/2010/02/16/AR2010021601302.html (containing the Carl Pope quote); and Matthew Walk, "U.S. Supports New Nuclear Reactors in Georgia," *New York Times*, February 17, 2010, http://www.nytimes.com/2010/02/17/business/energy-environment/17nukes.html. See also *The Future of Nuclear Power: An Interdisciplinary Study* (Cambridge, MA: MIT, 2003), http://web.mit.edu/nuclearpower/pdf/nuclearpower_full.pdf, and *Update of the MIT 2003 Future of Nuclear Power Study* (Cambridge, MA: MIT, 2009), http://web.mit.edu/nuclearpower/pdf/nuclearpower_update2009.pdf.

The claim that greater energy efficiency has offset three-quarters of increased demand is from the American Council for an Energy Efficient Future, *The Size of the Energy Efficiency Market: Generating a More Complete Picture* (Washington, DC: American Council for an Energy Efficient Future, June 16, 2010), http://www.aceef.org/pubs/e083.htm (last accessed June 30, 2010). The McKinsey estimates are from McKinsey & Company, *Unlocking Energy Efficiency in the U.S. Economy* (New York: McKinsey & Company, June 2009), http://www.mckinsey.com/clientservice/electricpowernaturalgas/downloads/US_energy_efficiency_full_report.pdf.

On the advantages of carbon and petroleum taxes over cap and trade for oil-importing nations and their disadvantages for oil exporters, see Jon Strand, "Taxes and Caps as Climate Policy Instruments with Domestic and Imported Fuels" in *U.S. Energy Tax Policy*, ed. Gilbert E. Metcalf (New York: Cambridge University Press, 2011) 233–259.

For a particularly pessimistic view of our ability to make the necessary technological, cultural, and political changes, see Saul Griffith's remarks in David Owen, "The Inventor's Dilemma," *The New Yorker* (May 17, 2010), 50.

For those who may not know, Scarlett O'Hara is the leading lady of Margaret Mitchell's Pulitzer Prize–winning 1936 novel *Gone with the Wind* (New York: Scribner, 1936), famous for saying: "I can't think about it now. If I do, I'll go crazy. I'll think about that tomorrow."

Index